PALAEOHISTORIA

PALAEOHISTORIA

ACTA ET COMMUNICATIONES
INSTITUTI BIO-ARCHAEOLOGICI
UNIVERSITATIS GRONINGANAE

23

A.A.BALKEMA/ROTTERDAM/1981

Editorial staff: Mette Bierma, J.W. Boersma, H.R. Roelink & J.D. van der Waals
Address: Biologisch-Archaeologisch Instituut, Poststraat 6, 9712 ER Groningen, Netherlands

ISBN 90 6191 520 1
© 1984 A.A.Balkema, P.O.Box 1675, 3000 BR Rotterdam, Netherlands
Distributed in USA & Canada by: A.A.Balkema Publishers, P.O.Box 230, Accord, MA 02018
Printed in the Netherlands

CONTENTS

A SITE OF THE HAMBURG TRADITION ON THE WADDEN ISLAND OF TEXEL (PROVINCE OF NORTH-HOLLAND, NETHERLANDS)

Dick Stapert

CONTENTS

1. INTRODUCTION

In recent years an extensive settlement research-project has been carried out by the Rijksdienst voor het Oudheidkundig Bodemonderzoek (R.O.B.) near the Beatrixlaan in Den Burg on Texel, involving the investigation of traces dating from the Middle Bronze Age until the Late Middle Ages (Woltering, 1975). In 1974, in the course of this research, a concentration of Upper Palaeolithic flint artefacts was found. On the basis of typological and geological evidence these finds can be placed in a late phase of the Hamburg tradition. They are described here below.

For the realization of this article I wish to thank the following persons: Mr. P.J. Woltering (R.O.B.), for the opportunity to study the material; Mr. H.J. Veenstra (Geologisch Instituut, Groningen) for the determination of the grain size distribution of two sand samples; Mr. J.M. Smit (B.A.I.) for inking my pencil drawings; Mr. F.W.E. Colly (B.A.I.) for the printing of several photos; Mette Bierma and Engelien Rondaan-Veger for typing the manuscript; Sheila van Gelder-Ottway for the translation into English. To all these people I am very grateful.

2. THE SITE, GEOLOGY

The coordinates of the site on the Topographical Map of the Netherlands (1:50,000, no. 9W – Den Helder) are c.: 114.95/563.72. For the situation on a map of the Netherlands see figure 1 (see also the maps in Woltering, 1975; the site lies on the border of his excavation trenches 131 and 136).

In contrast to the other Wadden Islands, the island of Texel has a Pleistocene core that is exposed at the surface. In different places there are outcrops of boulder-clay (the highest is the "Hoge Berg"), that form part of a series of hills in the Northern Netherlands, and that can be regarded as ice-pushed moraines of Saalian age (Ter Wee, 1962; 1973). The "Hoge Berg" reaches a height of about 15 m above

Fig. 1. Map of the Netherlands, on which the location of the Hamburg site on the island of Texel is indicated.

N.A.P. (Dutch Ordnance Level). On the flanks around the relatively high boulder-clay regions cover-sands occur, dating from the last part (Late Glacial) of the Weichselian. The site lies about 1 km to the north of the western part of the exposed boulder-clay region of the "Hoge Berg", within the area of cover-sand. It is possible, however, that during the Late Glacial sources of stones and flint (boulder-clay) were present at the surface also closer by, that became covered by cover-sand only after the Upper Palaeolithic occupation. From a map shown by Woltering (1975, fig. 5) it is evident that the site lies on the northwestern slope of a cover-sand ridge, that runs roughly SW-NE. For the entire area that he has excavated (c. 11.5 ha) Woltering has made a reconstruction of the original micro-relief (1975, fig. 7). From this it can be deduced that the site did not lie on the highest part of

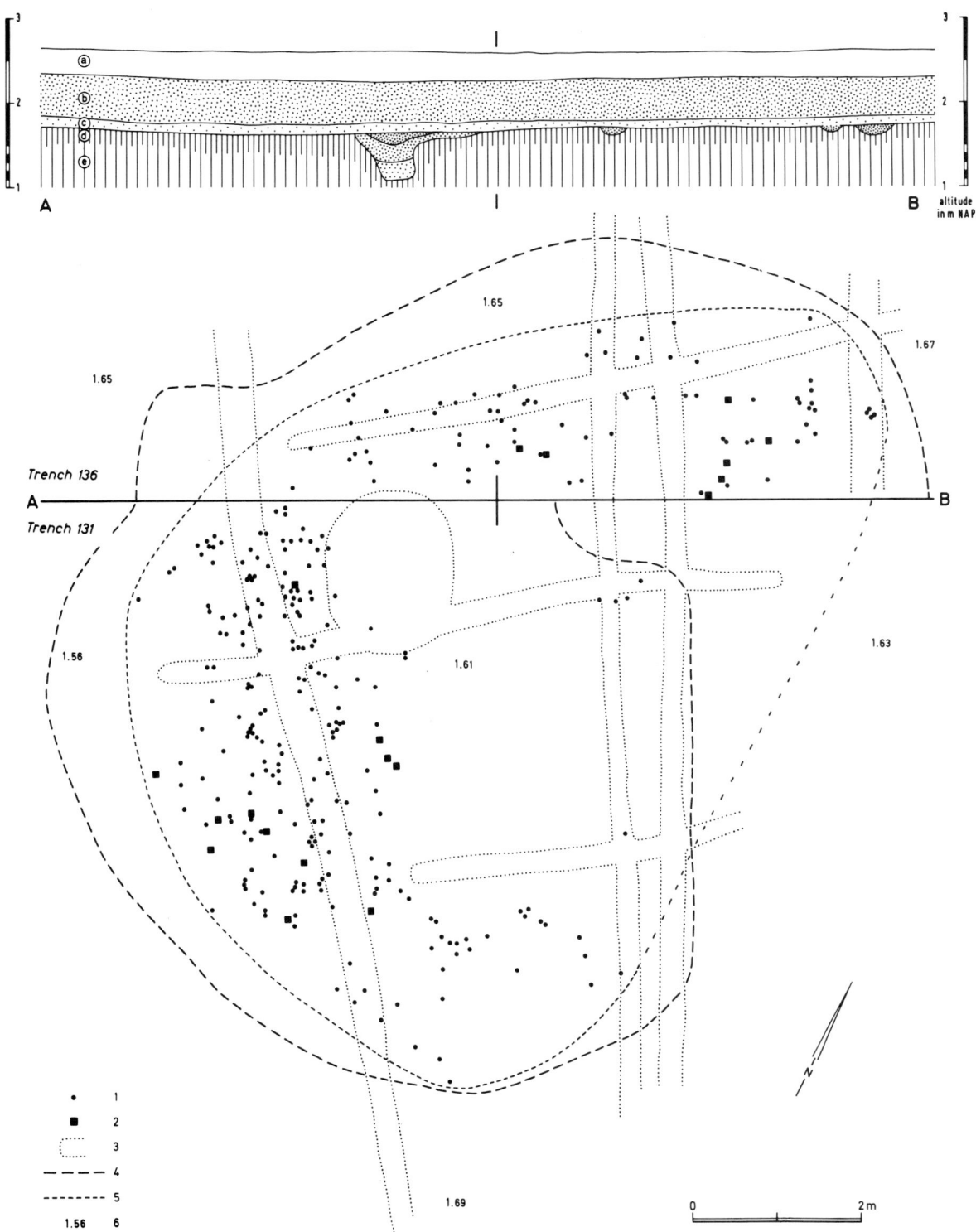

Fig. 2. Excavation plan of the horizontal distribution of the finds, and cross-section through the concentration. – Key: a. modern topsoil; b. *es* layer (Late Middle Ages); c. humous field layer (probably dating from Roman times); d. cover-sand, brownish in colour; e. cover-sand, yellow in colour; 1. flint artefacts; 2. stones; 3. ditches, dating from Roman times; 4. area that was searched for finds; 5. reconstruction of the most probable original shape of the concentration; 6. altitudes in m above N.A.P. (Dutch Ordnance Level). Redrawn by J.M. Smit (B.A.I.) after data provided by P.J. Woltering (R.O.B., Amersfoort).

the cover-sand ridge (c. 3.25 m + N.A.P.) that is situated c. 60 m to the SE of the site, but on a plateau-like projection of the northwestern flank of this ridge, about halfway up the slope.

Within a radius of 10 km around the site (the area necessary for hunters and gatherers to have at their disposal according to Sturdy, 1975) there were several relatively high boulder-clay outcrops, higher and lower cover-sand ridges, and undoubtedly also several smaller valleys and/or small lakes.

At that time the coast must have lain at a distance of 300 to 350 km to the north (see e.g. Veenstra, 1970).

The profile at the site of the flint concentration (fig. 2) can be described as follows (based on data of P.J. Woltering, R.O.B.; the surface lies at c. 2.60 m + N.A.P.):
a. 0–30 cm: modern topsoil, b. 30–80 cm: *es* layer (Late Middle Ages), c. 80–95 cm: humous field layer (probably dating from Roman times); d/e. 95–deeper than 150 cm: cover-sand (Twente Formation; Pleistocene-eolian); towards the top (d) brownish in colour, lower down (e) yellow.

The flints were present towards the top of the cover-sand, just under the humous field layer (c), and were distributed vertically over c. 5-10 cm, according to the excavators. From the profile description it is evident that the find level (at c. 1.60 m + N.A.P.) lies at the base of a podsol. In the cover-sand iron infiltration traces (brown) were still visible, but one cannot speak of a distinct B-horizon in the find level. The A- und B-horizons of the original podsol have evidently (largely) disappeared as a result of later human activities. We are most probably concerned here with a remnant of the Holocene heath-podsol, that is present almost everywhere in the cover-sand region at the top of the most recently formed cover-sand layer. It is not known for certain whether originally flints also occurred higher in the profile, e.g. in the B-horizon (that has now disappeared). No flints were found when a surface in level c (the humous field layer) was being cleaned. On the basis of this fact it could be proposed that originally the finds did not

Table 1. Grain size distribution of cover-sand samples 1 and 2.

sieve fractions	*percentages by weight*	
	sample 1	*sample 2*
> 710 mu	0.182	0.610
500 - 710 mu	0.437	1.666
425 - 500 mu	0.639	1.450
355 - 425 mu	1.075	1.964
300 - 355 mu	2.939	4.265
250 - 300 mu	5.814	7.286
212 - 250 mu	8.366	9.468
180 - 212 mu	15.416	16.603
150 - 180 mu	17.907	18.045
125 - 150 mu	17.066	15.108
106 - 125 mu	9.828	7.797
90 - 106 mu	8.847	6.346
63 - 90 mu	9.875	7.024
< 63 mu	1.609	2.369

occur substantially higher in the profile. In that case we could conclude that the flints were left behind before the entire cover-sand layer had been deposited, thus during (though presumably in the last part of) a stadial. From other excavations of Upper Palaeolithic flint concentrations in the cover-sand of the Northern Netherlands it is known, however, that the flints are often scattered vertically considerably more than 10 cm, up to c. 50 cm, as a result of various disturbances caused by burrowing animals and by roots (bioturbation). A vertical distribution of only 10 cm therefore seems improbable, so that in my opinion it must be assumed that originally the finds did indeed occur also higher up in the profile.

It is not known exactly in which geological period the cover-sand was deposited in this locality. Within the flint concentration no sand samples were taken, but after the excavation two samples were taken in the immediate vicinity by P.J. Woltering, at my request: *sample 1*, taken about 60 m to the S of the concentration, immediately under the soil disturbed by the excavation, and thus certainly from a slightly greater depth than the level at which the flints were present; *sample 2*, taken c. 50 m to the NE of the concentration just outside the excavation area, most probably from the same level as that of the finds (just below the culture layers, sand with slight iron infiltration), at about 1 m below the surface. Dr. H.J. Veenstra was so kind as to have the grain size distribution of both samples determined. The results are shown in table 1.

Ter Wee (1973) used a classification triangle for sands, in which 3 large fractions can be read off. For the 2 samples from Texel these fractions have the following values (the fraction limits are slightly different as a result of the sieves used, but this has no significant consequences for the purposes of comparison).

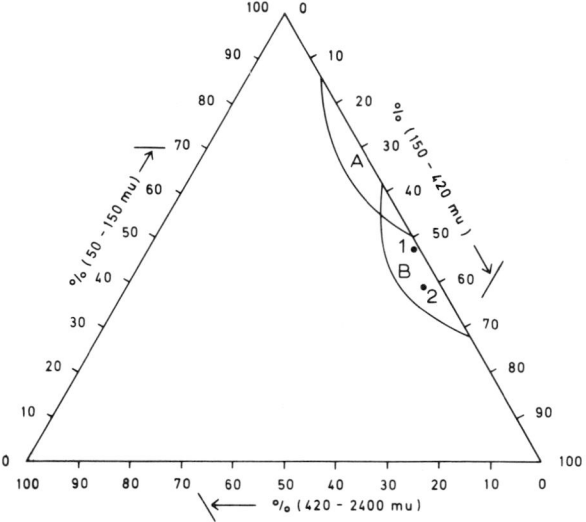

Fig. 3. Classification triangle for sands (after Ter Wee, 1973), in which the grain size distributions of the two samples from Texel are indicated. – Key: A. Older Cover-sand; B. Younger Cover-sand; 1. sample 1; 2. sample 2. Drawing by D. Stapert/ J.M. Smit (B.A.I.).

Table 2. Sieve fractions for the classification triangle (fig. 3).

sieve fractions	percentages by weight	
	sample 1	sample 2
50 - 150 mu	47.2	38.6
150 - 425 mu	51.5	57.6
> 425 mu	1.3	3.7

Figure 3 shows the classification triangle mentioned above. Indicated are the areas established by Ter Wee (1973) for A Older Cover-sand and B Younger Cover-sand. These areas show an overlap. Both samples from Texel fall in the middle of the area of Younger Cover-sand, outside the overlap. Sample 1 is somewhat finer than sample 2, which can be explained by the fact that the former originates from a somewhat greater depth than the latter. According to Van der Hammen *et al.* (1967) and Ter Wee (1966) Younger Cover-sand was deposited after the Bølling interstadial, and can be divided into Younger Cover-sand I and II. Younger Cover-sand I

was deposited during the stadial between the Bølling and Allerød interstadials (Early Dryas), and Younger Cover-sand II in the last stadial of the Weichselian (Late Dryas), after the Allerød interstadial. Younger Cover-sand I is somewhat finer than Younger Cover-sand II, and is furthermore characterized by the presence of thin loamy bands, that increase in number and thickness with increasing depth. In view of the fact that the "layer of Usselo" was not observed, it is most probable that we are here concerned with Younger Cover-sand I. Finds of the Hamburg tradition from after the Allerød interstadial are moreover from an archaeological point of view hardly to be expected.

To conclude, the most probable dating for the finds is as follows: the last part of the (short) stadial between the Bølling and Allerød interstadials (Early Dryas stadial). Naturally it must be pointed out that this dating can by no means be taken as definitive, but only as the most probable on the basis of the insufficient data that are available. It is noteworthy, however, that for at least 2 other sites of the Hamburg tradition in the Northern Netherlands a comparable dating has been established, namely Luttenberg (province of Overijssel, see Lanting & Mook, 1977) and Oldeholtwolde (province of Friesland, see Stapert, in prep.), while for several other Hamburg sites the same dating can be reasonably assumed. The typology of the material from Texel (see under 4) does not contradict this possible dating.

3. THE EXCAVATION, HORIZONTAL DISTRIBUTION OF THE FINDS

The excavation near the Beatrixlaan in Den Burg was carried out in 1967 and 1971-1975, on a terrain that was later built up with houses (see Woltering, 1975, fig. 9). The first flints were found in trench 131 (see Woltering, 1975, fig. 10), on August 7th, 1974.

In trench 131 first the recent topsoil and the *es* layer were removed mechanically. The surface that was thus exposed, surface 1, was localised in the field layer dating from Roman times (c in fig. 2). While this surface was being cleaned with a shovel no flints were noticed. Subsequently excavation proceeded mechanically down to surface 2: in the cover-sand just under the culture layers, i.e. in d in fig. 2 (at c. 1.60 m + N.A.P.). When this surface was being cleaned with a shovel the first flints came to light. Afterwards the whole of the surrounding area was searched intensively (also with trowels). The finds were registered on an excavation plan (with 2 symbols: one for flint, one for other sorts of stone), but not individually numbered. The sand was not sieved.

Part of the ground said to be removed from the area of the flint concentration in trench 131 was sieved; in this way a number of finds were recovered.

In trench 136 mechanical excavation proceeded directly down to surface 2. The area was searched for finds, but the dumped ground from this trench was not sieved.

Table 3. Numbers of finds, encountered in the two excavation trenches (131, 136) and sieved from loose soil.

	flints	stones	total
trench 136	76	7	83
trench 131	221	12	233
subtotal	297	19	316
sieved from loose soil (trench 131)	150	9	159
total	447	28	475

A total number of 475 finds were collected, of which about two-thirds were found during the cleaning of surface 2 (see table 3). In figure 2 is shown the horizontal distribution of the registered finds. A broken line (4) indicates the area that was searched for finds during the process of cleaning (with shovel and trowel) the surface just below the culture layers. A conspicuous feature is the area without any finds present in the eastern half of trench 131, that does not continue into trench 136. It must be assumed that here the machine removed most of the find-bearing level. The finds obtained by sieving the dumped ground (c. one-third of the total) would have originated mainly from this area. A thin broken line (5) in figure 2 indicates a reconstruction of the most probable original shape of the concentration: an oval of c. 10 x 8 m. The total number of finds recovered (475) is rather small for such a fairly wide concentration. It therefore seems probable that a considerable proportion of the finds are lost to us. This could also be evident from the fact that extremely few examples of the broken blades fit together (see under 4.3.2.1). Several reasons for this could be:
a. As a result of the mechanical excavation an unknown amount of the material was removed.
b. An unknown proportion of the finds could

have ended up in the later culture layers. This could not have been the case with very many finds, as during the cleaning of surface 1 in trench 131 no flints were noticed.

c. No sieving was done.

In view of the fact that the distribution plan shows a large hiatus (c. one-third), and that in addition an unknown (though probably considerable) proportion of finds is lost to us, it is clear that this plan cannot provide much further information. Moreover the finds have not been individually numbered, so it is not possible to study the distribution of specific groups within the material. For the rest, the plan shows no striking phenomena.

4. THE FLINT ARTEFACTS

4.1. General

The finds consist of 447 flint artefacts, in addition to 28 stones (see under 5). The flint used is of good quality. In fact this applies in general for sites of the Hamburg tradition, in contrast to many sites, for example, of the Tjonger (*Federmesser*) tradition. This may be partly connected with the more open vegetation during the Early Dryas stadial (in which many Dutch Hamburg sites can be dated), that made it possible to select good flint. During the Allerød interstadial (in which most of the Tjonger sites can be dated) it must have been much more difficult to collect good flint.

Frequently occurring within the material are grey to brown relatively fine-grained sorts of flint. These consist without exception of northern moraine-flint (including bryozoan flint). The origin of the flint is undoubtedly the nearby situated boulder-clay area of the "Hoge Berg".

In addition to a slight gloss patina that is generally present, brown and sometimes also white patina occur on the flint artefacts.

The flint material can be subdivided as shown in table 4, in which figures are also given concerning the occurrence of fractures and traces of burning. The high percentage of broken blades is conspicuous. An attempt was made to fit broken blades together. This was possible in only 2 cases, i.e. for 4 out of the 146 blade fragments (i.e. 2.7%). This almost certainly means that a large proportion of the original material has not been recovered

Table 4. Flint artefacts.

	no.	% of total	broken		burnt	
			no.	% per group	no.	% per group
blades	173	38.7	145	83.8	8	4.6
flakes	194	43.4	97	50.0	20	10.3
cores	12	2.7	–	–	2	16.7
blocks/nodules	8	1.8	–	–	0	0
splinters (< 1 cm)	33	7.4	–	–	6	18.2
subtotal non-tools	420	94.0	242	–	36	8.6
tools	27	6.0	21	77.8	1	3.7
total	447	100.0	263	–	37	8.3

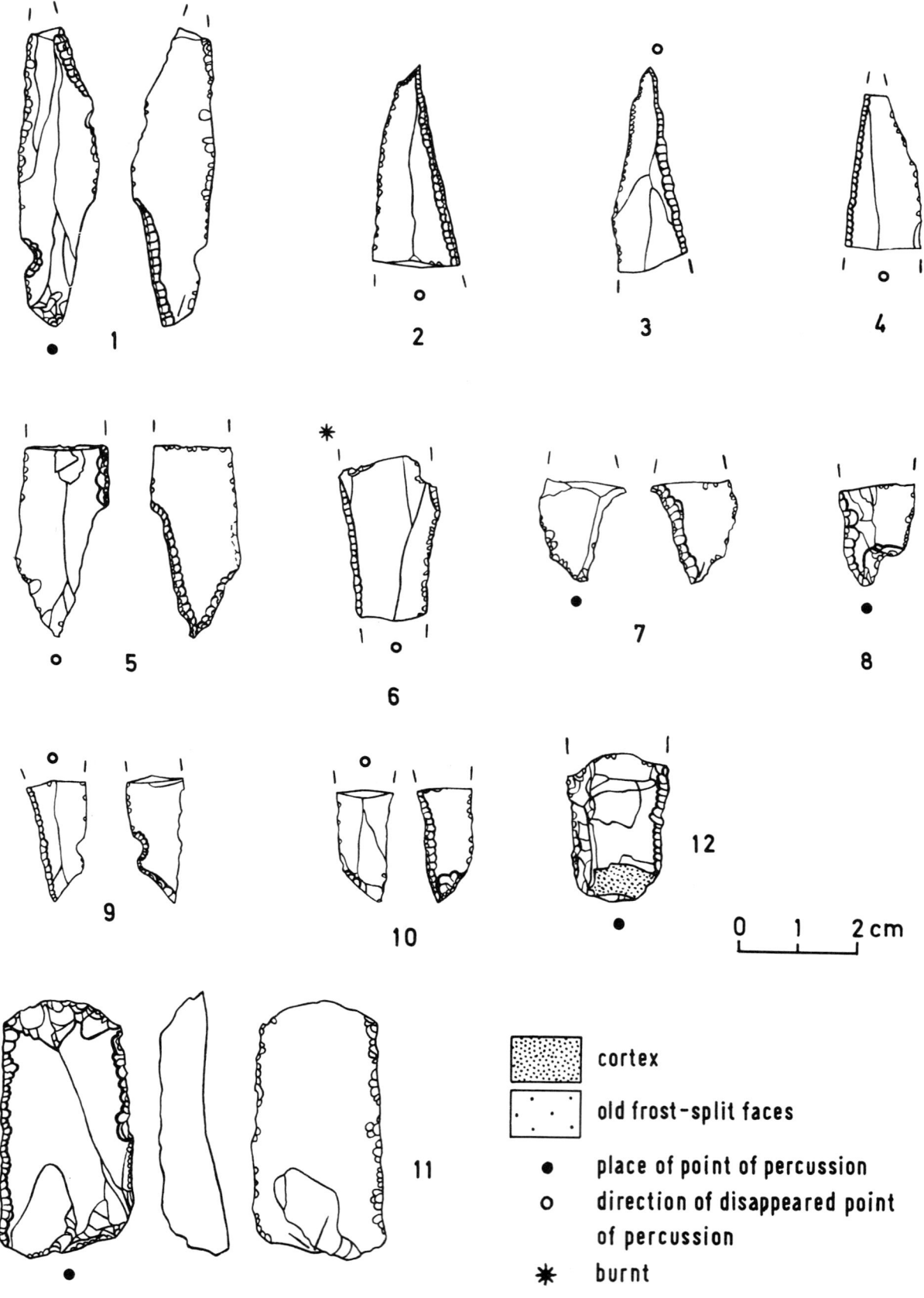

Table 5. Flint tools.

type	no.	perc.	burnt
shouldered points (mostly broken)	10	37.0	1
convex end scrapers	1	3.7	0
burins	2	7.4	0
borers	2	7.4	0
notched blades	5	18.5	0
(part.) retouched blades	5	18.5	0
(part.) retouched flakes	2	7.4	0
total	27	99.9	1

(see also under 2). Consequently no further attempts were made to fit the flints together. In view of this situation it must also be remembered that the relative numbers of e.g. different types of tools do not necessarily reflect the original proportions.

The numbers of specimens indicated as showing traces of burning must be regarded as minimum figures: it is not always possible to definitely ascertain former contact with fire.

4.2. Tools

The 27 tools (6% of all flints) can be subdivided as shown in table 5. Not included in the category of tools are 4 specimens, that could be described as "core scrapers" or "core burins". These specimens are classified as cores (see under 4.3.3.).

In the following, several remarks are made here and there about traces of use that are visible on some specimens. Here we are concerned with traces that are visible to the naked eye or with a stereomicroscope. The author is no specialist in this field, but is of the opinion that macroscopically observable phenomena should in any case be mentioned.

4.2.1. *Points and point fragments* (total 10; fig. 4, nos. 1-10)

Of the ten points 9 are fragmentary. Only one specimen (fig. 4, no. 1) is more or less complete, with only a small part of the top broken off — possibly as a result of being used as a projectile (the fracture surface is not recent). This is also possibly indicated by the relatively high degree of splintering along the non-retouched edges. Ventrally some scratches are also visible with the aid of a stereomicroscope, that run more or less parallel to the longitudinal axis of the point (fig. 5). Similar scratches also occur on at least one of the top fragments. Scratches of this kind can also indicate use as a projectile (see e.g. Odell, 1978).

Table 6. Lengths of point fragments.

top fragments	length
fig. 4, no. 2	3.3 cm
fig. 4, no. 3	3.4 cm
fig. 4, no. 4	2.6 cm
	(mean = 3.1 cm)

basal fragments	length
fig. 4, no. 5	3.2 cm
fig. 4, no. 7	1.7 cm
fig. 4, no. 8	1.7 cm
fig. 4, no. 9	2.0 cm
fig. 4, no. 10	1.8 cm
	(mean = 2.1 cm)

Fig. 4. Texel, flint tools. 1. shouldered point (tip anciently broken off); 2-10. point fragments; 11. blade end scraper with two retouched sides; 12. proximal blade fragment with one retouched side (possibly fragment of blade end scraper). Drawings by D. Stapert/J.M. Smit (B.A.I.).

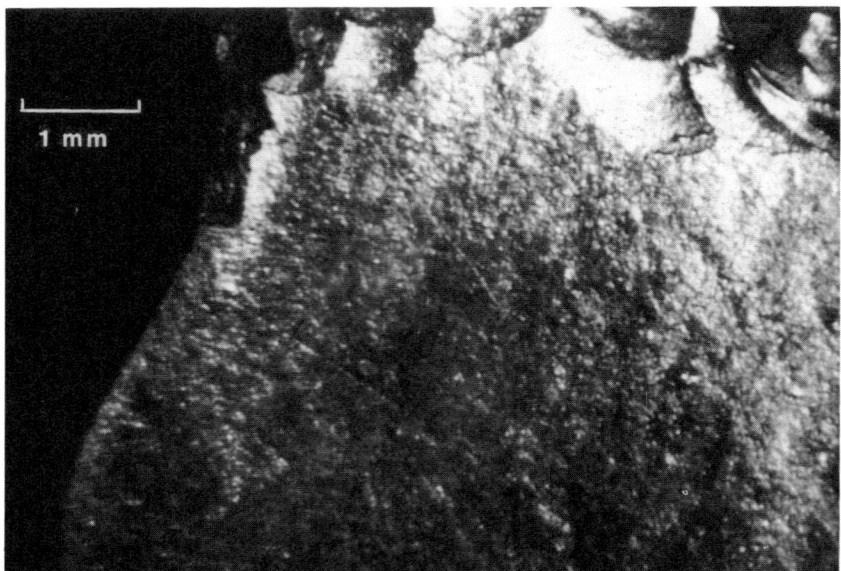

Fig. 5. Some fine scratches on the ventral face of the shouldered point (fig. 4, no. 1) near the fracture (to the left) at the tip of the point. The scratches run more or less parallel to the longitudinal axis of the point. Scale in mm. Photo made with a stereomicroscope by D. Stapert (B.A.I.).

The single complete specimen is a typical shouldered point (*point à cran*), as occurs in the Netherlands mainly within the Hamburg tradition.

The basal notch (shoulder) is retouched ventrally, the apical part dorsally. A conspicuous feature is the small dorsal notch at the base, opposite the (shoulder-) notch. This notch may have served to facilitate shafting of the point. A similar extra notch is also present on one of the basal fragments (fig. 4, no. 9). The original length of the point was c. 5.5 cm.

Of the 9 point fragments one is burnt (fig. 4, no. 6); this fragment is the middle part of a shouldered point.

For the other 8 fragments the lengths are given in table 6. The fractures are not recent.

None of the fragments fits together with any other. The numbers are of course too small to permit any statistically reliable conclusions to be drawn. There is a suggestion however that the fractures mostly occur at a little over 2 cm from the base, so the hypothesis could be proposed that these points were shafted over a distance of c. 2 cm. The two complete basal (shoulder-) notches within the material are 2.1 and 2.3 cm long respectively (fig. 4, nos. 1, 5), which does not contradict this proposition. The average lengths of the top and basal fragments are together 5.2 cm, which could have been approximately the average length of complete points.

It is understandable that basal fragments were left behind at a settlement. They may have been shafted, e.g. on to spears (in the Hamburg levels of Meiendorf and Stellmoor there are no arrows present, in contrast to the Ahrensburg levels), and could then have been replaced by new complete points during the time when the settlement was inhabited. The fact that top fragments were left behind is more difficult to understand, unless one assumes that these were transported to the camp site embedded in the flesh of game that had been killed. It is in any case interesting to note that within the material not a single completely intact point is present, and that basal fragments occur more frequently than top fragments.

It is reasonable to assume that one of the activities at this camp site was replacing broken or damaged points. The relatively high proportion of points among the tools may perhaps indicate that we are concerned here not with a

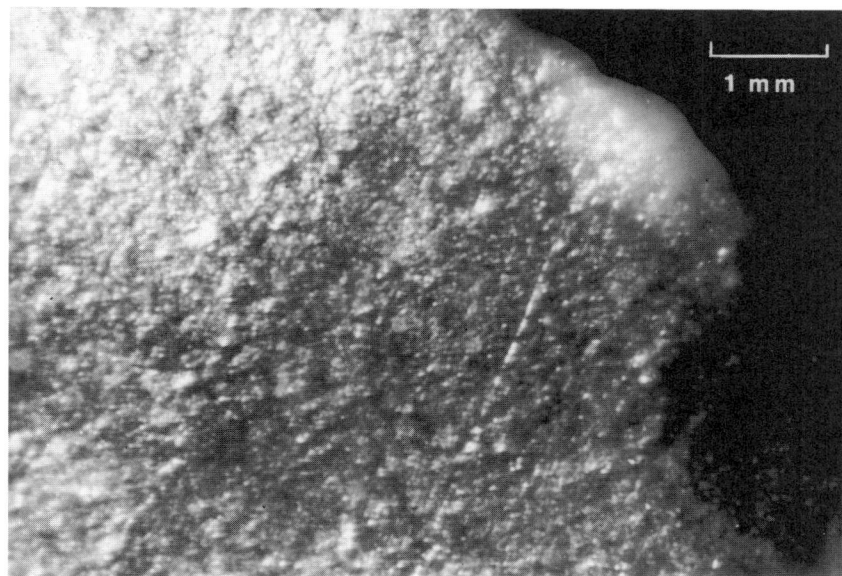

Fig. 6. The ventral face of the blade end scraper (fig. 4, no. 11) near the scraping edge. Visible are the roundedness of the scraper edge, and a scratch that runs c. perpendicular to the scraping edge. Scale in mm. Photo made with a stereomicroscope by D. Stapert (B.A.I.).

"base camp", but with a temporary camp of a group of hunters. Naturally this is all speculation, in view of the fact that we are not familiar with all the material from this site. (Compare however the material from Swalmen: Stapert, 1979).

Among the top fragments a noteworthy feature is the extra retouch on the apex of fig. 4, no. 2, that served to make the tip sharper.

Among the basal fragments especially the tang-like specimen no. 5 is worthy of note; also no. 9 has been finely finished off with an extra notch opposite the shoulder-notch. Especially these two specimens are suggestive of the possibility that we are concerned here with a typologically late phase of the Hamburg tradition (as postulated by Bohmers, 1947 and Tromnau, 1975), that is supposedly characterized by the presence of tanged points and backed pieces (e.g. *Federmesser*). Within the material there is also one, albeit atypical, backed blade present (fig. 7, no. 8; see under 4.2.6.). The material concerned is however of too small a quantity for a convincing "typological dating". As we have already seen, however (under 2) the scarce geological data are

not in contradiction with a "late" dating.

4.2.2. *Blade end scraper* (total no. 1; fig. 4, no. 11)

Only one (complete) blade end scraper is present. The scraping edge has been made distally on a fairly short blade. The scraper angle is c. 45°. The left edge of the blade is retouched completely, the right edge partly; the non-retouched part is naturally blunt. Blade end scrapers with retouched edges are a normal element within the Hamburg tradition and may be indicative of a relation with the Late Magdalenian.

With a stereomicroscope it can be seen that the working edge of the scraper is clearly rounded and glossy (fig. 6). These features indicate use in the processing of skins (see i.a. Keeley, 1980; Broadbent & Knuttson, 1975). Locally ventral scratches also occur close to the scraping edge; usually they make an angle of 30-80° with the scraping edge, but in addition there are a few scratches that run more or less parallel to the scraping edge.

One of the broken retouched blades may be a broken scraper (fig. 4, no. 12), but this is not certain (see under 4.2.6.).

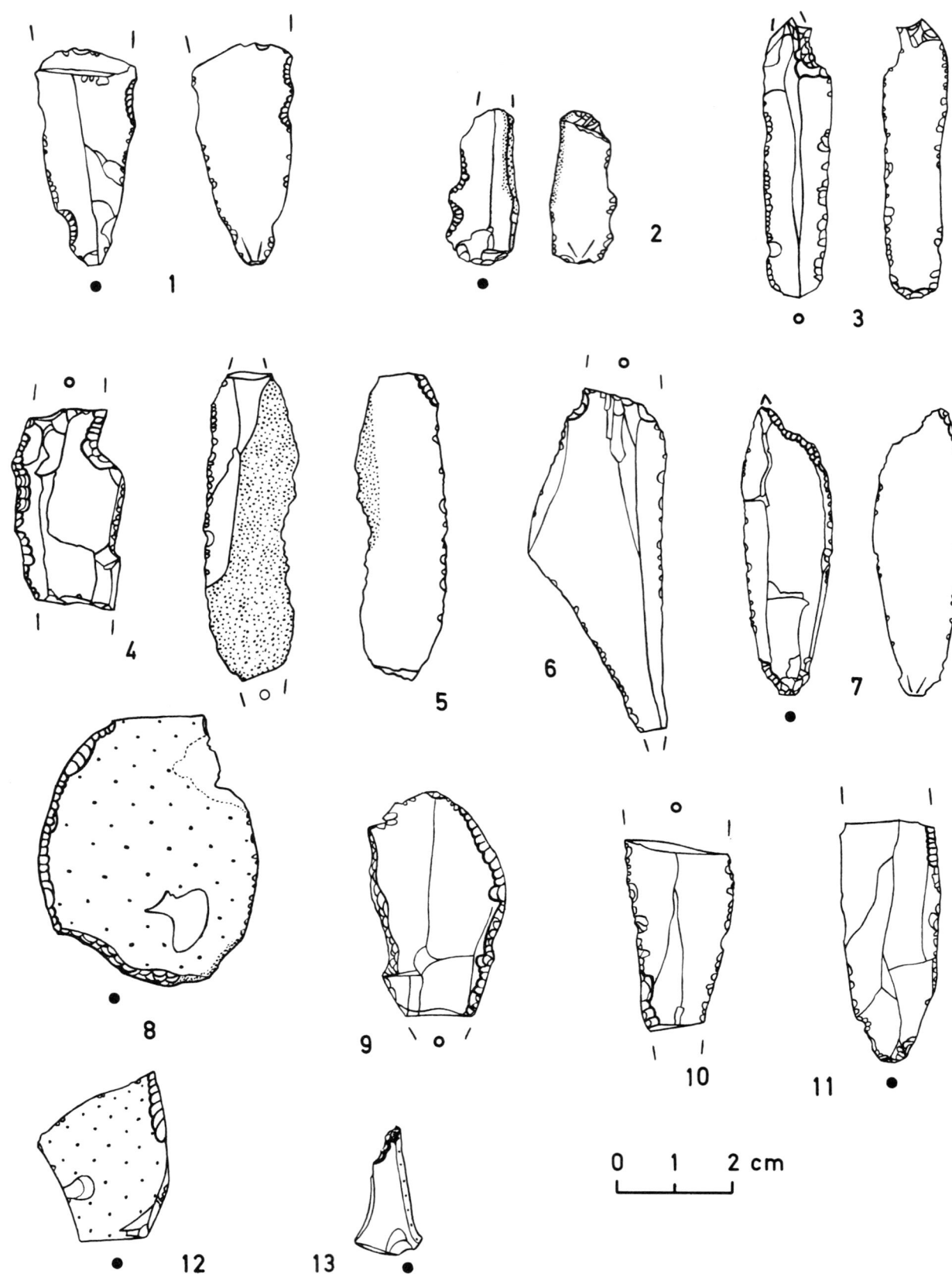

Table 7. Notched blades: number and position of notches.

	no.	dorsal	ventral
fig. 7, no. 1	4	2	2
fig. 7, no. 2	2	2	–
fig. 7, no. 3	1	1	–
fig. 7, no. 4	3	3	–
fig. 7, no. 5	1	1	–
total	11	9	2

4.2.3. *Notched blades* (total no. 5; fig. 7, nos. 1-5)

There are 5 blades with notches, that vary considerably in size and shape. Indicated in table 7 is the number of notches present on each specimen, and their position (dorsal or ventral).

All the specimens are broken, and in 2 cases only a medial blade fragment has been preserved. The average number of notches per specimen (2.2) is therefore almost certainly lower than the original number. The width of the notches (only complete cases) varies from 0.5-0.8 cm (average 0.64 cm, N=7). In 4 cases a fracture borders on to a notch, so one gets the impression that these fractures, in any case partly, could have something to do with the existence of the notches (see Dewez, 1970).

In some cases a certain amount of gloss was observed along the edge of the notch, on the surface from where retouch had been applied.

Fig. 7. Texel, flint tools. 1-4,6. notched blades; 7. borer (with inverse retouching; small tip anciently broken off); 13. small "borer" on flake; 5,9-11. (partially) retouched blade fragments; 8,12. retouched flakes. Drawings by D. Stapert/J.M. Smit (B.A.I).

4.2.4. *Borers* (total no. 2; fig. 7, nos. 6,12)

Within the material from Texel, there are no typical asymmetrical borers (*Zinken*) present. Of the two borers present one is a clear example, made from a blade (no. 6), while the other is an atypical specimen made out of a small flake (no. 12).

The first borer consists of a somewhat concave oblique truncation, with ventral retouch along the opposite edge. The very top part of the borer has been broken off (not recently, therefore presumably as a result of use). Near the top end of the borer the edges are somewhat blunted. Near the base of the blade (on the left) a small amount of retouch has been applied, while in addition both edges show ventrally and dorsally quite a lot of fine splintering, that can probably be regarded as the result of use.

The second implement that can be described as a borer is an atypical specimen. It consists of a small flake that has been sharpened into a point, with some additional retouch on the dorsal ridge.

4.2.5. *Burins* (total no. 2; fig. 8, nos. 1,2)

There are 2 burins present within the material: a very small dihedral burin made out of a frost-split fragment (no. 1), and an transverse angle burin made out of a broken core-preparation blade (no. 2). Burins on truncations, that generally occur frequently within find-complexes of the Hamburg tradition, are absent.

In the case of the dihedral burin the negative of a burin spall perpendicular to the longitudinal axis of the specimen has been used as a striking platform for 3 burin spalls parallel to the longitudinal axis (one primary, two secondary). The burin edge is 0.3 cm wide, and the burin angle is almost 90°.

One angle of the burin edge (indicated by a semicircle in fig. 8:1b) is clearly rounded (abraded), as can be seen already with the naked eye. This roundness, that can only be the result of intensive use, is not restricted to this angle itself, but is also clearly present over

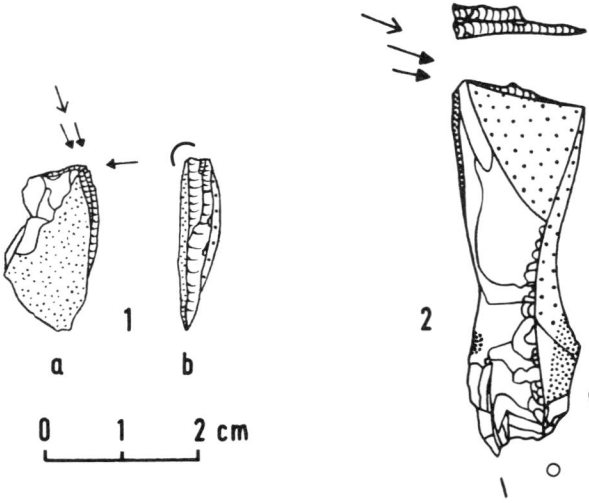

Fig. 8. Texel, burins. 1. small dihedral burin (face not shown is an old frost-fracture); 2. transverse angle burin. Drawings by D. Stapert/J.M. Smit (B.A.I.).

On the burin edge itself the roundedness is visible only over a very short distance, as is also the case along the edge of the burin facet downwards. The other angle of the burin edge (on the right in fig. 8:1 b) shows only minimal roundedness, that is present from the angle again mainly along the rim of the upper surface.

For these distinct traces of use to have originated, especially on the left angle of the burin edge, the tool must most probably have been manipulated in the following way. The implement would have been held in the left hand between thumb, fore- and middle finger, with the upper surface (of fig. 8:1a) directed downwards, in such a way that the burin edge pointed away from the user (fig. 10). Only one method of use appears possible, namely graving, and then by means of a movement towards the person. Only in this way could the upper edge have become rounded over a fairly long distance, and then to a lesser degree with increasing distance from the angle, and only in this way too could the scratches on the upper surface run parallel to the edge.

The edges of the negatives of the vertical burin spalls (in fig. 8:1 b) do not appear to have any distinct traces of use; only in two places

a distance of about 0.3 cm along the top rim (in fig. 8:1a, to the left), while right next to this edge on the top surface scratches are visible, that run more or less parallel to this edge (fig. 9).

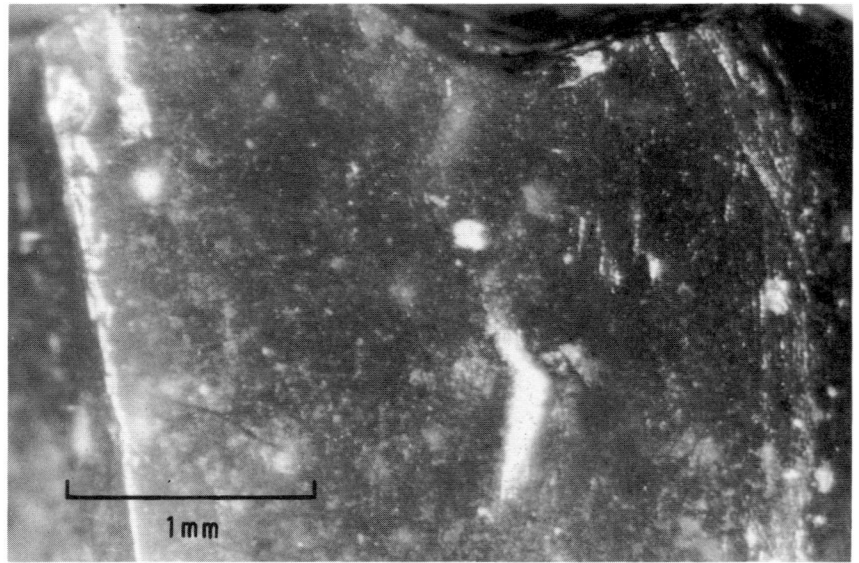

Fig. 9. Top surface of the dihedral burin just near the burin edge (at the top). The ridge on the right is heavily abraded, and scratches are visible on the top surface that run approximately parallel to this ridge. Scale in mm. Photo made with a stereomicroscope by D. Stapert (B.A.I.).

is slight splintering visible (on the right), but no roundedness or scratches can be seen. It is interesting that it can be shown that the probable function of this tool was graving with an angle of the burin edge (the traditionally assumed function of burins, see e.g. Rust, 1937), and not scraping with the edges of the burin facets (Bordes, 1965). Nevertheless the latter function is not improbable in some cases, e.g. some burins from the Late Ahrensburgian site of Gramsbergen (Stapert, 1979), where regular and distinct splintering occurs along the edges of the burin facets.

The second burin shows the negatives of 3 burin spalls (2 primary, 1 secondary), perpendicular to the longitudinal axis of the blade, struck off from a transverse rim (formed by cortex). The burin edge is c. 0.4 cm wide, the burin angle measures c. 85°. No distinct traces of use could be observed on this implement.

Within the material from Texel there are no clear burin spalls present.

4.2.6. *(Partially) retouched blades* (total no. 5; fig. 4, no. 12; fig. 7, nos. 8-10,13)

The only tool from Texel that could with some difficulty be called a backed blade (no. 8) is a fairly wide core-preparation blade, that is broken (distal fragment), and of which one edge is completely retouched. The other edge is not sharp, however, but consists of preparation negatives that are more or less transversely oriented, from which it is evident that this specimen is atypical.

In fig. 4, no. 12, a proximal fragment of a core-preparation blade is illustrated, of which the right edge is retouched; the left edge is not sharp, but consists of an almost transversely oriented face. It is very well possible that we are here concerned with a broken scraper. (With the blade end scrapers of the Hamburg tradition the sides are often retouched).

The three other specimens are larger blade fragments (two medial, one proximal) with partial retouching of one of the edges. Apart from the retouching these specimens show fine splintering along the non-retouched edges, both

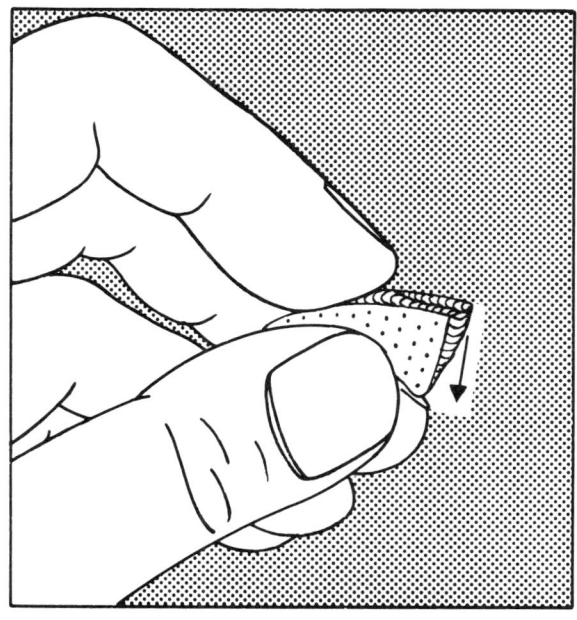

Fig. 10. Reconstruction of the most probable way of using the dihedral burin: graving while the implement is held in the left hand, with a movement towards the person. Drawing by D. Stapert/J.M. Smit (B.A.I.).

ventrally and dorsally, that could be the result of use.

4.2.7. *Retouched flakes* (total no. 2; fig. 7,11)

In both cases we are concerned with flakes that consist dorsally of (old) frost-split faces. The larger flake (no. 7) shows in addition features indicative of core-preparation, and is completely retouched along one convex edge. The other edge is sharp, so the specimen could have been used as a knife (some fine splintering is visible along the edge).

The smaller flake shows partial retouching along one edge, that ends distally in a point. The other edge is not sharp, but consists of a transversely oriented face. Fine splintering near the point suggests that this is the "working edge" of the tool, although the specimen cannot be described typologically as a borer, for example.

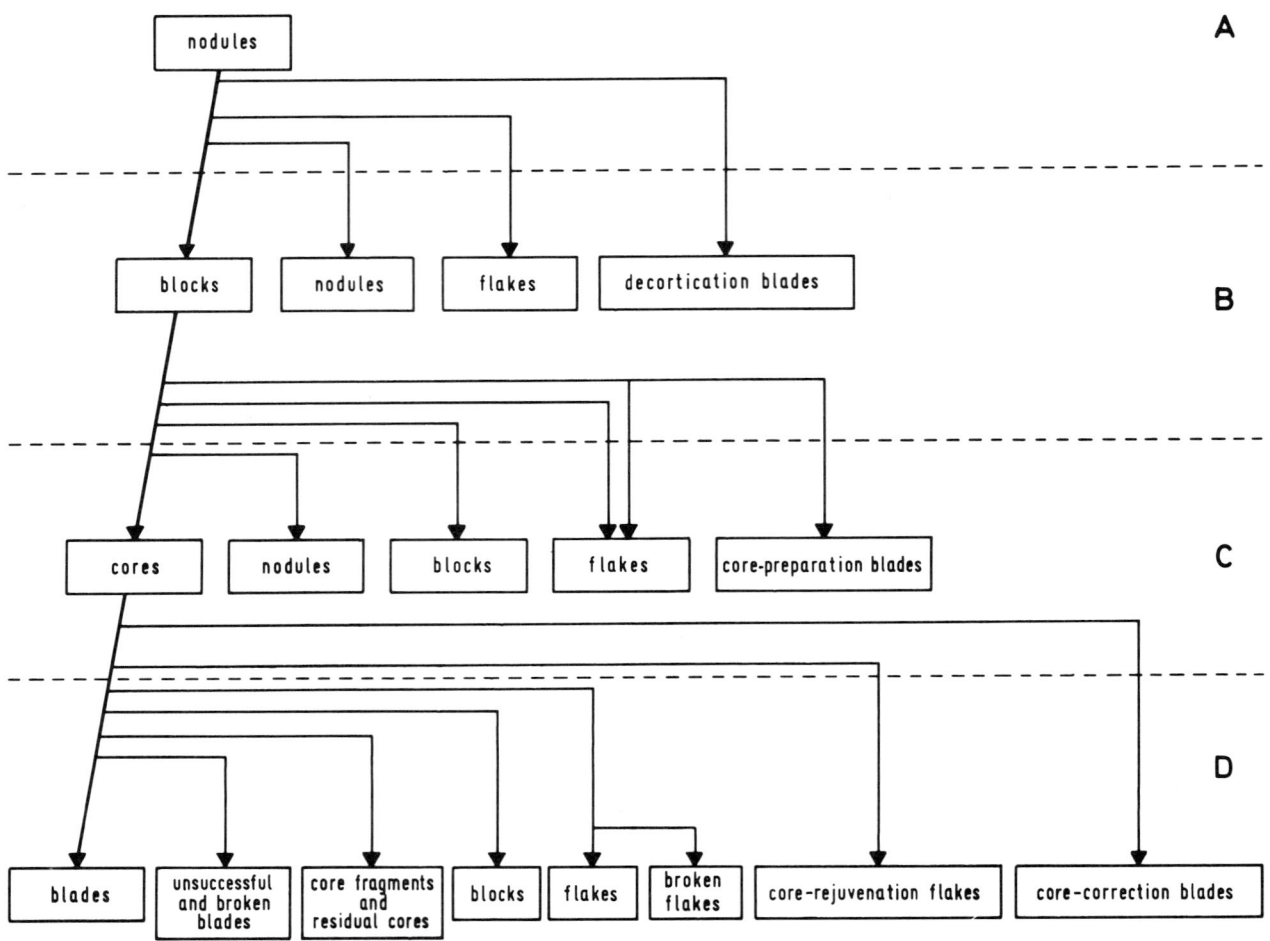

Fig. 11. Scheme illustrating the Upper Palaeolithic blade technology. A. selection of raw material; B. first testing and quartering of the nodules; C. core-preparation; D. blade production. Indicated are the products that result at each stage in the working. Drawing by D. Stapert/J.M. Smit (B.A.I.).

4.3. The artefacts other than tools

4.3.1. *General*

There is a total number of 420 flint artefacts that have not been modified further. For the numbers of several main categories within this material see table 4.

The flint material of Texel is clearly the product of a technology for the manufacture of blades, as is generally the case with Upper Palaeolithic sites. The blocks, flakes and splin-ters can be regarded for the most part as the waste products of this technology. The process is shown schematically in figure 11. The Upper Palaeolithic blade technology has already been described extensively in the literature (also experimentally), so it is not necessary to go into the subject in any detail here (see i.a. Bordes & Crabtree, 1969; Crabtree, 1972; Newcomer, 1975; Tixier, 1972).

The scheme is intended to indicate the diffe-rent stages in the process, and the products that result at each stage.

Table 8. Blades and blade fragments.

	total no.	not broken	proximal fragm.	medial fragm.	distal fragm.	burnt
normal blades	141	18	44	35	44	7
decortication blades	14	5	2	4	3	1
core-preparation blades	14	3	3	1	7	0
core-correction blades	1	1	0	0	0	0
plunging blades	3	1	0	0	2	0
total	173	28 (= 16.2 %)	49	40	56	8

4.3.2. *Blades and blade fragments* (total no. 173)

The blades can be subdivided as indicated in table 8.

4.3.2.1. *Normal blades* (total no. 141)

Only 18 normal blades are complete (of which 1 has been in contact with fire). Good long blades do not occur particularly frequently. The mean length is only 3.77 cm (S.D. 1.02), the mean width (measured in the middle) is 1.38 cm (S.D. 0.35), and the mean maximum thickness is 0.45 cm (S.D. 0.18). All complete normal blades show to a greater or lesser extent some fine splintering along the edges, that can be regarded as traces of use. Of the 123 blade fragments only two pairs fit together (medial/distal and proximal/distal) so it is clear that a large proportion of the material has not been recovered (see also under 3).

The blade fragments are divided into proximal (fragments with percussion bulb), medial (fragments bounded by 2 fractures) and distal. The numbers and several average measurements of the blade fragments are in table 9 (as a rule the burnt specimens were excluded when measurements were taken).

There are a number of conceivable ways in which blades could break, such as:
a. when being struck off from the core, b. as

Table 9. Numbers and average measurements of the blade fragments.

	no.	mean length	S.D.	mean width	S.D.	mean thickness	S.D.
proximal	44	2.50	0.90	1.19	0.31	0.38	0.15
medial	35	2.67	0.82	1.32	0.37	0.39	0.12
distal	44	3.24	0.88	1.18	0.35	0.41	0.17

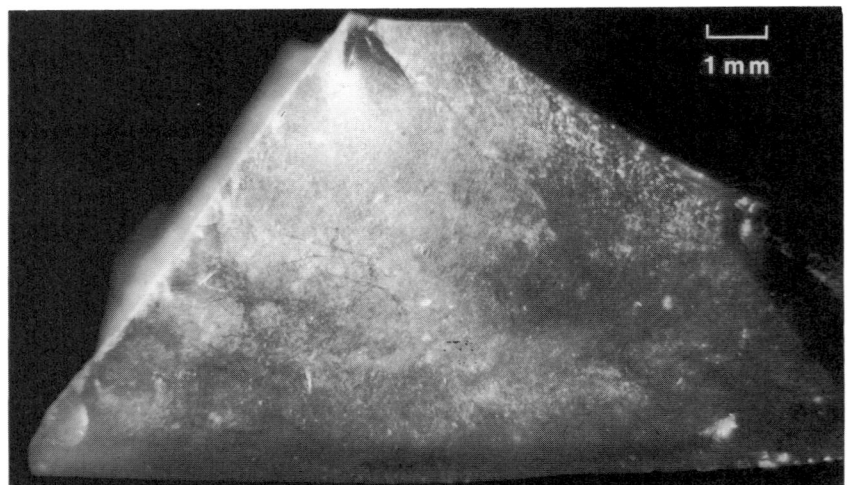

Fig. 12. The fracture surface of a proximal blade fragment. Dorsally a small percussion cone is visible, presumably caused by a blow with a hammer stone on a dorsal ridge of the blade. Scale in mm. Photo taken with a stereomicroscope by D. Stapert (B.A.I.).

a result of use, c. by being trodden underfoot, d. as a result of (secondary) frost-splitting, e. intentionally.

Secondary frost-splitting can with certainty be assumed to be one of the causes of fracture; this would then have occurred mainly during the Late Dryas period (the last 1000 years of the Weichselian), as is evident, for example, from the excavation results of the Hamburg site Oldeholtwolde (Stapert, in prep.).

A number of fractures within the blade material from Texel are strongly suggestive, however, of having originated intentionally. These fractures show a more or less distinct small percussion cone (fig. 12), and often also a "lip" opposite to it. These fractures mostly appear to have been created by a blow with a hammer stone on a dorsal ridge. In some cases fine splintering is visible near one of the fracture angles, so one gets the impression that the fracture was made with the aim of subsequently making use of the sharp but usually also strong fracture angles as "working edges" (possibly in a fashion analogous to the way in which burins were used).

There are also a number of cases in which a dorsal ridge is markedly damaged, sometimes over a distance of more than 1 cm, close to a fracture. In these cases the fractures may also be connected with a method of use (unknown to us) as a result of which the damaged ridges originated.

Assuming that a number of fractures originated intentionally (with the aim of using the fracture angles produced in this way), it does not seem illogical to expect that the lengths of the proximal and distal blade fragments thus produced do not show a normal distribution, but rather a bimodal one. It would have been attempted to keep the length of a blade fragment as long as possible in order to facilitate the manipulation of the "tool". In the frequency diagrams (fig. 13), in which numbers per length class of 0.5 cm are indicated, this expectation appears to be confirmed: the distributions of the proximal and distal fragments are more or less bimodal.

Also worthy of note is the fact that the proximal fragments are distinctly smaller than the distal ones. A hypothesis could be that for the production of the fractures there was a preference for a spot close to the proximal end of the blade, perhaps because the thickest part of the blade is usually present there, so that the fracture edges created will also be more massive.

We are concerned here with such small numbers, however, while moreover the other possible ways mentioned in which fractures formed cannot be excluded, that the hypotheses

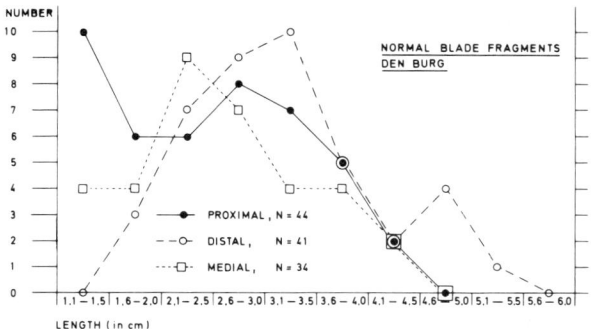

Fig. 13. Frequency diagrams of the lengths of the normal blade fragments, measured in classes of 0.5 cm. Drawing by D. Stapert / J. M. Smit (B.A.I.).

mentioned cannot be satisfactorily tested. The occurrence of intentional fractures seems to me to be certain, however, in view of the presence of a number of distinct percussion cones.

Finally it can be said that most of the blade fragments also show distinct splintering along the edges (edge damage), that can be explained as traces of use.

4.3.2.2. Decortication blades (total no. 14)

Decortication blades are blades of which the dorsal face consists for more than 3/4 of old faces (cortex and/or frost-split faces). Of these blades only about half show distinct splintering along the edges. There are 5 complete specimens (mean length 4.34 cm), 2 proximal, 4 medial and 3 distal fragments (mean length 3.82 cm). The mean width of all specimens is 1.38 cm (S.D. 0.52), the mean thickness 0.60 cm (S.D. 0.30).

4.3.2.3. Core-preparation blades (total no. 14)

Core-preparation blades are the first blades that are struck off from the core, in addition to the decortication blades. They are characterized by the occurrence of dorsal preparation, that may be one- or two-sided. This pre-

paration served to make an initial straight ridge over the whole length of the core, that can "lead" the blade during the process of splitting it off from the core (see Crabtree, 1972). Only three specimens are complete (lengths 3.8, 3.7 and 2.8 cm). In addition there are 3 proximal, 1 medial and 7 distal fragments. The average length of the broken pieces is 3.85 cm. This is more than the length of the complete normal blades; the same applies for the (broken) decortication blades. This is understandable, because these two types of blades originate at the beginning of the process shown in figure 11. Later the blades become gradually shorter, partly because the striking platform has to be renewed now and then. The mean width of all specimens is 1.21 cm (S.D. 0.37), the mean thickness 0.71 cm (S.D. 0.29). Core-preparation blades are thus generally longer, thicker and narrower than normal blades. For some tool-types, for which a relatively greater thickness is of importance, these properties make them more suitable than normal blades. Thus (at other sites) scrapers are often made out of core-preparation blades, as are *Zinken* too.

4.3.2.4. Core-correction blades, and plunging blades (total no. resp. 1,3)

Core-correction blades are usually thick blades, that were generally struck off for the purpose of removing the scar of a previous, unsuccessful blade. Here we are mainly concerned with blades with a "step fracture" or "hinge fracture" distally, that left behind a deep negative c. halfway across the core, as a result of which further normal exploitation of the core became impossible. This could be corrected by striking off a thick blade by means of a hammer stone, preferably from a striking platform situated opposite, so that the unwanted negative was removed at one stroke. One typical example is present, that is unbroken. The length is 5.7 cm, the width 2.1 cm and the thickness 1.1 cm.

In addition there are 3 so-called plunging blades, also known as *lames outre-passés* (see

Tixier, 1963). These are blades, of which the ventral face turned inwards when the blade was being struck off the core, so that sometimes a considerable part of the core (opposite the striking platform used) was removed. Often the distal part of such blades consists of part of an opposite striking platform. It is clear that the cores concerned generally became unusable as a result.

There are three distinct examples, of which 1 is complete (length 6.0 cm, width 2.1 cm, thickness 1.5 cm) and 2 distal fragments (lengths 5.0 and 4.4 cm, widths 2.2 and 1.6 cm, thicknesses 1.8 and 1.1 cm).

Both the core-correction blade as well as the three plunging blades appear to have been used nevertheless, in view of the presence of fine splintering along the edges.

4.3.3. *Cores and core fragments* (total no. 12)

Within the material from Texel 12 cores or core fragments are present. As already mentioned under 4.2., these include 4 specimens that are sometimes described in the literature as "core scrapers" or "core burins"; these specimens are classified here however as cores, and are described under 4.3.3.2. The other 8 specimens are described under 4.3.3.1.

4.3.3.1.   *"Normal" cores*

Out of the 8 specimens one is burnt. This core has one distinct striking platform. Two cores are fragmentary; while being used they fractured along (already existing) internal frost-cracks. Both these cores have one distinct striking platform.

Of the remaining 5 cores two appear to be pretty well "used up" (residual cores). One of these has 2 opposite striking platforms, the other one. Three cores are not used up – only a few blades have been removed from them, and still more blades could be struck off. Two have one distinct striking platform, the third has 2 striking platforms that are opposite each other.

The mean length of these 8 cores (measured in the striking direction of the blade negatives that are present) is 5.9 cm.

4.3.3.2.   *"Core scrapers" and "core burins"* (total no. 4; fig. 14, nos. 1-2; fig. 15, nos. 1-2)

This is a problematical category, that is sometimes included with the "tools". Negatives of "burin blows" on cores or core fragments can be regarded however as the scars of unsuccessful blades (e.g. blades that stopped too soon). "Retouching" along the edges of cores can often better be interpreted as preparation retouch, for example of striking platforms or, more often, for making an initial straight ridge on the core, to facilitate striking off the first blades (core-preparation blades). Such core-preparation blades do indeed occur within the material from Texel (see under 4.3.2.3.), so it is clear that this technique was applied here for the preparation of cores. The "core scrapers" and "core burins" are therefore included by some in the same category as cores, as is done here too. A solution to this problem could only be provided by an investigation as to the presence or absence of traces of use on any supposed "working edges" on cores.

Within the material from Texel 4 such specimens are present. They are discussed here separately, and illustrated (the specimens are illustrated in the same way as the tools).

The specimen in fig. 14, no. 1, can very well be regarded as a core with 2 striking platforms opposite each other; it is made out of a flat frost-split piece. Remnants of core-preparation are present along both long edges. One reason for possibly regarding this specimen as a "core burin" is the fact that there are a number of very small blade negatives present. But as stated above these could very well represent unsuccessful blades, and in my opinion this is most probably the case.

The second specimen (fig. 14, no. 2) is comparable with the previous one. Here too it is more likely that we are concerned with core-preparation rather than with scraper retouch, and with negatives of unsuccessful blades rather than with negatives of burin spalls. The core has one distinct striking platform.

In the case of the third specimen (fig. 15, no. 1) the "scraper retouch" can best be regarded as "reduction": removal of the little edges that stick out between the proximal extremities of the blade negatives. Similarly the negative of the "burin blow" can best be regarded as the result of a blade that turned out to be too short. The specimen can well be interpreted as a broken core, which probably fractured as a consequence of the presence of an internal frost-crack that was already present. It is not clear whether there was originally a second striking platform present.

The fourth specimen (fig. 15, no. 2) also shows "scraper retouch", that in my opinion can better be regarded as core-preparation. This core has one distinct striking platform.

The length of the 3 non-fragmentary cores are 6.5, 4.7 and 5.3 cm, respectively. Clear use traces were not observed, but only a stereomicroscope was used so this cannot be taken as definite proof for the non-tool character of these pieces.

4.3.4. *Flakes, splinters, nodules and blocks*

The remaining flint artefacts consist of 194 flakes or flake fragments, 33 splinters (smaller than 1 cm), 4 nodules and 4 blocks.

Among the flakes there are also specimens that are unsuccessful blades, but that cannot be included in the category of blades because they are too short or too irregular in shape. Moreover among the flake fragments a number of blade fragments will also be present that will not be recognizable as such on account of their being too fragmentary.

Among the splinters there are no distinct broken-off fragments of tools (such as borer tips, etc.).

Nodules are pieces of flint that have only natural surfaces (cortex, frost-split faces) on their exterior. These may be pieces of flint that were brought to the site with the aim of using them as cores, without this having

happened. The limited size of the 4 specimens from Texel argues against this however (max. lengths 5.8, 5.2, 6.0 and 3.0 cm — mean 5.0 cm; widths 3.5, 2.4, 4.4 and 1.2 cm — mean 2.9 cm). For three of the four specimens another possibility is conceivable, i.e. that when a piece of flint, for intended use as a core, was first tested and prepared, it fell apart along internal frost cracks that were already present. For the fourth (very small) specimen this is not possible, because the frost-split faces are all old (with wind gloss).

Blocks are pieces of flint with external surfaces that are natural for the most part, but that also show traces of working by man (without there being any distinct percussion bulbs). One of the 4 specimens (max. length 6.1 cm, width 2.3 cm) probably originated as the result of the splitting of a core along internal frost cracks; the other three are small (1.5-3 cm) fragments.

5. THE STONES

A total number of 28 stones were recovered during the excavation near the Beatrixlaan; 19 of these are indicated on the excavation plan (fig. 2). On this drawing no clear pattern is visible in the horizontal distribution of these stones.

The stones are generally rather small, the largest having a maximum length of 8.8 cm, while the smallest measures maximally 2.4 cm.

We are concerned here with angular lumps of rock, that without exception give the impression of having split off larger pieces. In table 10 the stones are roughly subdivided according to rock type, and a few measurements are given (the width is measured perpendicularly to the line of maximum length, and the thickness perpendicular to the plane in which length and width are measured).

Stones are found at many Upper Palaeolithic sites, in the Netherlands especially those belonging to the Hamburg tradition. As for the possibility of a "tent-ring", as had been postulated for some sites in other coun-

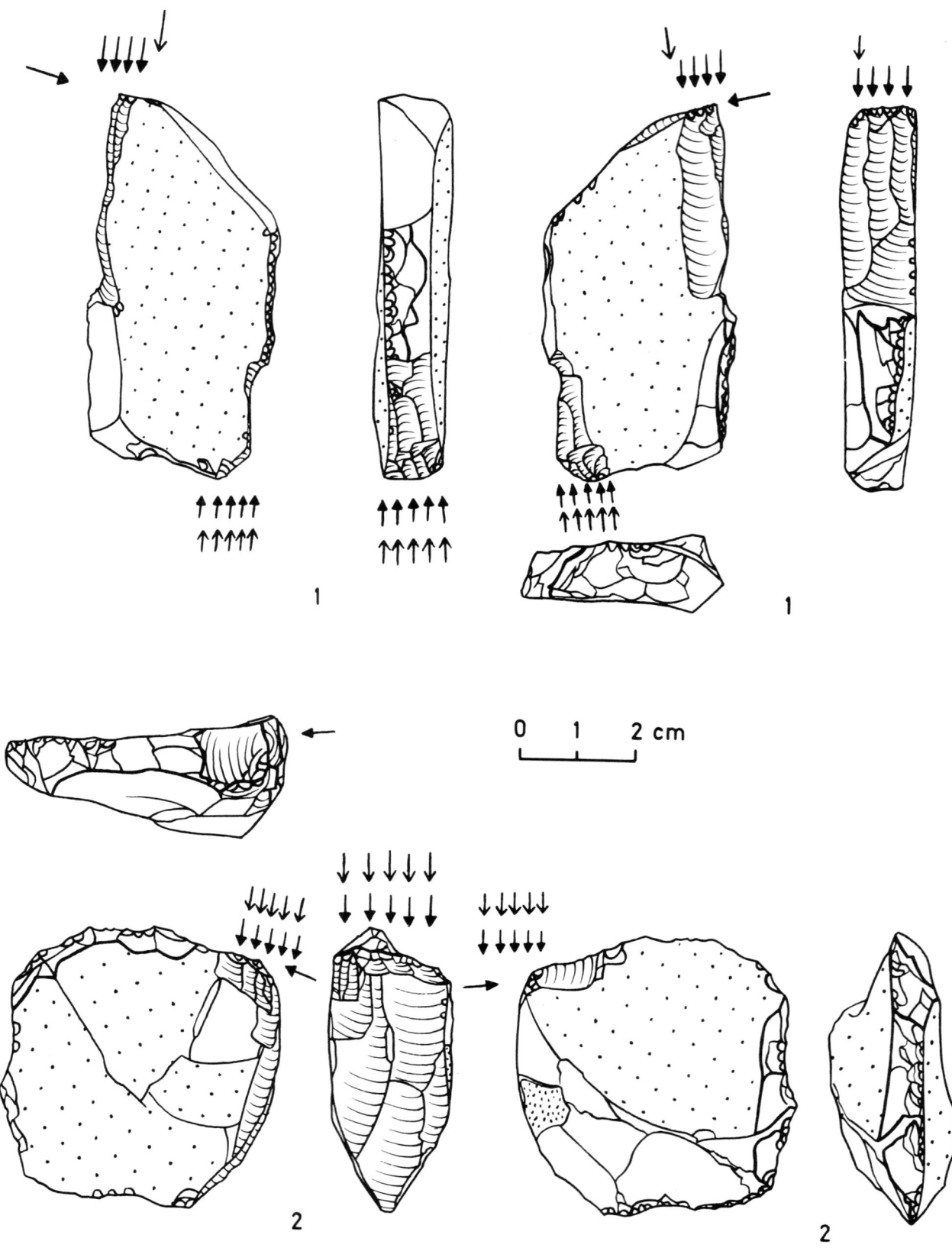

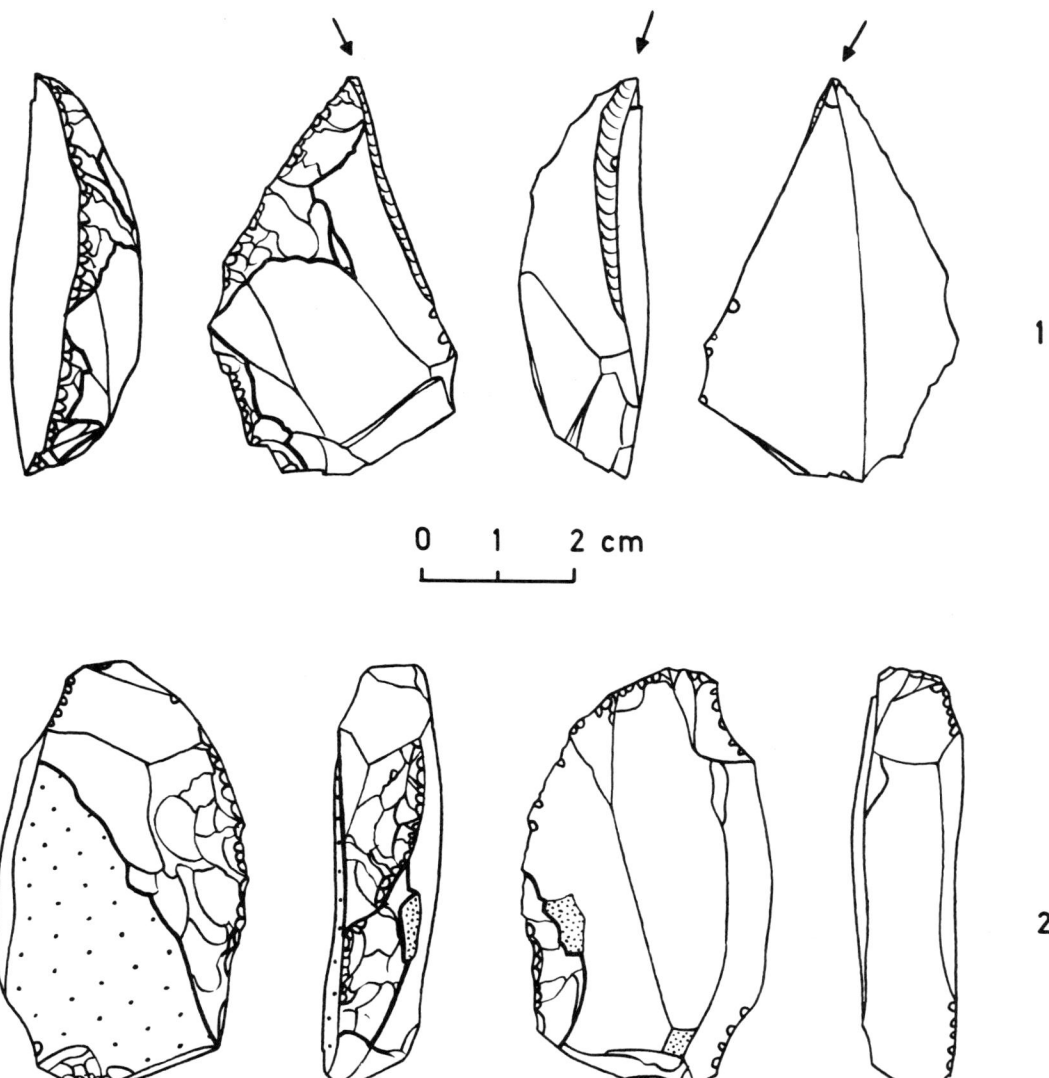

Fig. 15. Texel, "core tools". 1,2. cores with core-preparation retouch ("scraper retouch"), no. 1 also showing a scar of a small narrow blade ("burin spall"). Drawings by D.Stapert/J.M.Smit (B.A.I.).

tries, here there is no evidence of this as far as can be seen on the excavation plan, while moreover most of the stones appear to be too small for such a function.

At the Hamburg site of Oldeholtwolde a

Fig. 14. Texel, "core tools". 1,2. cores made of flat frost-split pieces of flint. Core-preparation retouch and short scars of unsuccessful blades give the impression of the pieces being "core scrapers/burins". Drawings by D. Stapert/J.M. Smit (B.A.I.).

large hearth was excavated, that was "paved" with stones (consisting e.g. of quartzites and granites), under which charcoal was present (Stapert, in prep.). These stones were generally flat, with an average thickness of about 2 cm. Although the average thickness of the stones from Texel is also about 2 cm, it cannot be said that we are certainly concerned here too with the remains of a hearth. In the first place, in the case of Texel flat stones occur far less consistently than is the case at Olde-

Table 10. Stones.

	max. length (cm)	width (cm)	thickness (cm)	weight (gr)
quartzite-like sandstone	6.7	5.8	0.8	44.7
	3.2	1.8	0.5	4.5
very coarse sandstone	5.8	3.2	1.6	44.9
	5.2	2.8	1.1	18.1
	3.8	2.3	1.6	11.2
	4.7	1.9	0.9	8.3
	2.7	1.7	0.7	2.5
rose-red granite	8.1	7.1	4.7	236.4
	8.8	7.8	2.6	136.5
	5.3	4.3	3.7	96.6
	5.8	4.7	2.8	85.2
	5.9	4.2	3.7	68.5
	5.9	4.4	2.5	57.0
	5.2	4.2	2.8	47.3
	3.6	2.8	1.5	12.1
	3.1	1.9	1.8	9.7
	3.6	2.5	1.1	8.4
white granite	5.3	5.1	2.4	61.4
	6.0	4.0	2.8	60.2
	6.0	4.3	2.9	59.4
	5.8	5.3	1.6	46.7
	4.5	4.5	1.7	32.5
	4.0	3.4	1.7	23.4
	4.7	2.8	1.4	19.7
	2.9	2.8	1.7	11.9
	4.0	2.1	1.4	8.8
	2.8	2.1	1.3	6.6
	2.4	1.9	0.9	4.6
			total weight	1227.1
mean	4.9	3.6	1.9	43.8

holtwolde. What is more important is that on the stones from Texel no distinct black coloration (carbonization) is visible, as occurs on the hearth-stones from Oldeholtwolde. Moreover the stones from Texel were not found in any concentration, as can be seen from the excavation plan. Nor was any charcoal found during the excavation. A few stones do appear to have some vague red coloration, however, while in addition some fracture surfaces could have originated as a result of heat. Therefore the possibility that there was a hearth at this site cannot be completely excluded, but the evidence is too slight for any positive conclusion to be drawn.

None of the stones shows in addition any clear traces of use, e.g. hammer stones, polishing stones, etc.

6. A FEW REMARKS CONCERNING THE DISTRIBUTION AND DATING OF THE HAMBURG TRADITION

The site on Texel is the most westerly one within the entire known distribution area of the Hamburg tradition (see e.g. *Karte 2* in: Tromnau, 1975). It is situated more than 50 km to the west of the dense concentration of sites on the sandy soils of the provinces of Drenthe and Friesland. (A few years ago a site was also discovered in the extreme SW of Friesland: Elzinga, 1972).

The location of the site indicates that in any case part of the North Sea (at that time for the most part dry land – the Late Glacial coastline would then have lain as far north as the northern tip of Denmark), to the west and north of Texel, must have formed part of the region that was occupied by the Hamburgian hunters.

As far as the west is concerned, it is still not yet clear to what extent there was any relation between the Hamburg tradition and some sites of the English "Late Upper Palaeolithic" (for an excellent recent survey see Campbell, 1977). Tool types such as shouldered points and *Zinken*, that are regarded as "indicator fossils" of the Hamburg tradition (actually incorrectly as a rule), occur in a number of English sites. Typical examples of both tool types are present for example within the material from the site of Gough's Cave (Campbell, 1977, vol. II, figs. 119-126), in addition to such tools as Creswell and Cheddar points, and also *Federmesser*. Notably shouldered points with a ventrally retouched shoulder (e.g. Campbell, 1977, vol. II, fig. 122,7) are strongly reminiscent of Hamburgian points. Also the notched blades that are present, long end scrapers, oblique truncations and truncation burins could well fit into the Hamburgian. In the Netherlands and Germany *Federmesser* (Tjonger points) and tanged points are found at late Hamburgian sites, but the Creswell and Cheddar points that are present in Gough's Cave would naturally constitute a problem if one were to ascribe

this site to the Hamburg tradition. But, as Campbell also mentions, it is very well possible that the large collection of artefacts from Gough's Cave represents a number of traditions. In Robin Hood's Cave too shouldered points have been found, also during the excavation by Campbell (1977, vol. II, figs. 152-153). It is even possible that shouldered points form the only type of point for one stratigraphical level (Campbell, 1977, vol. I, p. 174), but here we are concerned with very small numbers.

The open-air site of Hengistbury Head has also yielded typical, albeit rather large, shouldered points as well as tanged points, also of considerable size (Mace, 1959; Campbell, 1977 – for drawings see vol. II, figs. 157-165). The most typical shouldered and tanged points occur within the material excavated by Mace (site C1), and that from site A, while within the collection recovered by Campbell (site C2) especially Creswell points are conspicuous. Distinct *Zinken* are not present however.

The brief references made above to the English material serve to support the idea that it is not impossible that the Hamburg tradition is also represented in England, or in any case is related to a number of sites of the Late Upper Palaeolithic of Campbell. For further discussion on this problem it is important that the Dutch "Creswellian" sites be published (especially Zeijen, Siegerswoude II and Neer II), and in fact work is currently in progress with this aim in view. The impression that we get at the moment however is that these certainly do not appear to be less closely related to Hamburgian material than most of the Tjonger find complexes. In my opinion therefore it is not completely correct simply to classify the (Dutch) Creswellian together with the *Federmesser* tradition *s.l.* as proposed by Paddayya (1971), to my way of thinking this group of sites can better be considered separately, at least for the meantime.

As for the north: Texel is not the northernmost Hamburg site. If we disregard the isolated shouldered point, found near Ljerlev in Jutland, Denmark (Becker, 1970), the most northerly known site (according to Tromnau, 1975) is that at Ahrenshöft in Schleswig-Holstein (G.F.R., see Hinz, 1954). This could indicate that a strip of about 100 km to the north of the Dutch Wadden Islands belonged to the region of the Hamburg tradition. In that case the region frequented by the Hamburgian hunters would clearly not have extended as far as the coastline of that period. Moreover, even if that was indeed the case, we must assume that the sites in the Northern Netherlands lay too far from the coast to enable us to suppose that the people who used them (for a certain part of the year) travelled as far as the coast (in another part of the year). The distance (300-350 km) appears to be too great for this. The extent of the annual seasonal migration of most (sub-) recent hunters/gatherers is usually less than 100 km, and seldom more than 200 km (see also the discussion in Campbell, 1977, vol. I, p. 32 and vol. II, maps 5-8). For the meantime we must therefore assume that the Hamburgian hunters had an economy that was not coast-bound, and that they thus spent the whole year living inland.

Although Texel cannot serve as the best evidence, there are nevertheless also geological indications at this site of a dating *after* the Bølling interstadial (see under 2.). For a number of other Hamburgian sites such a dating appears to be certain, for example Oldeholtwolde and Luttenberg. At Oldeholtwolde the Hamburg finds were present in Younger Cover-sand I, that according to the geologists (e.g. Ter Wee, 1966) was deposited during the stadial between the Bølling and Allerød interstadials, a few decimetres below the "layer of Usselo" (Late Allerød soil) that has formed at the top of this cover-sand layer. Oldeholtwolde can therefore be dated to relatively shortly before the beginning of the Allerød interstadial. The finds from this site include tanged points and *Federmesser*, which corresponds to the expectation of both Bohmers (1947) as well as Tromnau (1975) for "Late" Hamburg sites. Moreover it is clear that the finds were left behind while the (eolian) sand

was still being deposited, thus during a cold (stadial) period. In my opinion most of the Dutch Hamburg sites, including Texel, are to be dated in this stadial. This is later than is generally assumed (Bokelmann, 1979, dates the Hamburg tradition in Northern Germany, for example, completely in the Bølling interstadial).

Therefore there most probably does not exist a "gap" in time between the appearance of the Hamburg tradition and that of the Tjonger (*Federmesser*) tradition.

7. SUMMARY

In this article is described the material from a small site of the Hamburg tradition on the Wadden island of Texel, in the extreme northwest of the Netherlands. The finds were present in a layer of cover-sand, and can be dated with some degree of certainty in the stadial between the Bølling and Allerød interstadials (Early Dryas). The material that was found includes, in addition to 28 mostly smaller stones, 447 flint artefacts. Among these are 27 specimens that can be described as "tools" (6%). What is remarkable here is the relatively large proportion of points (10: in all cases broken or damaged). For the other tools see table 5. The material recovered from this site is certainly only a part of what was once present: an unknown but probably considerable proportion of the original material has been lost to us. As appears from the excavation (carried out by the R.O.B., Amersfoort), the concentration was situated in an area measuring approximately 10 x 8 metres.

8. REFERENCES

BECKER, C.J., 1970. Eine Kerbspitze der Hamburger Stufe aus Jütland. In: K. Gripp, R. Schütrumpf &. H. Schwabedissen (Hrsg.), *Frühe Menschheit und Umwelt*, Tl. I: Archäologische Beiträge (= Fundamenta A/2). Köln, pp. 362-364.
BOHMERS, A., 1947. Jong-palaeolithicum en vroeg-mesolithicum. In: *Een kwart eeuw oudheidkundig bodemon-*

derzoek in Nederland; Gedenkboek A.E. van Giffen. Meppel, pp. 129-201.

BOKELMANN, K., 1979. Rentierjäger am Gletscherrand in Schleswig-Holstein? *Offa* 36, pp. 12-22.

BORDES, F., 1965. Utilisation possible des côtés des burins *Fundberichte aus Schwaben* N.F. 17 (= Festschrift G Riek), pp. 3-4.

BROADBENT, N.D. & K. KNUTSSON, 1975. An experimental analysis of quartz scrapers; results and applications. *Fornvännen* 70, pp. 113-128.

CAMPBELL, J.B., 1977. *The Upper Palaeolithic of Britain.* 2 vols. Oxford.

CRABTREE, D., 1972. *An introduction to flintworking.* Pocatello, Idaho.

DEWEZ, M., 1970. Contribution à la technique lithique du paléolithique supérieur final. *Bulletin de la Société Royale Belge d'Anthropologie et de Préhistoire* 81, pp. 39-59.

DOPPERT, J.W.Chr., G.H.J. RUEGG, C.J. VAN STAAL-DUINEN, W.H. ZAGWIJN & J.G. ZANDSTRA, 1975. Formaties van het Kwartair en Boven-Tertiair in Nederland. In: W.H. Zagwijn & C.J. van Staalduinen (eds.), *Toelichting bij geologische overzichtskaarten van Nederland.* Haarlem, pp. 11-56.

ELZINGA, G., 1972. Stichting het Fries Museum, jaarverslag 1971. *Jaarverslagen Fries Genootschap; Fries Museum 1971*, pp. 124-125.

HAMMEN, J. VAN DER, G.C. MAARLEVELD, J.C. VOGEL & W.H. ZAGWIJN, 1967. Stratigraphy, climatic succession and radiocarbon dating of the Last Glacial in the Netherlands. *Geologie en Mijnbouw* 46, pp. 79-95.

HINZ, H., 1954. *Vorgeschichte des nordfriesischen Festlandes.* Neumünster.

KEELEY, L.H. & M.H. NEWCOMER, 1977. Microwear analysis of experimental flint tools: a test case. *Journal of Archaeological Science* 4, pp. 29-62.

KEELEY, L.H., 1980. *Experimental determination of stone tool uses; a microwear analysis.* Chicago/London.

MACE, A., 1959. An Upper Palaeolithic open-site at Hengist-bury Head, Christchurch, Hants. *Proceedings of the Prehistoric Society* 25, pp. 233-259.

ODELL, G., 1978. Préliminaire d'une analyse fonctionelle des pointes microlithiques de Bergumermeer (Pays-Bas). *Bulletin de la Société Préhistorique Française* 75, pp. 37-49.

PADDAYYA, K., 1972. The Late Palaeolithic of the Netherlands – a review. *Helinium* 11, pp. 257-270.

RUST, A., 1937. *Das altsteinzeitliche Rentierjägerlager Meiendorf.* Neumünster.

STAPERT, D., 1979. Zwei Fundplätze vom Übergang zwischen Paläolithikum und Mesolithikum in Holland. *Archäologisches Korrespondenzblatt* 9, pp. 159-166.

STAPERT, D., in prep. A late Hamburgian site with hearth at Oldeholtwolde (prov. of Friesland, Netherlands). *Palaeohistoria.*

STURDY, D.A., 1975. Some reindeer economies in prehistoric Europe. In: E.S. Higgs (ed.), *Palaeoeconomy; being the second volume of Papers in Economic Prehistory*, Cambridge, pp. 55-95.

TIXIER, J., 1963. *Typologie de l'épipaléolithique du Maghreb.* Paris.

TIXIER, J., 1972. Obtention de lames par débitage "sous le pied". *Bulletin de la Société Préhistorique Française* 69, pp. 134-139.

TROMNAU, G., 1975. *Neue Ausgrabungen im Ahrensburger Tunneltal* (= Offa-Bücher Bd. 33). Neumünster.

VEENSTRA, H.J., 1970. Quaternary North Sea coasts. *Quaternaria* 12, pp. 169-184.

WEE, M.W. ter, 1962. The Saalian glaciation in the Netherlands. *Mededelingen Geologische Stichting* N.S. 15, pp. 57-77.

WEE, M.W. ter, 1966. *Toelichting bij de Geologische Kaart van Nederland 1:50.000, Blad Steenwijk Oost (16 O).* Haarlem.

WOLTERING, P.J., 1975. Occupation history of Texel, I. The excavations at Den Burg: Preliminary report. *Berichten van de Rijksdienst voor het Oudheidkundig Bodemonderzoek* 25, pp. 7-36.

EEN-SCHIPSLOOT – THE GEOLOGICAL-PALYNOLOGICAL INVESTIGATION OF A TJONGER SITE

W. A. Casparie and M. W. ter Wee*

CONTENTS

* Rijks Geologische Dienst, District Noord, Oosterwolde (Netherlands)

1. INTRODUCTION

In 1969-1971 P. Houtsma and J. Schilstra carried out the excavation of a site of the Tjonger tradition along a ditch known as the Schipsloot near Een, *gemeente* of Norg (province of Drenthe), fig. 1. The report on this excavation is published in this volume, see Houtsma, Roodenberg and Schilstra. The geology of this region has been studied in detail and two organic deposits have been analyzed palynologically, by the second and first author respectively. The aim of these investigations was to gain a better insight into the genesis of this region and into the geological and vegetational development at the time when the settlement was occupied, and also to provide an explanation for a number of curious basin-shaped structures and to obtain a more accurate dating for some of these phenomena. The interpretation of the data presented in this publication has been arrived at in close collaboration with the two excavators, in particular with P. Houtsma.

The drawings were made by W.J. Dijkema (B.A.I.) and M.A. Smakman (R.G.D.). The photos, except for those provided by the authors, were made by B. Klijnstra (R.G.D.). Mrs. G. Entjes-Nieborg (B.A.I.) typed the manuscript and Mrs. S.M. van Gelder-Ottway translated the article into English. We wish to express our thanks to all those concerned for their cooperation.

2. THE GEOLOGICAL SITUATION

The Tjonger site that concerns us here is situated near the watershed between the valley system of the Tjonger, that flows towards the south-west, and that of the Peizerdiep, that drains the area towards the north.

The primary watershed of these two valley systems became considerably narrower and lower-lying at the time when these valley systems originated. In the Pleniglacial period, when the sea-level was low in the last ice-age, the Tjonger valley increased in extent from the

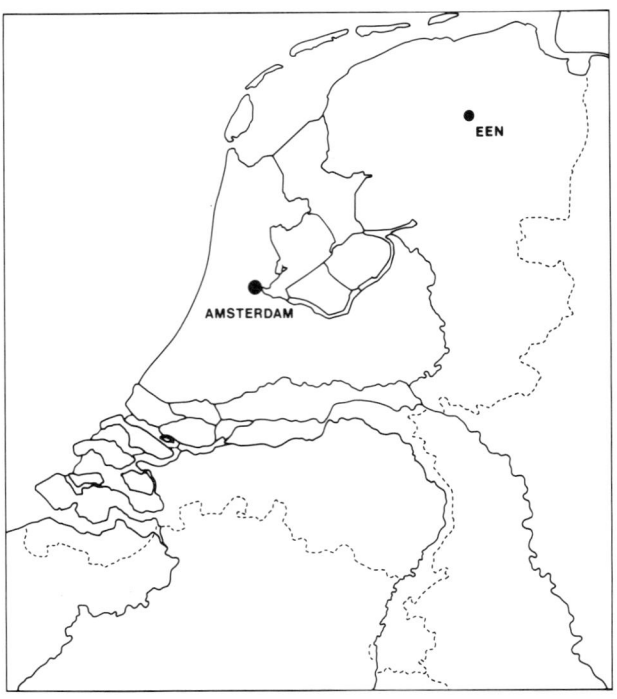

Fig. 1. Location of the Een excavation.

south-west as far as here as a result of headward erosion. From the north the valley system of the Peizerdiep also became more deeply extended back this far as a result of erosion, such that the two valley systems became interconnected and separated merely by a low, narrow watershed.

After this erosion phase, still during the Pleniglacial, both valley systems became filled up with stream sediments, i.e. moderately fine sands with layers of loam and gyttja, and some fine gravel. These sediments fill up the valley systems almost completely, so that on the geological map (fig. 2) it appears as though there is one single valley system; see also the stratigraphical table (below).

In the final phase of the last ice-age, the Late Glacial, the climate was cold but dry. Wind action had a severe effect on the landscape, that was sparsely vegetated at that time, and the sand on the surface was blown about by the wind. On the boulder-clay, on the pre-

Stratigraphic table for Een-Schipsloot.

		10,200	
	Late Dryas stadial		Younger Cover sand II
		11,000	
	Allerød interstadial		Layer of Usselo peat
Late Glacial (Late Weichselian)		11,800	
	Early Dryas stadial		Younger Cover sand I
		12,000	
	Bølling interstadial		
		13,000	
Pleniglacial (Middle Weichselian)		yrs B.P.	stream sediments

glacial sands and also to some extent in the stream valleys a layer of cover sand was deposited. The oldest cover sands here belong to the Younger Cover sand I and form a cover sand ridge in the watershed region (fig. 2). During the Allerød interstadial when the climate was less cold, a soil developed on top of the Younger Cover sand I, that undergoes a transition into a peat-gyttja layer (valley fill) in the low-lying part of the valley system (fig. 2).

The Allerød level has been definitely dated by means of the pollen-analytical investigation (see below) and the Tjonger tradition that has been investigated by Houtsma, Roodenberg and Schilstra.

The somewhat undulating Allerød horizon is an approximate reflection of the morphology of the former cover sand landscape. It is clear that the cover sand ridge in the Younger Cover sand I formed the watershed in Allerød time and that this cover sand ridge, situated high and dry, was an ideal passage for the reindeer-hunters. After the Allerød interstadial there followed the Late Dryas stadial, during which period the Younger Cover sand II was deposited. This cover sand lies like a blanket over the Younger Cover sand I and

accentuates the cover sand ridge already present and the watershed in this locality (fig. 2).

The Schipsloot was dug in historical times. This ditch served to facilitate drainage from the peat region in the upper reaches of the Tjonger, the Eenderveen, into the valley system of the Peizerdiep. As a result of the digging of this ditch the existing watershed was shifted to the beginning of the Schipsloot; about four kilometres south-west of the natural watershed.

3. PALYNOLOGICAL INVESTIGATION OF THE VALLEY FILL

In the low-lying part of the Allerød level the organic valley fill was sampled. The spot where sampling was done, to the west of the site, is indicated in fig. 2 by 1. At this spot the valley fill consists of a complex of thin layers of sandy peat, alternating with thin layers of sand, see fig. 3, that gives an overall picture of the fill. The section that was analyzed comes from the part on the left, near the arrow, where the deposit is richest in organic material. In fig. 4 a detail of the sandy peat layer is shown, just to the right of the

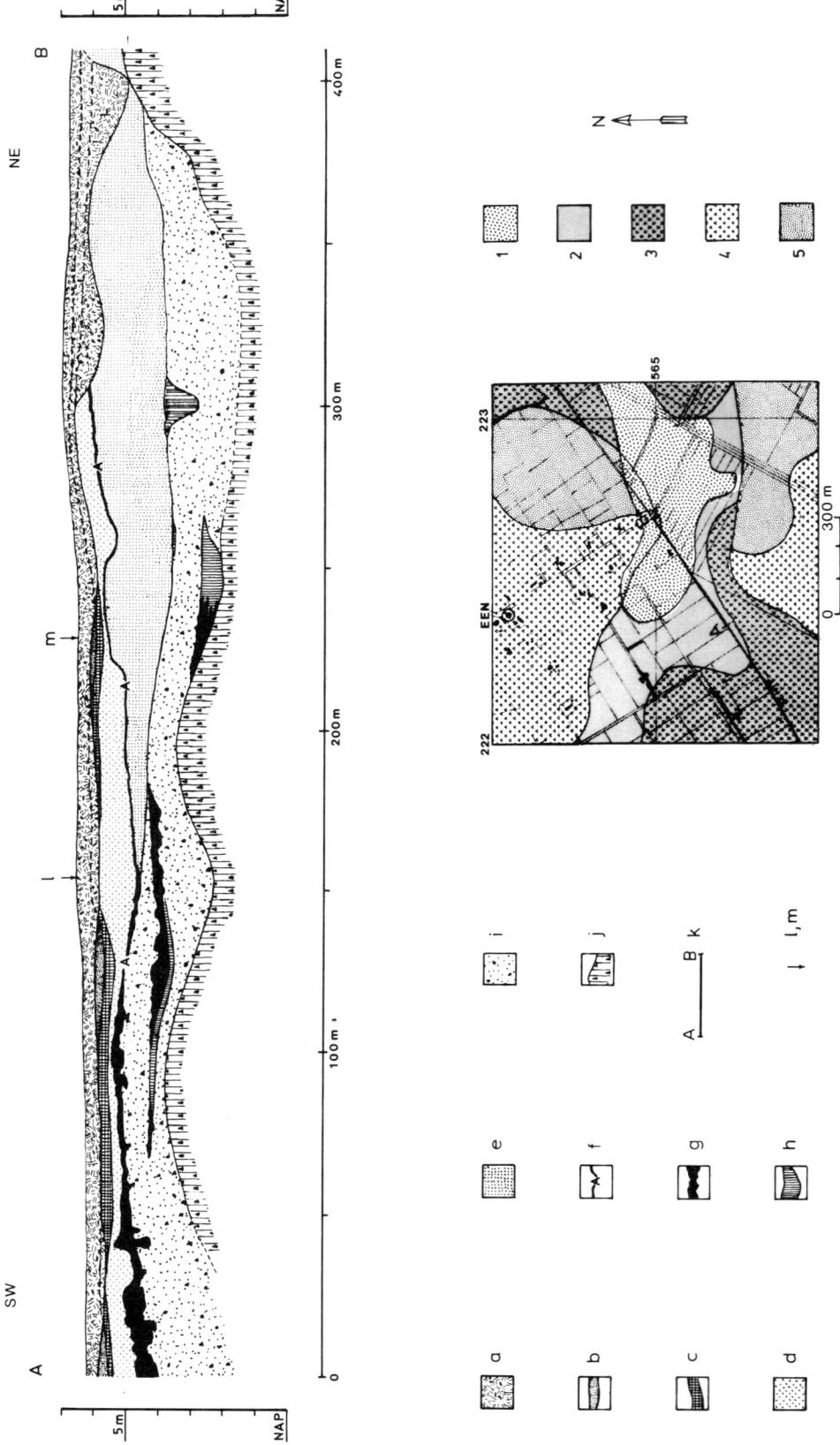

Fig. 2, below: *The area investigated*. Legend: 1. cover sand thicker than 2 m. – 2. stream deposits covered by cover sand thinner than 2 m. – 3. stream deposits covered by Holocene peat. – 4. boulder-clay (basal till) covered by cover sand thinner than 2 m. – 5. Preglacial sands covered by cover sand thinner than 2 m.

Fig. 2, above: *The geological section*. Legend: a. disturbed and/or transported soil. – b. A-horizon. – c. B-horizon. – d. Younger Cover sand I. – f. Allerød horizon. – g. gyttja. – h. fluvial loam. – i. fluvial sands. – j. boulder-clay (basal till). – k. situation of the cross-section (in the lower figure). – l. site of the analyzed peat section in the valley. – m. site of basin I.

spot where sampling was done. The (Late Dryas) frost crack, that originates from the top of the Tjonger culture layer, and the cryoturbate layering indicate a Pleniglacial or Late Glacial age of this deposit, that is c. 50 cm thick and that contains organic material particularly in its lower half. The pollen diagram, fig. 5, relates to this part.

Concerning the preparation of this pollen diagram, as well as of the pollen diagram shown in fig. 12 of the organic filling of one of the basins to be discussed in this article, the following can be said.

From both profiles samples, c. 1 cm thick, of peat or peaty material were prepared in the usual way and analyzed with the aid of a microscope. The results are set out in 'Iversen pollen diagrams'. As a basis for calculation purposes was taken the sum of pollen from trees, shrubs, wind-pollinated herbs, and also of *Empetrum* and the Ericaceae.

The wall from which the profiles were taken is the slope of the Schipsloot, that has an inclination of c. 45°. The scale-bars shown together with the pollen diagrams relate to this ditch slope; therefore they do not give the real thickness of the peat layers.

In the pollen diagram Een Schipsloot-A for the valley fill, fig. 5, the values for *Pinus* (c. 10%) and the fairly high *Betula* values (c. 50%) indicate a Late Glacial age, with *Pinus* being present already in this region. *Juniperus* occurs only sparingly, *Empetrum* is hardly present and also *Calluna* and *Erica* have only small percentages. This places the deposit in the *Betula* phase of the Allerød, before the beginning of the sharp increase in *Pinus*. The somewhat high herb percentages (c. 30-50%) for this period can be ascribed to local sedge vegetations. There is a relatively great abundance of herbs, but the heliophytes show no conspicuous maxima; thus *Artemisia* values are remarkably low. Of the herb types some occur on damp to very damp soils: *Filipendula, Sparganium, Menyanthes, Equisetum*. Several types are indicative of open water: *Batrachium, Myriophyllum, Pediastrum*. The last-mentioned shows no great maxima however. The indi-

cator species of extremely wet conditions are marked in the pollen diagram by means of an asterisk.

We are concerned here with a *Betula* forest, dating from the first half of the Allerød, in a fairly wet location, undoubtedly adjacent to or in the stream valley in question, of which the geological profile fig. 2 shows a cross-section. Between the stream and the forest there was a transitional peaty zone, probably of considerable extent, with sedges and *Sphagnum* present. The absence of extremely high maxima of *Pediastrum* suggests that there was relatively little open water.

The thin layers of sand that are present between the thin layers of peat are undoubtedly deposits of fluviatile origin, in view of the occurrence of secondary pollen such as *Quercus, Picea, Tilia, Acer* and *Taxus*. The spectra 1-5 show a considerable degree of mutual correspondence as against spectra 6-8, that show some degree of mutual correspondence, albeit to a lesser extent. This could indicate that peat formation and the sedimentation of sand took place at a considerably slower rate from spectrum 5 onwards, and that hiatuses are present in the upper part of the profile.

In the diagram a distinct *Pinus* phase with a conspicuous maximum is lacking. Nevertheless it is probable that spectra 6-8 in particular can be placed precisely at the beginning of such a phase; as a result of the conspicuously high (for the Allerød) Cyperaceae values in this section of the pollen diagram the onset of the *Pinus* expansion is, as it were, veiled.

The end of the formation of peat indicates that conditions in the valley system became drier, possibly as a result of the climate becoming drier. This situation favoured the expansion of *Pinus* and at the same time led to wind erosion taking place, for example as occurred in Southeast Drenthe during the *Pinus* phase of the Allerød (Casparie, 1972). In this valley system in any case wind erosion had a predominating effect; the level of the frost crack, that can be dated in a fairly early stage of the Younger Dryas (see fig. 3), is indica-

Fig. 3. The valley fill. The arrow indicates the spot where sampling was done.

tive of a sand deposit of c. 0.5 m after the *Betula* phase of the Allerød.

The fact that conditions became drier in the valley system could also be attributed to the disappearance of the permafrost during the warm Allerød phase, which resulted in a considerable lowering of the ground-water level and the pronounced compaction of the complex of sand and peat layers, as shown in fig. 3. The very shallow character of this valley system, i.e. in comparison to its width, does not exclude the possibility of peat formation on a frozen subsoil.

During the Late Dryas the deposits were affected by cryoturbation.

4. DESCRIPTION OF THE (COVER SAND) BASINS

In that part where the Schipsloot cuts through the cover sand ridge discussed above, in the course of the excavation of this Tjonger site nine round basins were found in the Younger Cover sand I, that were filled with Younger Cover sand II (Houtsma, Roodenberg & Schilstra, this volume). In addition an oval basin was found, that is included here in the category of the round type.

The level at which these basins begin is as a general rule, the uppermost part of the remnant of the culture layer containing the Tjon-

Fig. 4. Detail of the valley fill with the (Late Dryas) frost crack.

ger tools (fig. 6). The basins were already visible in part slightly above the culture level. The basins vary in depth from 0.15-1.00 m, while the diameter varies between 0.50 and 3.00 m. The largest, most spectacular basin, basin I, has the most complete and varied filling (figs. 6,7).

The wall of this basin against the Younger Cover sand I is very sharply defined and is partly accentuated by a thin secondary iron oxide band. At the base of the basin filling a coarse sandy layer is present, like a thick crust, and this continues also against the wall practically as far as the periphery of the basin. This sand is coarsest at the bottom, where the

stones, artefacts and wood also occur, and becomes increasingly finer higher up the slope. This crust-like sand-layer is finely layered, and its layering is more or less parallel to the boundary of the basin (fig. 7). There follows a light brown, fine-grained Hypnaceae-peat lens and from this there is a lateral transition into grey, very fine, possibly somewhat loamy sand with some rust coloration.

The rest of the filling of this basin consists of cover sand containing in the basal part one or possibly two thin irregular lenses with grey, very fine, loamy sand, partly accentuated by bands of rust. The cover sands in the basin correspond to the Younger Cover sand II out-

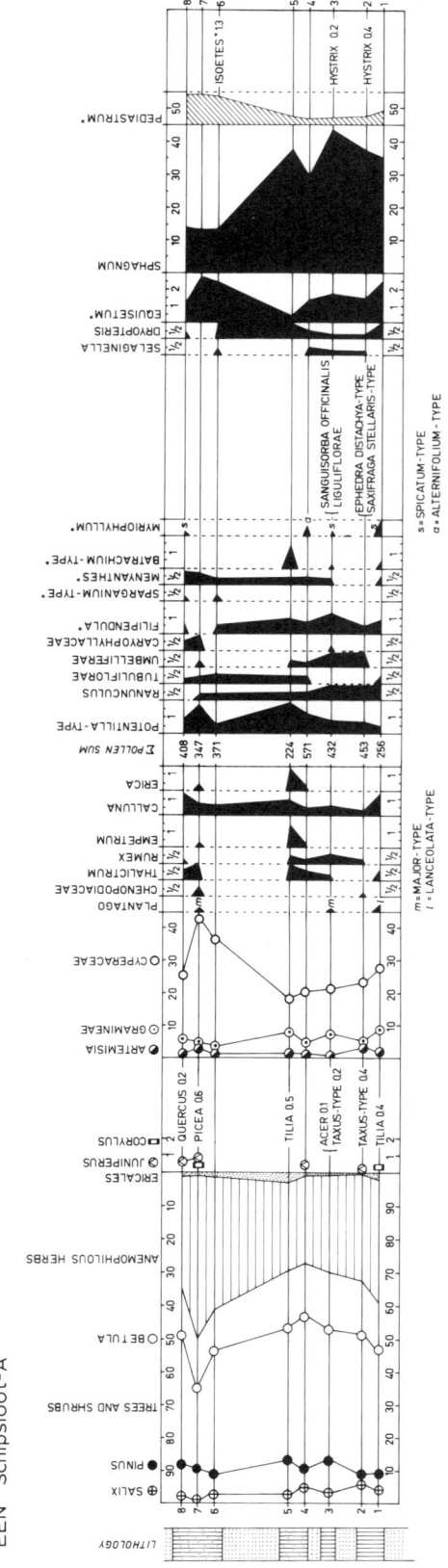

Fig. 5. Pollen diagram for Een Schipsloot-A (valley deposit). For legend relating to lithology: see fig. 12.

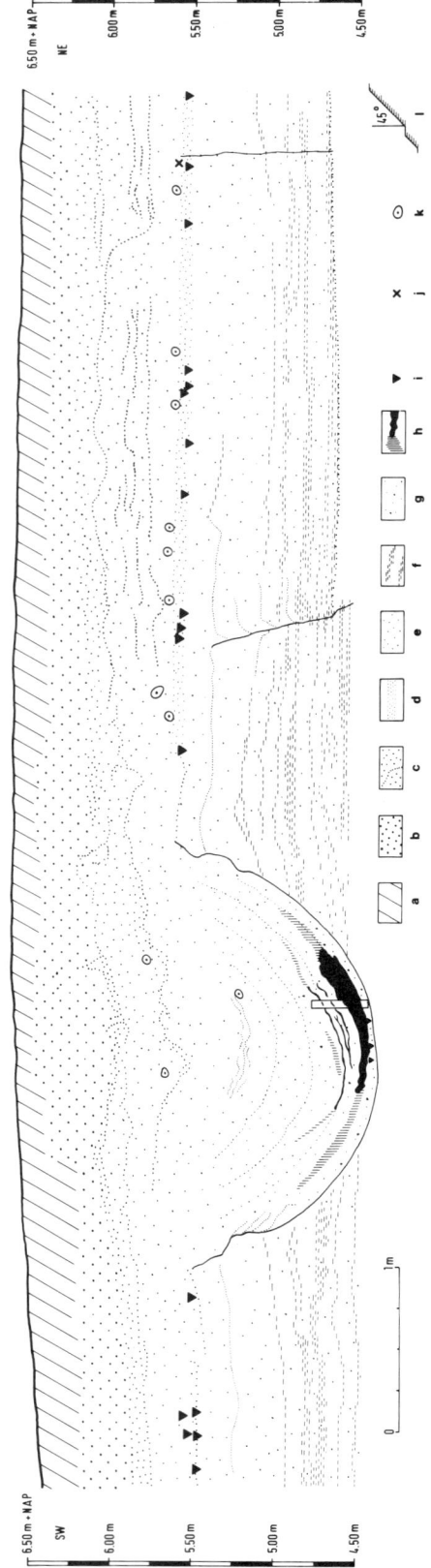

Fig. 6. *Cross-section through basin I.* Legend: a. disturbed and/or transported soil. – b. Younger Cover sand II, humous part of the podsol (A-horizon). – c. Younger Cover sand II, B-horizon with fibres. – d. Younger Cover sand I, slightly bleached layer (± level of the Tjonger tradition). – e. Younger Cover sand I. – f. loam and very loamy sand. – g. coarse sand, occasionally with fine gravel. – h. gyttja (left) and peat. – i. artefacts. – j. bone fragment. – k. fine gravel. – l. indication of the angle of inclination of this section. – Continuous lines: 1. frost cracks. – 2. contour of basin I.

side the basin, with which they form a whole. The layering in the basin is very much disturbed, however, and in some places the layering is more or less vertical. This basin I is the largest that has been found in the course of this excavation. The other basins show a similar structure; they are however much smaller and do not contain a Hypnaceae-peat layer, although in several basins there is a loam layer (see Houtsma, Roodenberg & Schilstra, this volume), that is possibly in the same position as the Hypnaceae-peat layer in basin I. Also the structure in the cover sand can vary; thus the filling of basin II (fig. 10) shows a pattern of layering of the cover sand that is parallel to the contour of the basin (figs. 8,9).

5. PALYNOLOGICAL INVESTIGATION OF THE PEAT FILL OF BASIN I

The location of this basin is indicated by m in fig. 2. The remarkable position of the peat layer is clearly visible in fig. 6. The exact spot where sampling was carried out is indicated here by means of a rectangle. As mentioned previously, under the peat layer a number of stones, several artefacts and a piece of wood were found. This wood, *Pinus*, came from a thick tree, in view of the pattern of annual rings. A C14-dating has been established for the peat: GrN-6341, 10,495±60 B.P. (Lanting & Mook, 1977). The precise thickness and the level of the dated sample are given in the pollen diagram, fig. 12. The diagram was prepared in the same way as that for the valley fill, see section 3 and fig. 5. Here too the indicator species of extremely wet conditions are marked by means of an asterisk.

Concerning the pollen diagram Een Schipsloot-B for the basin filling, fig. 12, the following can be said. The values for *Pinus* (almost 15%) and *Betula* (± 60%) and the herb pollen are indicative of a Late Glacial age for this peat layer. The occurrence of *Empetrum* shows that we are concerned here with the Late Dryas period. The C14-dating points to the same.

Pediastrum has high values; Gramineae and Cyperaceae on the other hand are relatively low. There is a not inconsiderable hydrophyte vegetation. In the pollen diagram there are nine indicator species of damp to very wet conditions. The peat most probably originated as floating bog vegetation in a depression with open water where there was no shade. In this vegetation Hypnaceae (not further identified) played an important role.

Also present are rather many types that are associated with dry soils, such as *Juniperus, Hippophaë, Urtica* (?), *Epilobium, Helianthemum* and *Sedum.* This shows that this basin was a very localized aquatic biotope in an otherwise relatively dry environment.

The regularity of the tree pollen curves and the low values for Gramineae and Cyperaceae make it probable that peat formation was able to continue for a relatively brief part of the Late Dryas. Here we are thinking in terms of decades rather than of centuries. In view of the C14-dating (10,495±60 B.P.) this would have been approximately in the middle of this period. The occurrence of the piece of *Pinus* wood, that probably came from the Allerød forest, indicates rather developments at the beginning of the Late Dryas, since a piece of wood lying on the surface does not remain preserved for a length of time of the order of centuries.

In view of the layered structure of the cover sand and this peat layer in the basin, we can assume that peat formation came to an end as a result of the basin drying up, so that the peat layer ended up at the bottom. The thin layers, indicated as peat, in the rest of the cover sand filling of the basin, see fig. 6, could point to brief wet periods during the time when the basin was becoming filled up. The analysis of these layers showed that these are thick illuvial iron-oxide horizons, not containing any pollen or other organic remains. Wet phases during the period of filling up with cover sand cannot be demonstrated in this way.

Fig. 7. Basin I, overall view. The puddle at the front of the basin indicates the deepest point. Here a lot of flint material was found.

6. GENESIS OF THE (COVER SAND) BASINS

The climatic conditions under which these basins originated and the time when this occurred have become clearly evident from the pollen-analytical investigation, the C14-dating of the Hypnaceae-peat, the geological phenomena and the archaeological study by Houtsma, Roodenberg and Schilstra, and others. The time at which these basins originated coincides, as one would expect, with a period in which the permafrost started to develop once again.

On the basis of the evidence provided below the time of origin of the permafrost, that resulted from a considerable decrease in temperature, can be dated more accurately. Many Late Glacial pollen diagrams show precisely at the transition from Allerød to Late Dryas a sharp decline of *Pinus*. From the pollen diagram for Waskemeer (Casparie & Van Zeist, 1960) it can be deduced that the decrease in pollen production of *Pinus* from 47% to 6% took place within 100 years and possibly even within 60-80 years. This is supported by various other pollen diagrams. The rapid rate of this decline indicated that not only did the rejuvenation of forest fail to take place, but that the flourishing of many pines came to an abrupt end, possibly with trees dying off altogether. We can therefore assume that the permafrost, in which the above-mentioned basins originated, started to form immediately at the beginning of the Late Dryas. As for the length of time that was necessary for a thickness of 1 m to be attained (that was required for the formation of basin I – see below), no direct information is available.

We assume that the origin of these basins must be regarded in association with the (developing) permafrost, although, we are not able to point out precisely how the two phenomena are related. Nevertheless we will mention a few possibilities here in further detail.

As a result of the downward extension of the permafrost, at the beginning of the Late Dryas, and possibly also on account of the inflow of water from higher areas, the ground-water in the aquiferous sand-layer above the hardly permeable boulder-clay (fig. 2) came to be under increasingly greater pressure. It is possible that the ground-water that was under pressure was able to find a way out via weak spots, such as cracks and fractures, and that an ice lens was able to develop next to the base of the permafrost or inside it, whereby a small pingo originated.

That the origin of such basins could be the result of human activities is an idea that has been suggested by Mr. Rinke Nolles of the Rijks Geologische Dienst (Geological Survey of the Netherlands). He has drawn our attention to the fact that these basins are only found in an archaeological context, notably in several sites of the Tjonger tradition (including Siegerswoude). It is quite remarkable, moreover, that in the course of this excavation no hearths were found. Now Mr. Nolles proposes that the Tjonger tradition people weakened the permafrost locally with their fires in such a way that the ground-water that was under pressure was able to escape. The actual hearth material (charcoal) could have become pulverized. During the excavation small fragments of charcoal were indeed found here and there.

From the photos and the lacquer peels it

Fig. 8. Basin II, overall view. The cover sand layering here follows the curvature of the basin. At the bottom of the basin several artefacts are present. The dark-coloured layer near the top is the B-horizon of the podsol profile; the darker layers towards the bottom are iron-oxide illuvial horizons.

can clearly be seen that the boundary of the basins is very sharply defined and that the adjacent finely layered cover sand is in no way disturbed. This leads to the conclusion that this contact surface originated as a result of water erosion, while the adjacent layered cover sand was in a frozen state (fig. 13). This view is supported by a local undermining of part of the wall of the basin (fig. 14).

Whether the upward escaping water, in the event of the situation suggested above, crys-

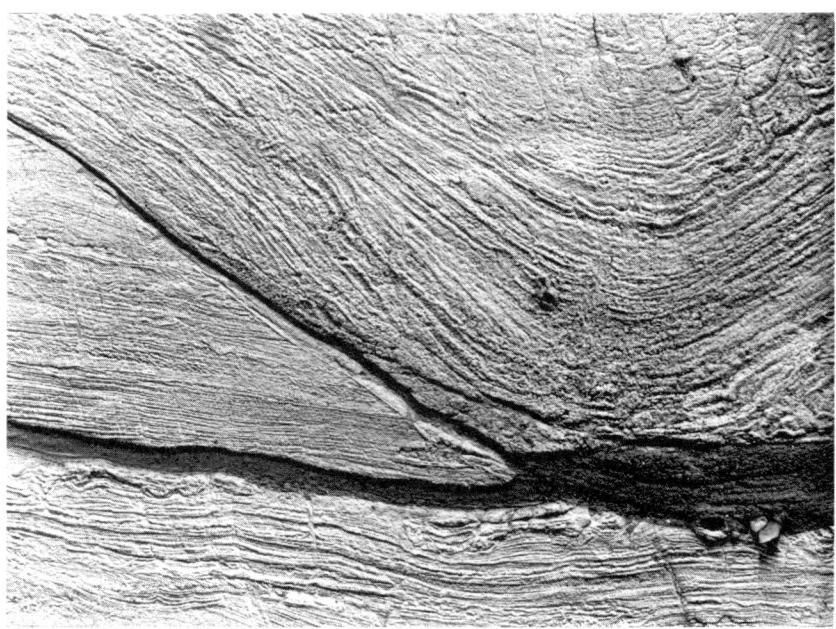

Fig. 9 a-b. Basin II, detail. The coarser sand with artefacts at the bottom.

tallized into an ice-lens and thus resulted in the formation of a pingo, a hillock with an ice-core, or whether the escaping ground-water washed out a basin it is impossible to say.

That water played a role at the time when these basins originated is also evident from the crust-like sand layer against the boundary of the basin (fig. 15). This residue layer, that shows fine parallel layering, consists of the

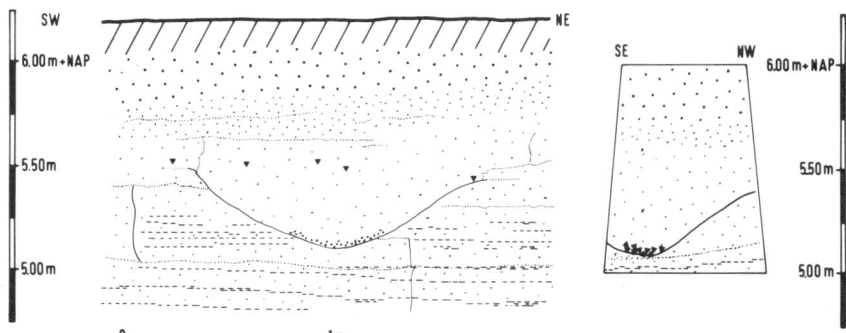

Fig. 10. Sections through basin II, left: parallel to the Schipsloot; right: at right angle to the Schipsloot. Legend: see fig. 6.

coarser sand grains of the cover sand complex, several glacial stones, artefacts and (in the case of basin I) wood. In fact this material represents the coarsest fraction of the cover sand plus culture layer at the spot where each basin occurs. Perhaps this situation could have arisen when at each spot a hearth melted the permafrost, the ground-water gushed out under pressure and washed out a small pond, in which this material became sedimented. How this material became washed out we do not know.

As stated previously, the peat originated as floating bog, undoubtedly in open water. In view of the bowl-like shape of the basin the soil must still have been frozen. The peat layer is situated remarkably deep in the (dried up) basin, namely immediately above the crust-like sand layer. During the process of peat formation there was therefore no sedimentation of mineral material of any significance. The almost sand-free character of the peat indicates that during the process of peat formation there was little or no transport of wind-

Fig. 11. Basin IV, horizontal cross-section. Above right two finds next to the periphery of the depression. Measuring staff: 5 cm.

blown sand taking place. It can be excluded that the peat-forming vegetation hindered sedimentation in the basin.

While peat formation was taking place the basin may still have been filled to a considerable extent with ice, so that there may have been open water of very limited depth in which peat formation occurred. The (virtual) absence of slumping phenomena along the wall of basin I is indicative of the permanent presence of ice in the basin.

The occurrence of the conspicuous crust-like sand layer, that consists partly of coarse sand, is not conclusively explained in this way. This applies particularly to the layering, that continues practically as far as the periphery of the basin, as already mentioned in section 4.

The ice lens finally melted completely, and the water sank away downwards fairly rapidly, so that the peat layer ended up on top of the coarser layered sand of the crust-like layer. With regard to the dating of this development the following can be said. The smoothness of the curves of many pollen types, see fig. 12, such as *Salix, Pinus, Betula, Juniperus* and *Artemisia*, suggests that we are not concerned here with a process of peat formation lasting several centuries, as already stated in section 5. The emptying of the basin can therefore probably be dated only somewhat later than the age obtained for the peat layer, 10.495 ±60 B.P. It must be remembered that this C14-dating does not indicate the time of emptying, and that we cannot say how long a period of time elapsed after this dating before the basin became dry.

The fact that the water in the basin sank away fairly rapidly is attributable in our opinion to the disappearance of the permafrost, as the result of a climatic improvement. For the rest the basin subsequently became filled up with cover sand. Towards the bottom of the cover sand filling there are one or more thin loamy sand-lenses present, that could indicate repetition of the process, possibly as a result of very slight improvements in climate. But subsequently the basin became filled up

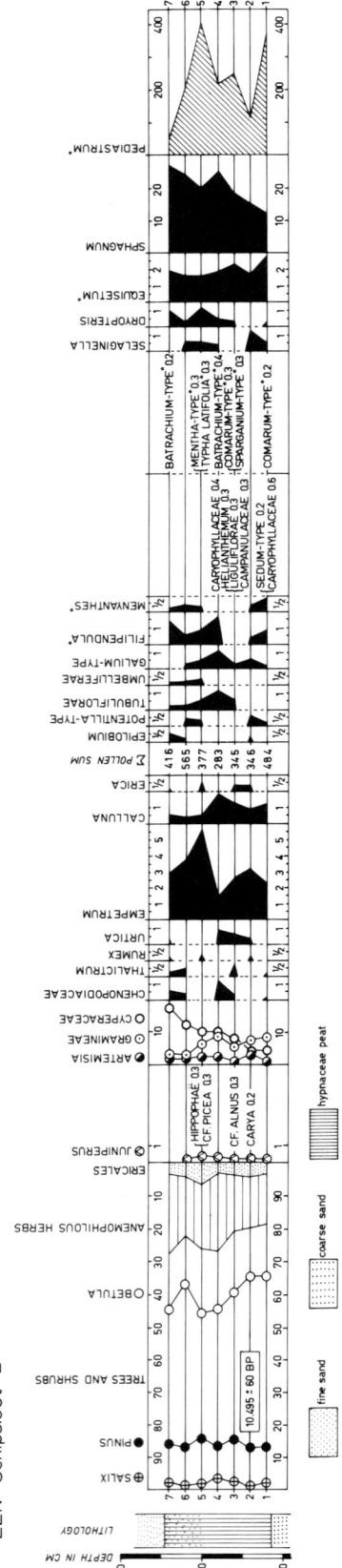

Fig. 12. Pollen diagram for Een Schipsloot-B (filling of basin I).

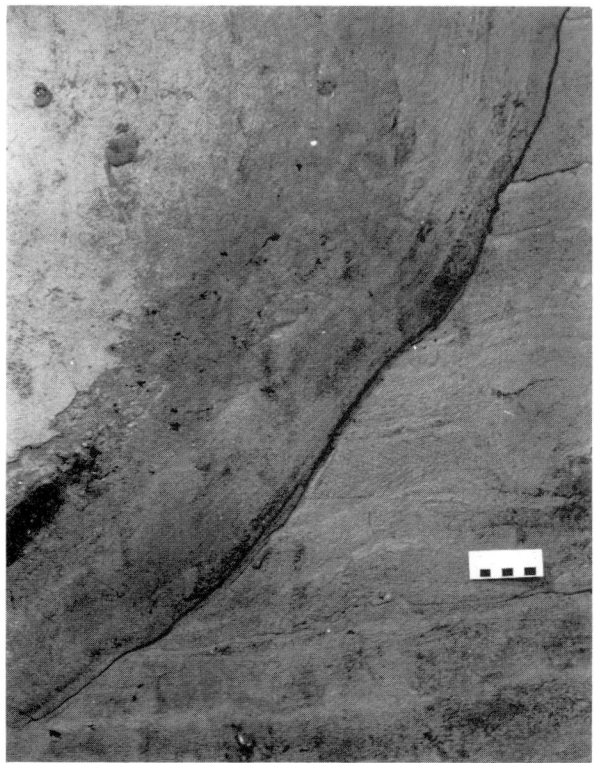

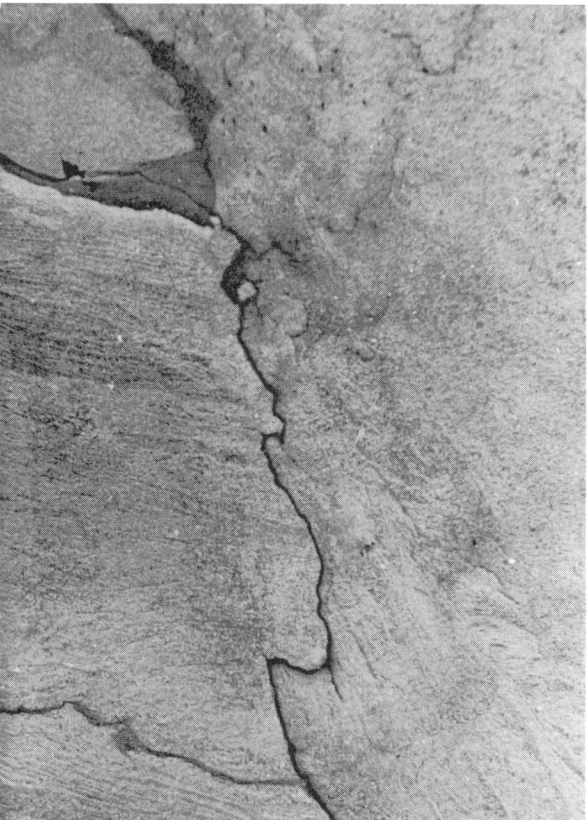

Fig. 13. Basin I, detail, showing the sharply defined base of the basin filling. This base is accentuated here by an iron-oxide horizon.

Fig. 14. Basin I, detail of the upper part of the wall, where locally undercutting had taken place (photo 1/3 of actual size).

Fig. 15. Basin I, detail, showing at upper right the edge of the peat lens and underneath it the finely-layered, crust-like layer of coarse sand. The base of the basin filling is very sharply defined against the horizontally layered cover-sand.

with cover sand, Younger Cover sand II; outside the basins too this cover sand was deposited in a series of layers.

In the small basins the layering of this cover sand follows the curvature of the wall (figs. 8,9), while in the large and fairly deep basin I (fig. 7) these finely layered sands are disturbed in the central part of the basin.

The depth of the basins varies from 0.15-1.00 m. Assuming that the depth of the basin indicates approximately the thickness of the permafrost at the time when the basin originated, the question arises whether we are concerned here with considerable differences in thickness in a developing permafrost, or with differences in time. In the latter case a relative chronology could be established on the basis of depth.

7. SUMMARY

In the very narrow watershed region of the valley systems of the Tjonger and Peizerdiep, that originated during the Pleniglacial, remains have been found of the Tjonger tradition. The site lies at a relatively high point of the somewhat undulating Allerød horizon, developed as a soil over the Younger Cover sand I. At the lower spots of this horizon a thin, sandy peat layer formed in the *Betula* phase of the Allerød.

At the beginning of the Late Dryas a permafrost was formed here; possibly in connection with this, round basins originated in the frozen soil, with a diameter of 1-3 m and a depth of 0.15-1.00 m. The upward movement of ground-water under pressure probably played a role in this. In the basins open water

was present; in the largest basin (basin I) peat formation took place. This is dated to 10,495 ± 60 B.P. (GrN-6341).

Precisely how the basins originated cannot be satisfactorily explained; it is possible that hearths of the Tjonger people weakened the permafrost to such an extent that the process of upward movement of ground-water was set in motion as a result. At the bottom of the basin filling a coarse sandy layer is present, like a crust, that continues against the wall of the basin practically as far as its periphery. The boundary between the wall of the basin and the crust-like sand layer is very sharply defined. Ultimately the basins dried up, after which they became filled up with Younger Cover sand II. This occurred still during the Late Dryas, possibly in connection with the disappearance of the permafrost.

8. REFERENCES

CASPARIE, W.A., 1972. Bog development in Southeastern Drenthe (the Netherlands). *Vegetatio* 25, pp. 1-271.

CASPARIE, W.A. & W. VAN ZEIST, 1960. A Late-Glacial lake deposit near Waskemeer (Prov. of Friesland). *Acta Bot. Neerl.* 9, pp. 191-196.

Geologische overzichtskaarten van Nederland, 1:600.000, 1975. Rijks Geologische Dienst, Haarlem.

HAMMEN, T. VAN DER & T.A. WIJMSTRA, 1971. The Upper Quaternary of the Dinkel valley. *Med. Rijks Geol. Dienst* N.S. 22, pp. 55-213.

HOUTSMA, P., J.J. ROODENBERG & J. SCHILSTRA, 1981. A site of the Tjonger tradition along the Schipsloot at Een (gemeente of Norg, province of Drenthe, the Netherlands). *Palaeohistoria* 23, pp. 45-74.

LANTING, J.N. & W.G. MOOK, 1977. *The Pre- and Protohistory of the Netherlands in terms of radiocarbon dates*. Groningen.

MAARLEVELD, G.C. & J.C. VAN DEN TOORN, 1955. Pseudo-sölle in Noord Nederland. *Tijdsch. Kon. Ned. Aardr. Gen.* 72, pp. 344-360.

A SITE OF THE TJONGER TRADITION ALONG THE SCHIPSLOOT AT EEN (*GEMEENTE* OF NORG, PROVINCE OF DRENTHE, THE NETHERLANDS)

P.Houtsma,* J.J.Roodenberg** and J.Schilstra***

CONTENTS

* Kroezestraat 4, 8434 NN Waskemeer, Netherlands
** Nederlands Historisch-Archaeologisch Instituut, Istiklâl Caddesi 393, Istanbul-Beyoğlu, Turkey
*** Ger. Terborchlaan 28, Lochem, Netherlands

1. INTRODUCTION

In the autumn of 1970 the ditch known as the
Schipsloot, to the south of Een (*gemeente* of
Norg), was widened in order to improve its
drainage capacity. At the highest point of a
cover sand ridge, intersected by the Schipsloot, P. Houtsma and J. Schilstra found in the
slope a few flint artefacts, in the sand, with
distinctly Upper Palaeolithic features. The coordinates of the site are: Topographical map
of the Netherlands, sheet 12A (Norg), 564.
925/222.550. For location on a map of the
Netherlands, see fig. 1, for the local topographical situation, fig. 2.

From further observations made in September 1970 it became evident that the find-
bearing level lay in the undisturbed C-horizon
of a podsol. This level was interrupted by a
basin-shaped depression I, filled with peat,
loam and sand (figs. 3 and 11; and Casparie
& ter Wee, this volume: fig. 2). Under the peat,
at the bottom of this basin, an accumulation of
flint artefacts was present. Artefacts were also
found in heaps of recently dug-out sand that
had been dumped on the north side of the
Schipsloot.

In view of these circumstances Prof. H.T.
Waterbolk (Biologisch-Archaeologisch Instituut) decided to make a more detailed investigation of the site, by means of an excavation.
The excavation was carried out by the first and
third author, in the periods February-October
1971 and February-May 1972. Those provi-
ding assistance included Messrs. H. ter Haagha,
R. Jansen and R. Nolles.

Two lacquer peels of the largest basin (I)
were made under the supervision of Mr. A. Meij-
er (B.A.I.), who also saw to it that the topsoil
was removed from four trenches marked out
by the excavators. These trenches were each
2.50 m wide and 20 m long.

The pollen analysis of some peat profiles was
carried out by Dr. W.A. Casparie (B.A.I., see
Casparie & ter Wee, this volume). A C14-da-
ting for the peat from basin I was determined
by Prof. W.G. Mook (Laboratorium voor Iso-
topenfysica, Groningen). Drs. M.W. ter Wee
(Rijks Geologische Dienst, northern district)

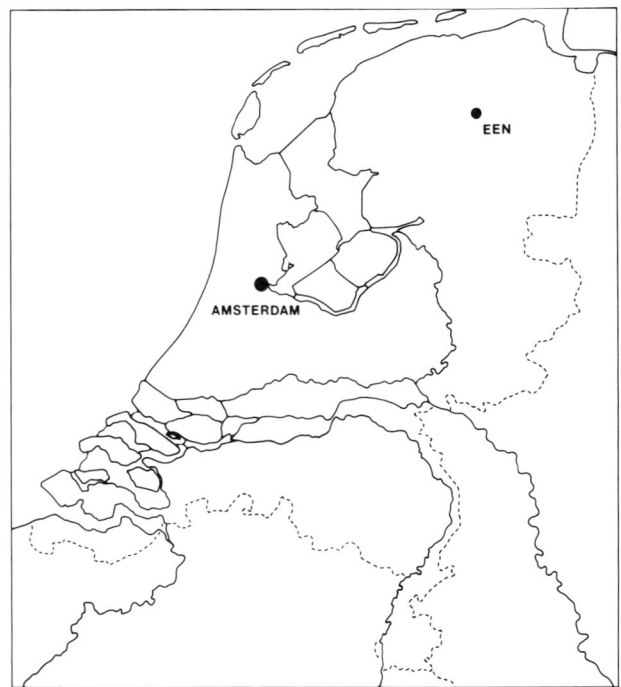

Fig. 1. Map of the Netherlands showing the location of Een.

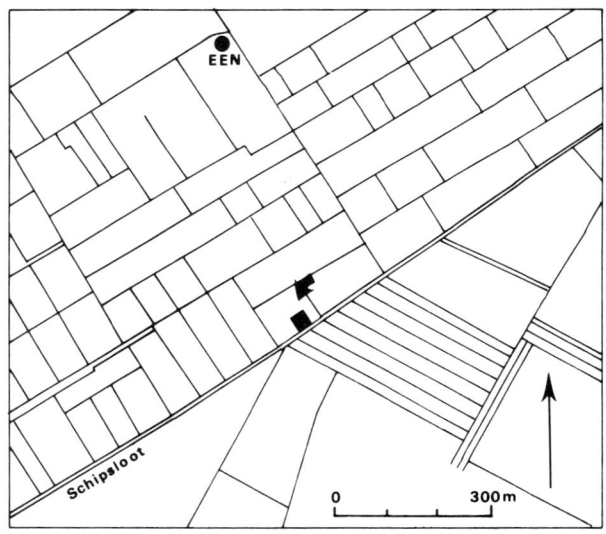

Fig. 2. Location map.

was kind enough to undertake the geological
investigation of the surroundings of the site
(see Casparie & ter Wee, this volume).

The owner of the parcel of land in which the
flint concentration was found, Mr. R. Vunde-
rink of Rijswijk, kindly gave permission to
carry out the excavation on his property. The

local water board "Noordenveld" gave permission to continue the excavation in the upper slope of the Schipsloot.

The authors wish to express their gratitude to all those who contributed in any way to the success of the excavation and to this publication,* and to all those who have shown interest in their investigations.

2. THE EXCAVATION

The investigation began, soon after the discovery, on September 12th 1970 with the drawing of the profile exposed in the north wall of the Schipsloot, in which a cross-section of basin I was visible. The part that was drawn was 20 m long, and the most important part of this is shown in fig. 3. (The inclination of the ditch slope was ca. 45°.) On September 17th 1970 a peat profile was sampled by Dr. W.A. Casparie, from the filling of basin I, that was subsequently covered over, and from peat from the adjacent valley. The actual excavation began in February 1971 with the investigation of the first trench (20 x 2 m) immediately to the north of, and parallel to, the Schipsloot. All finds made in the culture level were drawn on a scale of 1:10. The tools, blades, cores and larger stones that were recognized during the excavation were numbered individually. A total surface area of ca. 400 m² was excavated (see the distribution map, fig. 5, in pocket). On account of the presence of an earthen wall on the NW side and of the Schipsloot on the SE

* Ir. P. Offringa of Waskemeer measured in the horizontal position of the culture level with respect to N.A.P. (Dutch Ordnance Level). Drs. O.H. Harsema (B.A.I. and archaeological department of the Provinciaal Museum in Assen) provided technical assistance and arranged for a work-hut from the Koninklijke Nederlandsche Heidemaatschappij. This was placed on land belonging to Mr. A. Huizinga.
Sections 1 - 4 of this article were written by P.H. and J.S., sections 5.2.2, 5.2.3 and 6 by P.H., and the remaining sections by Drs. J.J. Roodenberg; Drs. D. Stapert provided assistance in editing the text. Mrs. Sh. van Gelder-Ottway translated the text into English. The drawings were made by Messrs. W.J. Dijkema and J.M. Smit (B.A.I.). Mrs. E. Rondaan-Veger (B.A.I.) typed the manuscript. The photos 13-17 were made by W.A. Casparie and the first author.

side, it was not possible to ascertain how far the flint concentration extended in those directions. The ground was carefully removed with a spade, and a hand-sieve was used where necessary.

Between the various excavation trenches profile walls 0.50 m thick were initially left intact. In these profile walls small pieces of coloured plastic were inserted to indicate the level of all artefacts found at a horizontal distance of 0.30 m or less from these profile walls, in order to give a picture of the course of the find-bearing level in profile. The culture level was 3-7 cm thick, and showed little variation in height (except for the basins, see under 3). Only a gradual slope (downwards) from north to south (about 0.20 m over a length of 15 m, i.e. about 1.3%) could be shown. In an east-west direction hardly any differences in height could be observed. At basin I the culture level was present at ca. 5.50 m +N.A.P.

In the sand immediately above and in the culture level some larger stones and a few smaller stones (up to 8 cm) are present, in addition to the flint artefacts. From a stratigraphical point of view the find-bearing level is situated at the boundary between deposits that are to be regarded as Younger Cover sand I and II. The "layer of Usselo", that one would expect to find at this level, was not visible (see also Casparie & ter Wee, this volume).

3. THE BASINS

In addition to the basin (I) found already in the NW wall of the Schipsloot, another 9 similar basins were found in the course of the excavation (numbered II-VII, VIII A, B, C). Drawings and photos were made of cross-sections through these basins, also a number of lacquer peels. These basins are described here below (see also Casparie & ter Wee, this volume). See fig. 5 for the location of these basins within the excavation area.

Basin I

Basin I is situated in the 1 m² squares D-I,

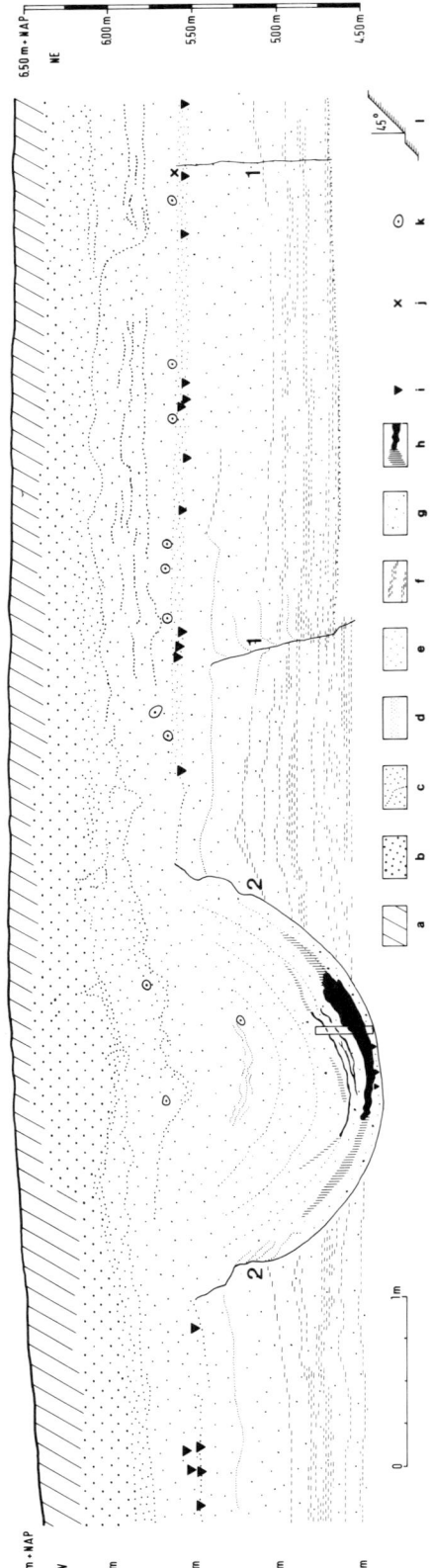

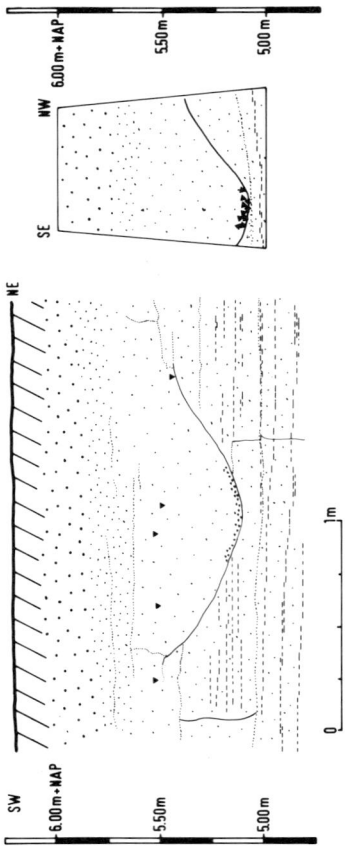

Fig. 3. SW-NE profile through basin I. Legend: a) disturbed and/or transported soil; b) Younger Cover sand II, A-horizon of the podzol; c) Younger Cover sand II, B-horizon of the podzol, with fibres; d) Younger Cover sand I, slightly paler layer (± level of the Tjonger finds); e) Younger Cover sand I; f) loam and very loamy sand; g) coarse sand, sometimes with fine gravel; h) gyttje (left) and peat; i) artefacts; j) bone fragment; k) fine gravel; l) indication of the angle of inclination of this section; Continuous lines: 1. frost cracks; 2. contour of basin I.

Fig. 4. Sections through basin II. Left: SW-NE profile with artefacts above the basin. Right: SE-NW profile with artefacts at the bottom. For legend see. fig. 3.

D-II, E-I, E-II, F-I, F-II (figs. 3, 5, 11 and 12; and Casparie & ter Wee, this volume: figs. 2 and 7). Lacquer peels and drawings were made of the SW-NE and the SE-NW cross-sections. Horizontally the culture level (outside the basin) was present at ± 5.50 m +N.A.P. As the basin was situated in the oblique ditch-slope, only part of it was preserved; the rest had been removed by the dragline.

Close to the basin a proportion of the artefacts was present a few centimetres higher than in the culture level, a short distance away, so that it seemed as though the basin had been surrounded by a small earthen wall. This was best visible on the east side; on the west side the sand was more or less disturbed by decayed tree roots.

On the inner slope of the basin, there were few artefacts, while at the bottom of the basin artefacts were relatively abundant. The depth of the basin was ± 0.90 m. The flint material had evidently slipped down the slope, and was found in a layer of coarse sand that showed up clearly against the almost horizontally layered fine sand outside the basin. The filling of the basin consisted of fairly loose coarse sand, alternating with thin layers of loam, with here and there a small piece of charcoal or small pebble, and at the bottom brown peat. For this peat a C14-dating was determined: GrN-6341: 10,495 ± 60 B.P. (= 8545 B.C.). Among the artefacts that were present in the coarse sand under the peat was a backed blade. As these tools are characteristic of the Tjonger tradition, this shows that the finds from the bottom of the basin belong to the same tradition as the finds from the culture level, in which similar tools are also present.

We have tried to ascertain the dimensions of the entire basin by reconstruction. The basin must have had an east-west diameter of 2.70-3.00 m, a north-south diameter of 2.50-2.70 m and a depth of 0.90 m. So the basin was not exactly round.

Basin II

Basin II is situated in the squares C-2, C-3,

D-2, D-3 (figs. 4, 5 and 13; and Casparie & ter Wee, this volume: fig. 9a). Lacquer peels of the SW-NE cross-section and of part of the SE-NW cross-section were made, as well as a drawing of the perimeter of the basin, seen in a horizontal plane at 5.45 m +N.A.P., and two profile drawings (see fig. 4).

In the horizontal plane both basin I and basin II became visible at the culture level, that was present near basin II at ± 5.45 m +N.A.P.

At this level the east-west diameter was ± 1.70 m, the north-south diameter ± 1.50 m. The depth was ± 0.45 m. The artefacts, apart from a few that had sunk down along the slope of the basin, were present at two levels: firstly at the top, at the culture level proper, that here extended over the basin undisturbed, and secondly right at the bottom. This could be an indication of the presence of (at least) two occupation phases. In contrast to basin I no loam or peat was present.

In the case of basin I it was conceivable that this basin had originated shortly after the occupation period and that the artefacts had slid down together from the culture level. The basin had been filled with water, in which plant and animal life developed, and later became filled by cover sand and loess-like material.

In the case of basin II it appears more likely that this basin became filled up during the occupation period, or between two different occupation phases. Evidently there was insufficient time for peat formation. During the first occupation phase artefacts could have ended up at the bottom, after which the basin became filled up with sediment, and in a second occupation phase new artefacts would then have come to lie above the basin, that had already been filled up. This second phase can be dated somewhere in the first centuries of the Late Dryas period, immediately after the formation of the basins.

Basin III

Basin III is situated in the squares J-I, J-1, K-I, K-1 (figs. 14-15). A lacquer peel was made of the NE-SW cross-section through basin III.

This profile was formed by the southern wall of trench I, that happened to run precisely through the middle of basin III. On the lacquer peel it can clearly be seen that the basin shows up already at 5.60 m +N.A.P., while the culture level begins only at 5.50 m +N.A.P. This basin thus came into being after the occupation period. How much later it cannot be said, as we do not know the rate at which these 0.10 m of sand were deposited. A frost crack ran through the basin (this was also the case with basin VIII-C). We could follow this crack horizontally over the entire breadth (15 m) of the excavation (see fig. 5). In square J-4 of trench 2 we found in the culture layer a piece of granite (no. 833a and no. 833b) more than 0.30 m long, 0.10 m wide and 0.07 m thick. This piece of granite lay transversely over the frost crack, and was broken in the middle precisely at that point (figs. 16-17). This illustrates the tremendous forces that are exerted in the formation of such cracks, that may extend to a length of tens of metres, and are sometimes more than a metre deep.

We assume that the formation of such deep and long cracks took place for the last time in the Late Dryas period. The frost cracks are secondary with respect to the basins III and VIII-C. If our assumption concerning the dating of these cracks is correct, then it follows that these two basins must have been formed at any rate during the Late Glacial, which is confirmed by the C14-dating of the peat in basin I.

A lacquer peel was also made of the horizontal plane at 5.50 m +N.A.P. This is the height of the culture level. At this level the east-west diameter was 1.30 m, the north-south diameter 1.20 m and the depth 0.30 m. Artefacts were present only on the inner slope of the basin and at the bottom.

Finally another lacquer peel was made of the NW-SE cross-section of basin III from 5.50 m +N.A.P. to 4.75 m +N.A.P.

The loam layer in this basin, that is characteristic of some of these basins, showed "worm-holes" with which we are familiar elsewhere in the "layer of Usselo".

Basin IV

Basin IV is situated in the squares Q-3 and R-3. Lacquer peels of NE-SW and NW-SE cross-sections. Drawing of horizontal plane 5.48 m +N.A.P. (the culture level). East-west diameter at this level 0.85 m, north-south diameter 0.65 m, depth 0.30 m.

As in the case of basin III here too the artefacts have sunk downwards. A humous loam layer was present here also. It was not clearly visible at which level the basin had originated. It was our impression that it was approximately at the culture level; however, it may have been slightly higher.

Basin V

Basin V is situated in the squares P-3 and Q-3. This basin was noticed too late: hence no lacquer peels. This small basin was situated right next to basin IV. No artefacts were present at the bottom of this basin.

Approximate dimensions at the culture level (5.50 m +N.A.P.): east-west 0.56 m, north-south 0.50 m, depth 0.20 m.

Basin VI

Basin VI is situated in the squares G-2, G-3, H-2, H-3. This basin was noticed too late: hence no lacquer peels. The periphery is drawn horizontally at 5.33 m +N.A.P., i.e. ± 0.15 m below the culture level.

At 5.33 m +N.A.P. the dimensions were as follows: east-west diameter ± 0.80 m, north-south diameter ± 0.70 m, depth ± 0.20 m.

At the bottom an accumulation of 45 artefacts was found. No loam layer was found in this basin.

Basin VII

Basin VII is situated in the squares D-7, D-8, E-7, E-8. This basin begins to show up in the NW-SE profile (lacquer peel) at 5.60 m +N.A.P., i.e. about 0.10 m above the culture level, that is present here at about 5.47 m +N.A.P.

The east-west diameter is unknown, the north-south diameter is ± 0.90 m. The depth is ± 0.25 m. The artefacts have sunk down about 0.07 m. No loam layer was present. The basin must have originated after the occupation period.

Basin VIII-A

Basin VIII-A is situated in the squares U-5, U-6, V-5, V-6. East-west diameter ± 0.90 m, north-south diameter ± 0.70 m. Depth ± 0.20 m. This basin is clearly younger than the arte-facts; the culture layer continues undisturbed under the basin.

Basin VIII-B

Basin VIII-B is situated in the squares D-13, D-14, E-13, E-14. No lacquer peels; this basin was noticed too late. This basin was visible 0.05-0.10 m above the culture level and there-fore originated after the occupation period.

The maximum width was about 1.50 m, the maximum length 2.30 m. In contrast to the other basins this one was not oval but pear-shaped. A loam layer was clearly visible. The depth was approximately 0.15 m. There were few artefacts; these had sunk slighty down-wards.

Basin VIII-C

Basin VIII-C is situated in the squares R-13, R-14, S-13, S-14. No lacquer peels. This basin was visible at the culture level (5.56 m +N.A.P.). East-west diameter ± 1.50 m, north-south diameter ± 1.70 m, depth c. 0.30 m.

At the bottom of the basin there were more than a hundred artefacts. Two (secondary) frost cracks ran through the basin. Once again loam had been deposited.

As for the age of this basin the same applies as for basin III: in any case not younger than the Late Dryas period.

Although basins I, II and VIII-C only became clearly visible at the culture level, this does not mean that they could not have originated a couple of centimetres higher. This was often difficult to ascertian accurately. With the exception of basin II, that must have originated during the occupation period, all the other basins could have been formed after the occupation period. This can be said with cer-tainty for basins III, VII, VIII-A and VIII-B.

With regard to the time of origin, the Late Dryas period is the only possibility. For the extensive frost cracks, that run right through the excavation, begin at the same level as most of the basins, i.e. 0.05-0.07 m above the cul-ture level.

It is evident that these basins are not artifi-cial pits or anything similar, but rather a peri-glacial phenomenon. Their origin is perhaps connected with man-made hearths.

4. THE HORIZONTAL DISTRIBUTION OF THE FINDS

In fig. 5 the finds are mapped insofar as they were classified during the excavation. In addi-tion the basins and frost cracks described above are indicated in this figure. In the culture level scattered fragments of charcoal were found (too few for a C14-dating), but no hearth(s). The question arises whether the material may have been moved secondarily by geological pro-cesses. The answer could be that this is pro-bably hardly the case, at least in a horizontal direction. The presence of some heavy pieces of granite (brought to the spot by Palaeolithic people) in the middle of the flint concentration is not indicative of shifting caused by any such external process. It is possible, however, that vertical shifting may well have taken place.

Finally, it is fortunate that in the deposits above the culture level there were no finds da-ting from the Mesolithic or Neolithic, so it is certain that all the mapped finds are of Upper Palaeolithic age.

Seeing that there are indications that we may be concerned here with two occupation phases, there seems to be little sense in trying to analyze the distribution map in detail.

5. THE FLINT ARTEFACTS

5.1. Artefacts other than tools

Given below is a description of the 8785 flint artefacts that were collected during the excavation. The typological nomenclature is for the most part taken from Schwabedissen (1954), Bohmers (1956) and Taute (1963).

The overall composition of the material is as follows:

	number	% of total
tools (see 5.2.)	871	10.1
artefacts other than tools	7914	89.9

		% of artefacts other than tools
flakes	4665	59.0
blades	1279	16.2
flake cores	228	2.9
blade cores	183	2.3
core preparation pieces and core renewal pieces	406	5.1
blocks	982	12.4
burin-spalls	170	2.2
"microburin" (notch remnant)	1	0.1

The flint used varies in colour from light grey to dark brown, and is of local origin (namely from the Saalian boulder-clay). The quality of the flint is not especially good, as is also evident from the relatively small number of blades. Generally speaking this is fairly characteristic of sites associated with the Tjonger traditon; this may have something to do with the vegetation cover that predominated during the Allerød interstadial, in which most of the sites of this tradition can be dated.

5.2. The tools

5.2.1. *Overview*

The tools (871 in total) can be divided as follows:

	number	% of tools
points (and probable point fragments)	50	5.7
backed blades	51	5.9
triangles	2	0.2
rectangle	1	0.1
borers	37	4.2
scrapers	406	46.7
combination tools (scraper/burin)	18	2.1
burins	161	18.5
obliquely truncated blades	6	0.7
tools with notches	35	4.0
retouched blades, flakes, blocks	104	11.9

5.2.2 *Points* (total no.: 50; figs. 6-8)

The fact that the term "points" is used here below does not mean that it is clear in all cases that these tools were used as projectiles. They could also have been used, for example, as cutting tools; for many of these pieces show use-"retouch" along the non-retouched edge.

The following types of points are present:

A. Gravettian points, total no.: 3.

These points are at least 50 mm long, slender, and with a straight or almost straight retouched back. The apex is situated on or close to the longitudinal axis of the point. Base-retouch may or may not be present.

B. Châtelperronian points, total no.: 3.

These points are at least 50 mm long, and relatively broad (at least 15 mm); the retouched edge is markedly curved. Base-retouch may or may not be present.

C. Tjonger points, total no.: 26.

These points have the same shape as the Châtelperronian points, but are narrower and usually smaller than 50 mm. Also included in this group are those points that on account of their shape could be called Gravettian points, but that are smaller than 50 mm. Base-retouch may or may not be present. Günther (1973: 21) classifies Gravettian points and Tjonger points (*Federmesser*) jointly as *Rückenspitzen*.

D. Azilian points, total no.: 1.

Like Tjonger points, but semilunar in shape, and with two distinct pointed ends. Schwabedissen (1954: 9) calls them *Halbmondmesser*, Taute (1968: 15, fig. 2:7) *Halbrundmeser* (see also Schwabedissen, 1954: 24, fig. 12: f-i).

E. Creswellian points, total no.: 2.

Like Tjonger points, but then with one oblique truncation, with the nick closer to the apex than to the non-pointed end of the artefact. Schwabedissen and Taute call this type: *Dreieckmesser Typ Kent*. If the nick is in the middle then Schwabedissen speaks of *Dreieckmesser Typ Petersfels*. If the nick is below the middle (i.e. closer to the non-pointed end of the artefact) then one could call it a shouldered point (see F) (see Paddayya, 1973: 204, fig. 15:12).

With the types described here above under A-E usually one edge is completely retouched, but sometimes part is left unretouched. From the overall external appearance of the artefact it can be seen in that case whether or not it is a long B-point (see G).

F. Shouldered points, total no.: 6 (of which only 1 is complete).

For a definition see Bohmers (1956), and here above (see E). The shouldered points from Een-Schipsloot are not very reminiscent of those of the Hamburgian tradition, although they somewhat resemble the points

of the Havelte type, that represent a kind of tanged points. From three fragments it is clear that one edge is retouched entirely; which shows their backed point character, that is typical of the *Federmesser* tradition. These artefacts are in a way shouldered Tjonger points (see Schwabedissen, 1954: 49 and fig. 65:4 and 6; Bohmers, unpublished drawings: Neer II; and Mace 1959: 246, fig. 4:19).

G. Long B-points, total no.: 2.

The term long B-points is used here for blades with an oblique truncation, if the truncation makes an angle of 45° or less with the longitudinal axis of the artefact. If the angle is greater than 45°, then we ascribe the tool to the category of obliquely truncated blades (see 5.2.10.). Perhaps it is preferable to use the term B-point only for microliths, as Bohmers originally did (Bohmers & Wouters, 1956: 29 and fig. 6:28-36). Taute (1968: 183) uses the name *Zonhovenspitze mit oder ohne Basisretusche* exclusively for microlithic types, and avoids the term B-point.

The artefacts referred to here as long B-points are of little value as far as dating is concerned, for they also occur within the Hamburgian and the Creswellian tradition. Moreover the borderline between long B-points and obliquely truncated blades is arbitrary.

H. A-points, total no.: 1.

A-points are microlithic points, of which one edge is retouched entirely (Bohmers & Wouters, 1956: 29 and fig. 6:19-27). It is remarkable that in the case of the one example from Een the basal two-thirds part is retouched dorsally, while the topmost third, that forms the point, is retouched ventrally. According to Dr. R.R. Newell (pers. comm.) this does not occur in the Mesolithic.

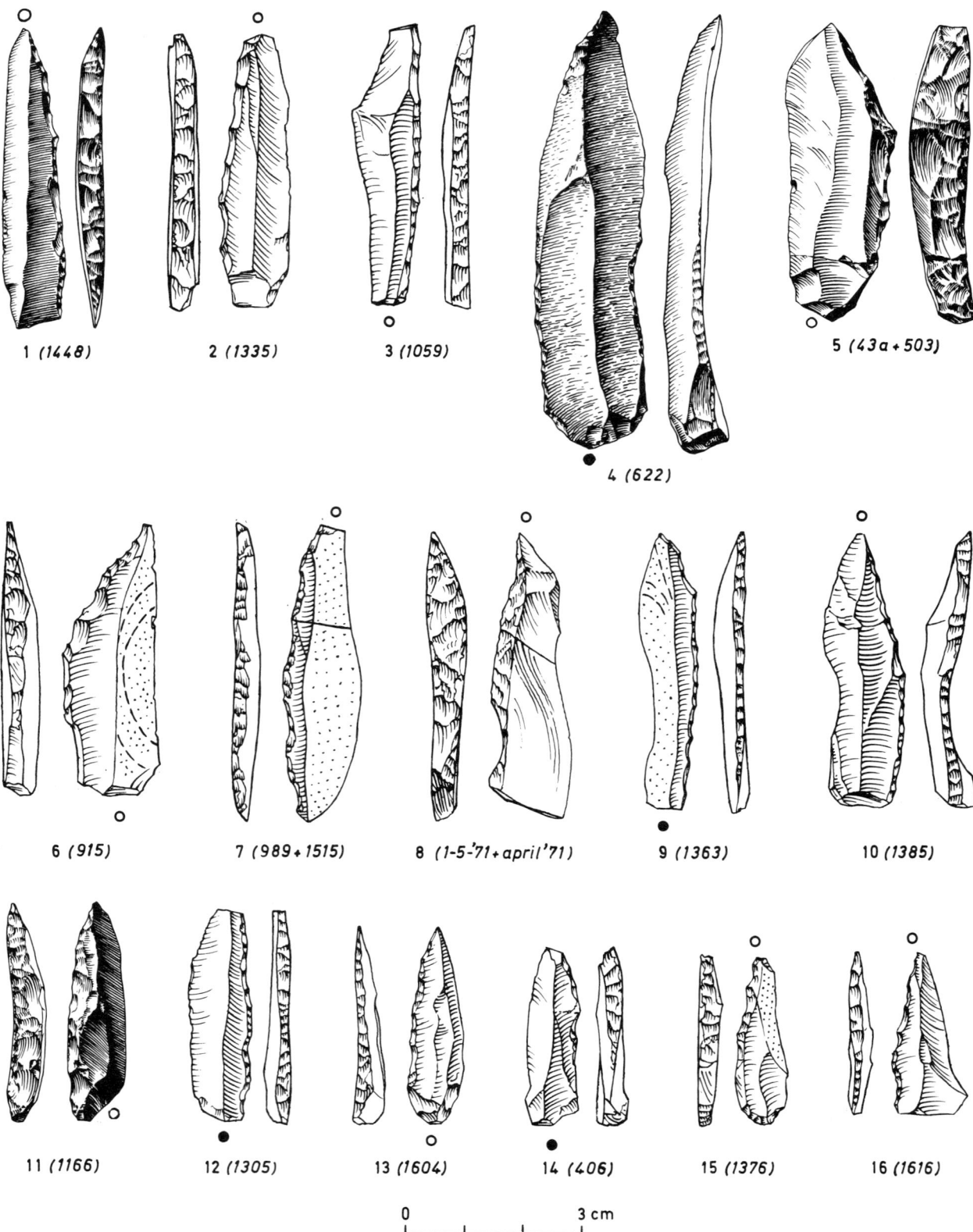

1 *(1448)*   2 *(1335)*   3 *(1059)*   4 *(622)*   5 *(43a + 503)*

6 *(915)*   7 *(989 + 1515)*   8 *(1-5-'71 + april '71)*   9 *(1363)*   10 *(1385)*

11 *(1166)*   12 *(1305)*   13 *(1604)*   14 *(406)*   15 *(1376)*   16 *(1616)*

0 ___ 3 cm

Fig. 6. Flint artefacts: 1-3 Gravettian points, 4-6 Châtelperronian points, 7-16 Tjonger points (the numbers between brackets indicate the inventory numbers).

I. Point fragments (not attributable to any particular type), total no.: 6.

Here below the points of each type are described briefly (the numbers in italics between brackets indicate the inventory numbers). Retouch has been applied dorsally unless mentioned otherwise.

A. Gravettian points

1. Fig. 6:1 (*1448*). Complete, length 51 mm, breadth 11 mm, thickness 5 mm. Backing on the right. Along the left edge some fine retouch is present (utilized?). The top is retouched ventrally, and is situated proximally.
2. Fig. 6:2 (*1335*). Top broken off, length 45 mm, breadth 12 mm, thickness 6 mm. The backing is on the left. The artefact is burnt (*craquelé*). The entire length must originally have been c. 50 mm. Narrower at the base on account of counter notch. The top is situated proximally.
3. Fig. 6:3 (*1059*). Length 49 mm, breadth 12 mm, thickness 4 mm. The backing is on the right, the top is broken off. The entire length must have been c. 52 mm. Base-retouch (the percussion bulb has disappeared as a result of retouching). The ventral surface is slightly damaged (recently) 1 cm below the top.

B. Châtelperronian points

1. Fig. 6:4 (*622*). Length 75 mm, breadth 18 mm, thickness 9 mm. The point is made out of a blade, of which the left edge already had a more or less curved shape, and the distal top of which has been further retouched into a point (both ventrally and dorsally). In addition part of the left edge has not been backed, while the retouch begins again 35 mm above the base. Of the right edge only the lowermost 33 mm is blunted. Use-"retouch" on the right edge, at the top, and the heavy percussion bulb make it improbable that the artefact was used as a point. The broad striking platform remnant gives the impression that the blade came from a core with a prepared percussion surface, like Levallois cores that were used for making blades; occasionally blades of this type are encountered in the Upper Palaeolithic.
2. Fig. 6:5 (*43a + 503*). Length 51 mm, breadth 18 mm, thickness 10 mm. The backing is on the right, with the lowermost part dorsally retouched and the uppermost part ventrally and dorsally retouched (this was necessary on account of the 10-mm-thick back). The distal top is missing. The point originally must have measured ± 57 mm in length. Use-"retouch" is present on the left edge. The percussion bulb is missing. The artefact shows wind-gloss on the dorsal surface.
3. Fig. 6:6 (*915*). Length 46 mm, breadth 15 mm, thickness 7 mm. The backing is on the left. The base (with percussion bulb) is missing. The right edge is utilized. Because the base is broken off we classify this point as a Châtelperronian point; the original length would have been more than 50 mm.

C. Tjonger points (*Federmesser*)

1. Fig. 6:7 (*989 + 1515*). Length 50 mm, breadth 13 mm, thickness 5 mm. The backing is on the left. The top (proximal) is broken off. The original length is estimated at c. 55 mm. The back shows a slight nick on the upper part (cf. Creswellian points).
2. Fig. 6:8 (*1-5-'71 + april '71*). Length 49 mm, breadth 13 mm, thickness 7 mm. The backing is on the left, and retouch is present ventrally and dorsally. Of the base 2 mm is blunted. The top is proximal. The "ventral" surface shows white patina. This point is not made out of a blade, but out of a naturally frost-split fragment.
3. Fig. 6:9 (*1363*). Length 47 mm, breadth 10 mm, thickness 4 mm. The backing is on the right. The top is distal. The back shows a small nick towards the top. The flat percus-

sion bulb is still present. The top lies on the longitudinal axis, as with the Gravettian points. The artefact has a slight patination.

4. Fig. 6:10 (*1385*). Length 46 mm, breadth 13 mm, thickness 6 mm. The backing is on the right. The top is proximal and lies, as with no. 3, on the longitudinal axis. Burnt (*craquelé*).

5. Fig. 6:11 (*1166*). Length 37 mm, breadth 11 mm, thickness 5 mm. The backing is on the left. The percussion bulb has disappeared at the base as a result of retouching. The distal top lies on the longitudinal axis.

6. Fig. 6:12 (*1305*). Length 35 mm, breadth 10 mm, thickness 4 mm. The backing is on the right. The top is missing; the original length must have been c. 39 mm. The percussion bulb is present at the base. Use-"retouch" is present on the left edge. The top must have lain on the longitudinal axis.

7. Fig. 6:13 (*1604*). Length 33 mm, breadth 9 mm, thickness 5.5 mm. The backing is on the left. The top is distal, and lies on the longitudinal axis. At the round, thick base use-"retouch" is present, like that which is sometimes found on scraper edges; the basal end may therefore have been used as a scraper; in any case this part is too thick to permit shafting for use as an arrowhead.

8. Fig. 6:14 (*406*). Lenght 30 mm, breadth 9 mm, thickness 6 mm. The top (distal) is broken off. The original length must have been about 32 mm. The backing is on the right. Retouch is absent over the basal 6 mm. The top lies on the longitudinal axis. The thick base would make shafting as an arrowhead impossible. Perhaps an unfinished item.

9. Fig. 6:15 (*1376*). Lenght 27.5 mm, breadth 9 mm, thickness 4 mm. The backing is on the left. The top, of which c. 1 mm is missing, is proximal and lies on the longitudinal axis. The lowermost 7 mm of the back are not retouched. The base is retouched all round and shows traces of use; the artefact may have served as a small scraper. The top is also slightly blunted on the right edge, and may therefore have served as a small borer.

10. Fig. 6:16 (*1616*). Length 27 mm, breadth 12 mm, thickness 4 mm. The backing is on the left. The top (proximal) lies on the longitudinal axis. The right edge is utilized.

11. Fig. 7:17 (*142*). Length 31 mm, breadth 11 mm, thickness 4 mm. The backing is on right. The top lies on the longitudinal axis, and is probably proximal. Fine use-"retouch" is present on the left edge near to the base.

12. Fig. 7:18 (*1399*). Length 44 mm, breadth 11 mm, thickness 4 mm. The backing is on the left. The top distal. Traces of use are present on the right edge. The base becomes slightly narrower on the right edge. The percussion bulb is present.

13. Fig. 7:19 (*912*). Length 44 mm, breadth 10 mm, thickness 4 mm. The backing is on the left. The base is slightly narrower on the left edge. The top is distal. The percussion bulb is present. Because two-fifths of the back is unretouched, the artefact could also be classified as a long B-point (see also Schwabedissen, 1954: table 1: 6-8; Bohmers, 1956: fig. 2:17; Movius *et al.*, 1968: 46 and fig. 27:9 and 10).

14. Fig. 7:20 (*1590*). Length 41 mm, breadth 9 mm, thickness 5 mm. The backing is on the right, except for the lowermost 4 mm, where retouch is absent. The uppermost 6 mm are retouched ventrally. The top is distal; the percussion bulb is present.

15. Fig. 7:21 (*120 + 121*). Length 40 mm, breadth 10 mm, thickness 4 mm. The backing is on the right. The top is distal; the percussion bulb is present, but a small part of it has disappeared as a result of retouch.

16. Fig. 7:22 (*166*). Length 39 mm, breadth 9 mm, thickness 4 mm. The backing is on the left. The distal top has broken off. The original total length must have been c. 44 mm. The percussion bulb is present. Towards the base the retouch bends round to the right, so the base is narrower.

17. Fig. 7:23 (*171*). Length 37 mm, breadth 12 mm, thickness 3 mm. The backing is on the left, and the edge is almost straight. The right edge is strongly curved, and to-

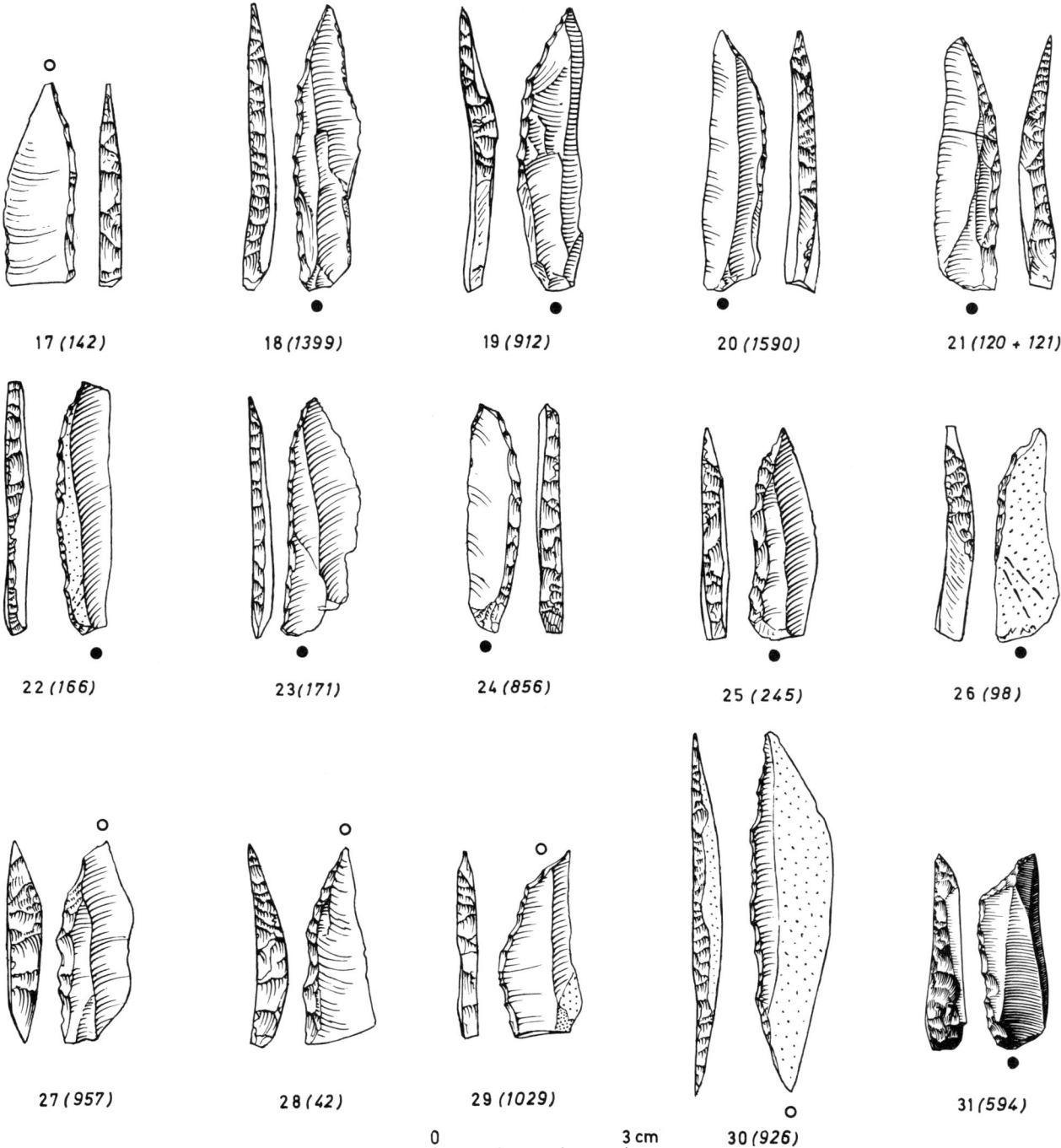

Fig. 7. Flint artefacts: 17-29 Tjonger points, 30 Azilian point, 31 Creswellian point (the numbers between brackets indicate the inventory numbers).

wards the base becoming gradually narrower (this has not been done intentionally, i.e. as a result of retouch). The top is distal; a flat percussion bulb is present.

18. Fig. 7:24 (*856*). Length 35 mm, breadth 8 mm, thickness 3,5 mm. The backing is on the right. The top is distal; the percussion bulb is present. The base becomes slightly

nàrrower on each edge. The top is concavely retouched on both edges, so the top slightly resembles a small borer. The left edge shows use-"retouch"; the dorsal face of the artefact shows white patina.

19. Fig. 7:25 (*245*). Length 33 mm, breadth 10 mm, thickness 4 mm. The backing is on the left edge; that is rather strongly curved, as well as the right edge. The top lies on the longitudinal axis, distally. Towards the base the retouch bends round to the right, so the base becomes slightly narrower. Base-retouch present. The right edge is utilized.

20. Fig. 7:26 (*98*). Length 32 mm, breadth 11 mm, thickness 5 mm. The backing is on the right. The top has broken off, and was situated distally. The artefact must originally have been c. 33 mm long. The back is retouched ventrally. Percussion bulb present. The artefact is made out of a flake. Only the top two-thirds of the back is retouched (here the same applies as to no. 13).

21. Fig. 7:27 (*957*). Length 31,5 mm, breadth 11 mm, thickness 6 mm. The backing is on the left edge, that is strongly curved; the proximal top is retouched ventrally and dorsally.

22. *939* (not illustrated). Length 29 mm, breadth 10 mm, thickness 7 mm. The backing is on the right. The distal top is retouched ventrally and dorsally, and has broken off. The original length would have been c. 31 mm. The left edge is marginally retouched almost from the base over a distance of 27 mm. Percussion bulb present. The base is formed by a triangular striking-platform remnant, of which the sides measure 9 mm and the base 7 mm in length; in view of this thickness shafting is unlikely. The artefact shows a white patina (stippled here and there).

23. Fig. 7:28 (*42*). Length 29 mm, breadth 11 mm, thickness 5 mm. The backing is on the left. Top proximal. Back strongly curved. The artefact is probably broken and the original length cannot be estimated.

24. Fig. 7:29 (*1029*). Length 28 mm, breadth 12 mm, thickness 3 mm. The backing is on the left edge, that is strongly curved. Of the back the top 17 mm is dorsally retouched, the following 4 mm is ventrally retouched and the rest is again dorsally retouched. The top is proximal. The artefact is probably broken and the original length cannot be estimated.

25. *304* (not illustrated). Length 27 mm, breadth 10 mm, thickness 4 mm. The back runs straight, and is on the left edge; the right edge is strongly curved. Burnt (*craquele*). The percussion bulb is no longer recognizable. Top distal.

26. *1537* (not illustrated). Length 21 mm, breadth 9 mm, thickness 4 mm. The backing is on the left. Towards the base the retouch bends round to the right, so that the base becomes narrower. Top distal; percussion bulb present.

D. Azilian point

1. Fig. 7:30 (*926*). Length 55,5 mm, breadth 13 mm, thickness 5,5 mm. The backing is on the left. Very fine use-"retouch" is present on the lowermost 20 mm of the right edge.

E. Creswellian points

2. Fig. 7:31 (*594*). Length 31 mm, breadth 10 mm, thickness 6 mm. The backing is on the left. Top distal; percussion bulb present. Shafting unlikely, on account of the thickness of the percussion bulb. Base more or less retouched all round, on the left dorsally, on the right ventrally.

2. Fig. 8:32 (*426*). Length 24 mm, breadth 9 mm, thickness 5,5 mm. Backing on the right. Overall shape dumpy. Top distal; percussion bulb absent. Back retouched ventrally and dorsally at the base, and ventrally at the top.

F. Shouldered points

1. Fig. 8:33 (*1671*). Length 43 mm, breadth 25 mm, thickness 11 mm. This artefact is broken, and the distal top is missing. The en-

tire point must originally have measured c. 62 mm in length. On account of its massive size this artefact can hardly be ascribed to the shouldered points. Nor can we classify it as a Lyngby point; the Lyngby points belong to the tanged points. The retouch on the left edge of the base is insignificant here compared with the heavy retouch on the right edge, so we cannot speak of a tang. Taute (1968:11) comments on this matter: *,,Als Stielspitzen sollten ferner nur solche Stücke bezeichnet werden, deren Schaftangeln einigermassen gleichwertige Retuschierung beider Kanten aufweisen, da anderfalls die Unterscheidung zwischen Stiel- und Kerbspitzen verwischt wird''.*

2. Fig. 8:34 (*524*). Length 32,5 mm, breadth 14 mm, thickness 6 mm. Backing on the left, retouched both ventrally and dorsally. Shoulder on the right. Base missing. The percussion bulb must have been at the base. Top distal.

3. Fig. 8:35 (*41*). Length 32 mm, breadth 10 mm, thickness 4 mm. Complete artefact. Backing on the left, over the entire length. Base-retouch present. Shoulder on the right. Percussion bulb at the base.

4. Fig. 8:36 (*1611*). Length 29 mm, breadth 10 mm, thickness 4 mm. Backing on the left, over the entire length, the topmost 18 mm dorsally, the lowermost 11 mm ventrally and dorsally. The distal top is formed on the right edge by a kind of burin facet and two flat flake-negatives on the ventral surface. Shoulder on the right. Basal part (with percussion bulb) is missing. Since the back is quite strongly curved on the left edge, and the shoulder-retouch on the right edge is rather straight and does not therefore constitute a notch, the artefact could perhaps better be classified as a tanged point.

5. Fig. 8:37 (*691*). Length 22 mm, breadth 9 mm, thickness 3 mm. The (distal) top part of the artefact is missing. The backing, on the left, is over its entire length (insofar as it is present). Flat percussion bulb at the base. The right edge of the blade has two ventral notches opposite the shoulder notch. The 2-mm-long part between the two notches shows three very fine dorsal retouch strokes. This artefact is very reminiscent of some points of the Hamburgian tradition from Duurswoude-IV (see Bohmers & Houtsma, 1961: 139, fig. IV:1, and 145, fig. IX:3 and 7; in all three cases the shoulder has been made on the left and the so-called hafting notches have been made on the right, ventrally). In addition there are six artefacts from Duurswoude-IV (145, fig. IX:1, 2, 4, 8, 9, 11) where the shoulder has been made on the right and the hafting notches have been made on the left edge, dorsally.

6. *819* (not illustrated). Length 21 mm, breadth 13 mm, thickness 6 mm. Broken, only basal part present. Backing on the right. Shoulder on the left.

G. Long B-points

1. Fig. 8:38 (*457*). Length 51,5 mm, breadth 11 mm, thickness 4 mm. Backing on the left over a length of 20 mm. Top distal. Percussion bulb at the base. We classify this as a long B-point, because the back is only over a small part of the blade. In terms of shape, however, it is more like a well-proportioned Gravettian point (see also no. 13, fig. 7:19, Movius *et al.*, 1968: 46).

2. Fig. 8:39 (*1593*). Length 42 mm, breadth 10 mm, thickness 4 mm. Backing on the left at the top by means of fine retouch. The tool could also be regarded as a retouched blade, but the narrowed base and the removal of part of the percussion bulb could be indicative of shafting, so the term point seems to be preferable here.

H. A.-point

1. Fig. 8:40 (*1037*). Length 24 mm, breadth 7 mm, thickness 2 mm. Backing on the right, at the top end 8 mm ventrally, at the bottom 16 mm dorsally. Percussion bulb (at the base) has disappeared as a result of retouch.

5.2.3 *Backed blades* (total no.: 51; fig. 8)

The broken specimens could be fragments of

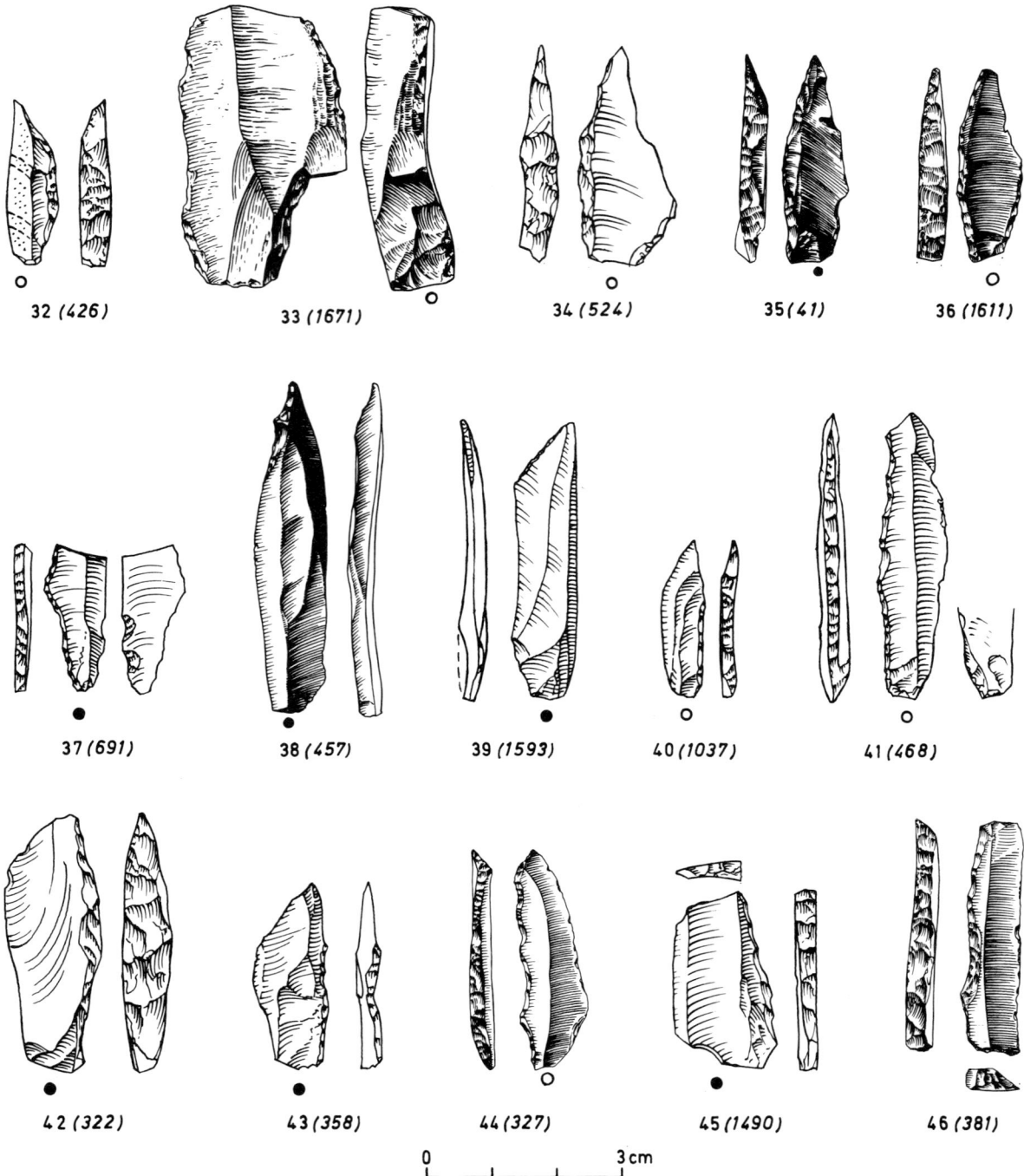

Fig. 8. Flint artefacts: 32 Creswellian point, 33-37 shouldered points, 38-39 long B-points, 40 A-point, 41-46 backed blades (the numbers between brackets indicate the inventory numbers).

points. A few specimens from Een-Schipsloot are not made out of blades, but out of flakes. We distinguish the following types:

A. Only one edge backed, no end-retouch

(called *Rückenmesser* by Schwabedissen and Taute), 45 specimens.

B. Both edges backed (called *parallelseitiges Messer* by Schwabedissen and Taute), 3 fragments. (These could also be fragments of so-called Kremser or Font-Yves points; see Bohmers, 1947; Garrod, 1926).

C. Right or left edge backed plus one of the ends (called *Messer mit returschiertem Ende* by Schwabedissen and Taute), 2 specimens.

D. Right or left edge backed plus both ends (called *Rechteckmesserchen* by Schwabedissen and Taute), 1 specimen. The broken specimens are not described in any further detail.

Backed blades, type A

1. Fig. 8:41 (*468*). Length 43 mm, breadth 10 mm, thickness 4 mm. Backing on the left. Percussion bulb present. Use-"retouch" present on right edge.
2. Fig. 8:42 (*322*). Length 40 mm, breadth 15 mm, thickness 8 mm. Backing on the right. RA-burin at the base. Along the left edge, that is strongly curved, use-traces are present.
3. Fig. 8:43 (*358*). Length 28 mm, breadth 11 mm, thickness 4 mm. Backing on the right, only the lowermost 18 mm retouched. Flat percussion bulb at the base. Left edge strongly curved; over a span of 3 mm fine retouch ventrally at 15 mm from the base.

Backed blades, type C

1. Fig. 8:44 (*327*). Length 34 mm, breadth 10,5 mm, thickness 3 mm. Backing on the left, more or less undulating, bending round to the right near the top and forming there an oblique truncation. Top distal. Right edge utilized.
2. Fig. 8:45 (*1490*). Length 27 mm, breadth 15 mm, thickness 3 mm. Backing on the right. Flat percussion bulb at the base. End-retouch is almost perpendicular to the back on the right. Left edge utilized.

Backed blades, type D

1. Fig. 8:46 (*381*). Length 36 mm, breadth 9 mm, thickness 4 mm. Backing on the left

and on both ends. Percussion bulb has disappeared as a result of retouch. Right edge utilized.

The six artefacts described here above, of the types A, C and D, are all complete; all the other specimens are broken.

5.2.4. *Triangles, rectangle* (total no.: 3; fig. 9)

Both of the triangles are made out of flakes. In the case of one specimen the 3 edges all show rather shallow retouch; one of the corners is unworked and still shows a fragment of cortex. The length of this specimen is 38 mm, the thickness 8 mm.

The other triangle (fig. 9:54) has one unretouched edge, that shows fine retouch near the corners. The two other edges are steeply retouched. One of the corners has broken off. The length is c. 45 mm, the thickness 8 mm.

The only rectangle has four more or less perpendicular edges, of which three are retouched semi-abruptly, the fourth edge is a fracture surface. The cross-section of this specimen, in contrast to that of the backed blades, is trapeziform. The breadth is 16 mm, the thickness 4 mm.

5.2.5. *Borers* (total no.: 37, fig. 9:51-52)

Among the tools 37 borers are present, of which 16 are broken, either at the apex or at the base.

From close inspection it appears that 4 specimens are made out of blades, 2 out of blocks and the rest out of core-preparation pieces or core-renewal pieces.

The apex of one borer is oriented almost at right angles to the longitudinal axis; all the other borers have an apex with almost the same orientation as the longitudinal axis. The apexes are generally fairly blunt, and triangular in cross-section; some show fine retouch applied from the extremity.

One specimen, that is made out of a block, has an apex that has been produced by making two deep notches. Slight traces of use are present on the otherwise unretouched apex.

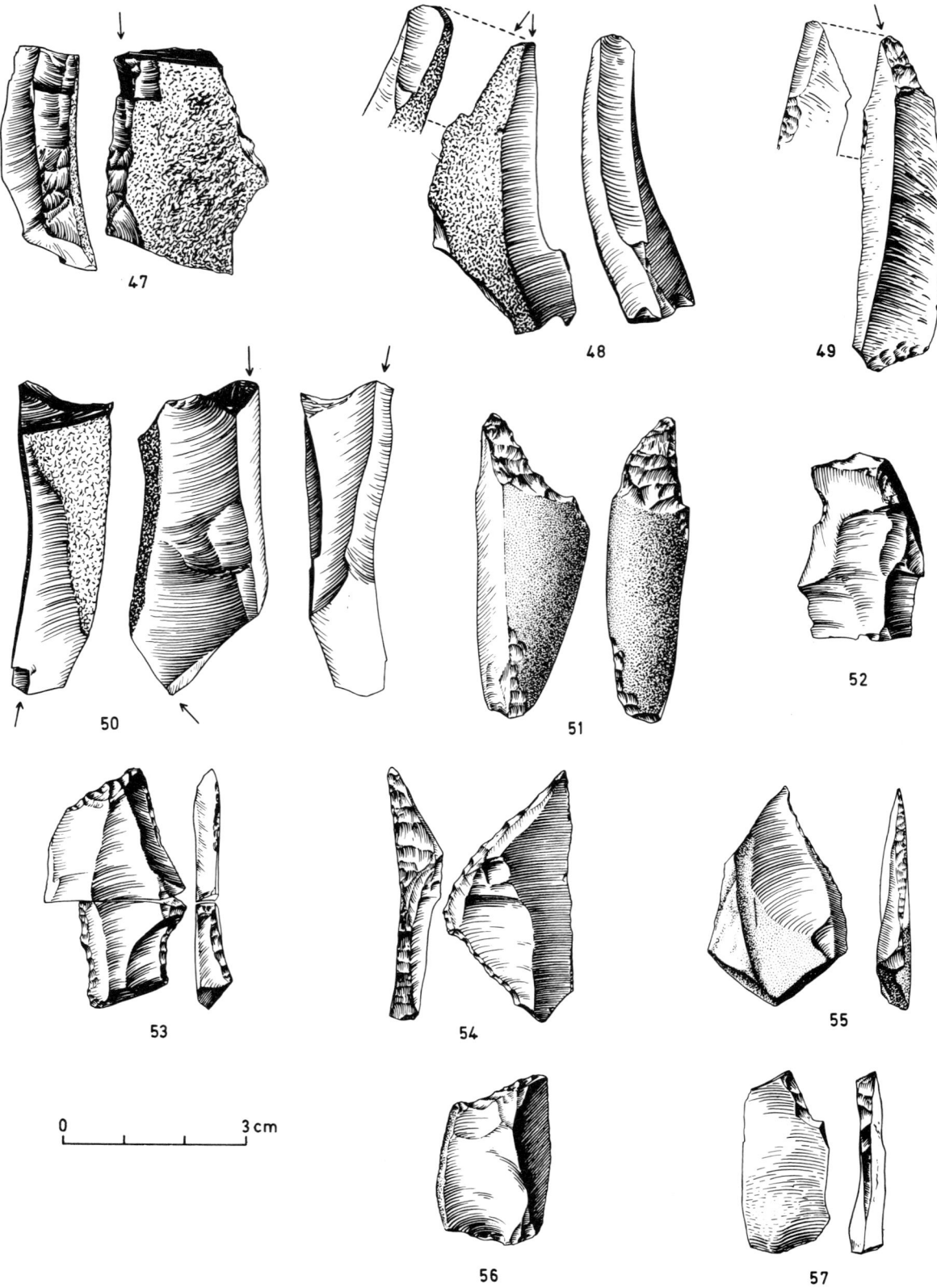

Fig. 9. Flint artefacts: 47 A-burin, 48 AA-burin, 49 RA-burin, 50 multiple burin, 51-52 borers, 53 disregarded, 54 triangle, 55 flake with retouch, 56 truncated blade, 57 notch remnant ("microburin").

Seventeen borer-apexes are retouched on one edge. Of these 12 show ventral retouch, and the rest dorsal retouch, with the exception of one specimen, that is worked both dorsally and ventrally. This is a large borer with a thick apex, that is triangular in cross-section, with some fine, longitudinally running retouch present on the lower edge.

Twenty borers show retouch on both edges of the apex, either dorsally (18 specimens), ventrally (2 specimens), or alternating, i.e. one edge ventrally and the other edge dorsally (6 specimens). Dimensions: length varying from 24-61 mm, breadth from 7-40 mm and thickness from 3-5 mm.

For the sake of completeness it should be mentioned that according to the terminology currently employed in Germany 5 of the borers described here above can be ascribed to the *Zinken*.

5.2.6. *Scrapers* (total no.: 406; fig. 10:58-69)

A scraper is an implement with a working edge that is not sharp and that is usually retouched dorsally. The angles, made by the retouch of this working edge and the ventral edge vary in the material concerned here from 30 to 75°.

This morphologically rather heterogeneous group of tools fall into distinct categories if one distinguishes the original basic material out of which they were made.

We distinguish primary and secondary basic material. We apply the term primary basic material (A) to flakes and blades that served primarily as blanks for the production of tools — in this case scrapers. We use the term secondary for material that was initially intended for some purposes other than the preparation of tools, such as cores and blocks (B), and by-products such as core-preparation pieces and core-renewal pieces (C), but that was evidently considered suitable for the production of scrapers.

A. Scrapers made out of primary basic material (total no.: 268)

The scrapers made out of flakes and blades can be further subdivided according to the place where the working edge has been made: end scrapers, side scrapers and round scrapers.

a. Round scrapers (8 specimens) — the working edge runs more or less uninterrupted all round; the shape is more or less circular. The diameter varies from 16-26 mm (average 20,13 mm), the thickness from 5-10 mm (average 7 mm), fig. 10:58.

b. Side scrapers (1 specimen) of the type consisting of side scrapers made out of flakes/ blades only one specimen has been found; length 33 mm, breadth 19 mm. The working edge, that is convex, runs parallel to the longitudinal axis, fig. 10:59.

c. End scrapers (259 specimens) — the working edge is situated at the distal extremity, perpendicular to the longitudinal axis. This group consists of 2 subtypes:

Double scrapers (28 specimens) — working edges have been prepared at both the proximal and the distal extremity. Length 18-30 mm; breadth 11-28 mm, thickness 5-11 mm, average length 24 mm, average breadth 19 mm, average thickness 8 mm, fig. 10:61.

Single scrapers (231 specimens) — the working edge is situated at the distal extremity of the flake or blade. This category includes a large number of broken pieces (127 specimens). On the basis of the shape of the working edge the single scrapers can be divided into: symmetrically convex (69 specimens) fig. 10:62-63, asymmetrically convex (24 specimens) fig. 10:65-66, straight (5 specimens) fig. 10:67, and denticulated (6 specimens) fig. 10:64. This classification, that is simple compared to those of others (Movius *et al.*, 1968), can be applied at a glance and without any danger of mistakes being made.

B. Scrapers made out of blocks or cores (total no.: 31; fig. 10:68-69)

Scrapers made out of blocks/frost-split fragments (25 specimens) form, together with those made out of cores/core fragments (6

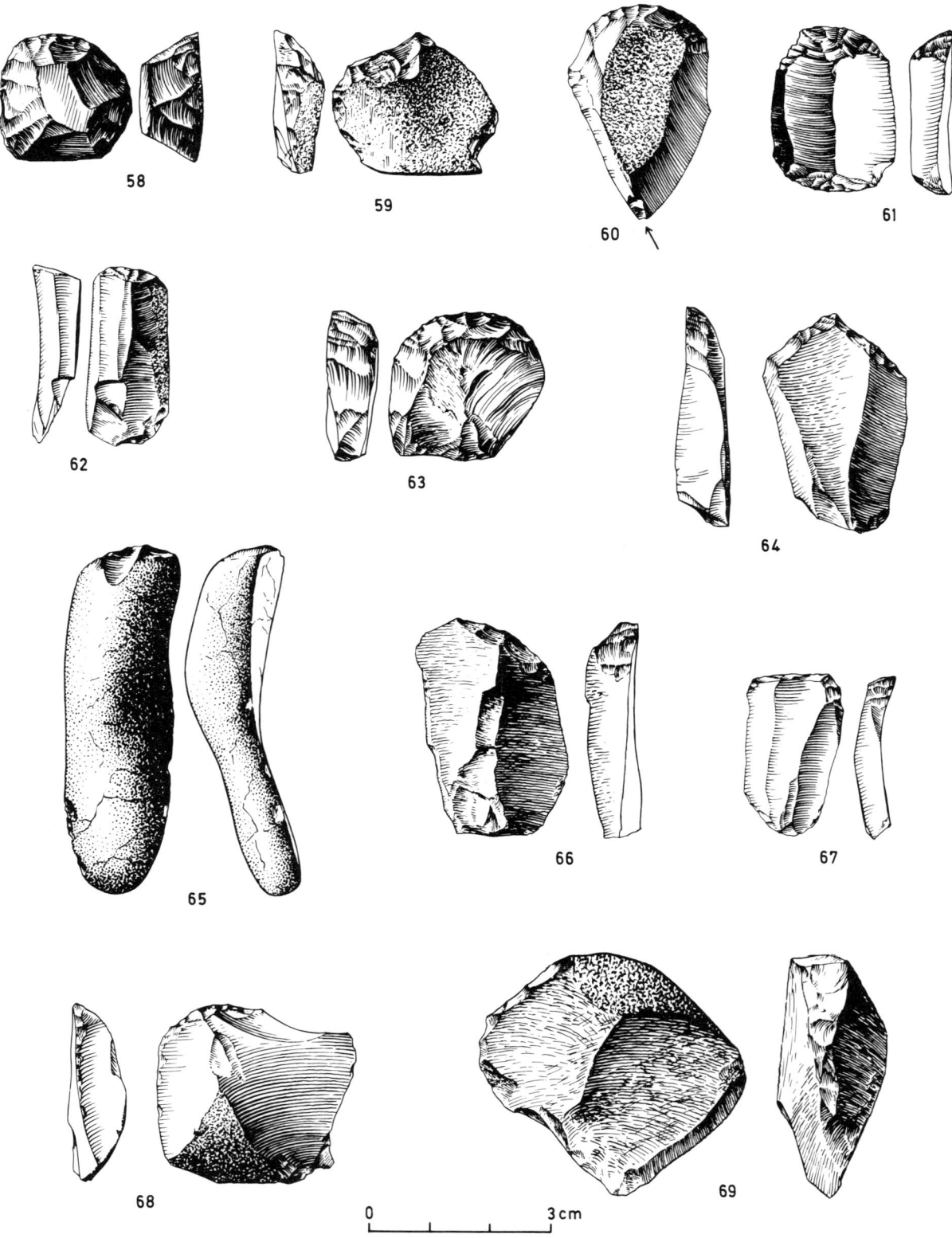

Fig. 10. Flint artefacts. Primary basic material: 58 round scraper, 59 side scraper, 60 combination burin/end scraper, 61 double end scraper, 62-63 end scrapers, symmetrically convex, 64 end scraper, denticulated, 65-66 end scrapers, asymmetrically convex, 67 end scraper, straight. Secondary basic material: 68 end scraper, 69 side scraper.

specimens), a category of highly irregular tools. It is not possible to make a subdivision into specific types. There are no morphological similarities, neither in terms of shape, nor of situation of the working edge, nor of size.

The working edge, that has usually been prepared in a rudimentary fashion, is of varying form: it may be convex, concave, straight, or denticulated. The scraper angles also vary considerably.

Among the tools these scrapers can in fact be regarded as occasional tools, that were made to be used only once, out of pieces of flint that just happened to be available. The maximum length varies from 22-56 mm with an average value of 36,57 mm; the thickness varies from 6-13 mm.

C. Scrapers made out of core-preparation pieces or core-renewal pieces (total no.: 107)

This group can be divided into side scrapers and end scrapers, while the end scrapers can be further subdivided into combination, double and single scrapers. The single scrapers fall into 4 subgroups according to the shape of the scraper edge. Without dealing with all these various subtypes extensively, as we have done already with the scrapers made out of flakes and blades, we give some data here below.
a. Side scrapers (5 specimens, 1 of which is broken). The side scraper, insafar as it has been prepared on core-renewal pieces, is usually situated on the rim of the striking platform remnant, that formed part of the striking platform of the core, before the scraper was removed. The scraper angles are therefore rather steep (between 75 and 90°). The length varies from 25-46 mm.
b. End scrapers (102 specimens):
Combination side/end scrapers (2 specimens), length 30 and 35 mm.
Double end scrapers (6 specimens), length varying from 24-36 mm, average length 27,33 mm.
Single end scrapers (94 specimens): broken

(21 specimens), complete (73 specimens); length varying from 14-58 mm, average length 30,03 mm. According to the shape of the scraper edge the complete specimens can be further subdivided: symmetrically convex (58 specimens), asymmetrically convex (2 specimens), denticulated (8 specimens), straight (5 specimens).

The broken scrapers vary in length from 15-40 mm, average length 20,1 mm. The complete scrapers vary in length from 11-56 mm, average length 27,6 mm.

The original scraper angles (before the tool was used and/or additional retouch was applied) were measured in classes of 15° and then only for the single scrapers (total no.:211). The measurements show a normal distribution, with the modus in the 45-60° class (50.7% of the material). The angles vary from 30-75°. If the scraper angles had been measured close to the working edge (i.e. mainly angles that arose after wear and the application of additionel retouch) then higher values would have been obtained.

5.2.7. *Burins* (total no.: 161)

Among the tools, after the scrapers the burins form the largest type-group, with 161 specimens.

The subdivision of the burins is based on the classification of De Sonneville-Bordes and Perrot (1956) and the study of Movius (1968) that is expanded upon here below.

Bohmers (1956) straight away made use of this classification and conceived Dutch equivalents for the French terms:
a. *burin dièdre* (Dutch: *tweevlaksteker, AA-steker*); dihedral burins.
b. *burin sur troncature* (Dutch: *afgeknotte middensteker, RA-steker*); truncation burins.
c. *burin sur cassure/sur plan naturel* (Dutch: *steker op gebroken einde of natuurlijk vlak, A-steker*); break burins.

The term *troncature* is in fact used in the literature rather loosely. Often there is not so much a truncation as rather simply the occurrence of marginal retouch, that does not influence the shape of the tool.

We share Bohmers' scepticism regarding a subdivision according to the location of the burin edge with respect to the longitudinal axis of the tool such as lateral and median burins (Movius *et al.*, 1968:34). This method, that is often applied to Late Palaeolithic material, is rather unreliable, in view of our experience with the collection from Een-Schipsloot, because the longitudinal axis cannot always be determined accurately.

A. Multiple burins (26 specimens, fig. 9:50)

There are 22 tools with burins at both ends, that can be divided as follows:
4 combinations of RA- and RA-burins;
6 combinations of RA- and AA-burins;
5 combinations of RA- and A-burins;
3 combinations of AA- and AA-burins;
2 combinations of AA- and A-burins;
2 combinations of A and A-burins.
Four tools have two burins at one end, namely:
3 specimens with 2 RA-burins;
1 specimen with 2 A-burins.

B. RA-burins (68 specimens, fig. 9:49)

42.88% were made out of flakes/blades, 10.72% out of blocks/core pieces and 34.48% out of core-preparation/renewal pieces.

The length values (in the case of blocks the length is measured parallel to the burin facet) vary from 17-61 mm, the average length is 35.5%; 56% lies between 25 and 40 mm. The burin angle was measured with the aid of a protractor with a level (Movius *et al.*, 1968:16). 74.1% have a burin angle between 60 and 80°. Also the width of the burin edge was noted. This varies considerably, from 1-11 mm. The reason for this wide variation undoubtedly lies in the different basic material that was used for making the burins. The most frequently occurring values fall between 3 and 4 mm (60%).

C. A-burins (43 specimens, fig. 9:47)

The basic material consists for 19.26% out of flakes/blades, for 53.5% out of blocks/core pieces and for 19.26% out of core-preparation/renewal pieces. The length varies from 20-54 mm, the average length is 37,03 mm, 52.63% of the total has a length between 30 and 40 mm. Of the burin angles 51.8% fall between 70 and 80°. The width of the burin edge shows the highest frequency between 4 and 5 mm (44.3%).

D. AA-burins (24 specimens, fig. 9:48)

Flakes/blades 25.86%, blocks/core pieces 38.79%, core-preparation/renewal pieces 34.48%. Length varies from 23-68 mm, with the highest frequency between 25 and 45 mm (79.17%), the average length is 38,79 mm. Burin angles between 50 and 70° form 51.8% of the total. A striking feature however is the high percentage of burin angles of 90°, namely 33.3%! 58.3% of the burin edges have a width of 6-8 mm.

A remarkable feature is the great difference in basic material between RA-burins and A-burins. Only a small proportion of RA-burins (10.72%) are made out of blocks and core pieces. Yet these coarse irregular pieces determine the character of more than half (53.50%) of the A-burins. Taking into account the fact that this type was made without any previous preparation of the striking platform, and by means of only one burin blow, it can be said that the A-burin is the most rudimentary tool among the burins.

The RA-burin, most usually made out of primary basic material and characterized by a carefully prepared striking platform, is a more sophisticated tool. Moreover the frequency distribution of the widths of the burin edges shows that 74% of the RA-burins have a burin edge 3-5 mm wide. This relatively large number of narrow burin edges, compared with the other types, is possibly indicative of a specific function.

A frequency distribution of the burin angles shows a remarkable feature of the AA-burins. While the RA- and the A-burin angles have a normal distribution, with a modus at 70°, the distribution of AA-burin angles is bimodal and has a modus at 90°. The sharp decline after 90° can be explained by the technical impossibility of removing a burin spall that would form an angle of more than 90° with the striking platform.

5.2.8. *Burin spalls* (total no.: 170)

These are not included in the category of tools, but are discussed here. Of the total number of 170 burin spalls 44 show retouch on the back. This retouch was applied to the blade prior to effecting the burin blow, with the intention of influencing the splitting of the burin spall (Bosinski & Hahn, 1973: 135).

Eleven of these burin spalls with retouch show retouch at the distal end; this means in fact that this retouch determined the lenght of the spall. This retouch at the distal end is called termination retouch. The other 8 specimens show retouch in the middle of the back.

Ninety-eight burin spalls are primary: these are flakes with a triangular cross-section. The other 42 are secondary; these show the negative of a previous spall. As Bosinski and Hahn already ascertained, it is mainly the primary burin spalls that are retouched; there are only 3 secondary burin spalls with retouch as against 14 primary ones.

The ratio of burin spalls to burin edges is 170:205. Here it must be pointed out that for each burin edge one burin spall is reckoned, thus including those that have more negatives. Although the number thus obtained, 205, is on the low side, the ratio does not seem to be unsatisfactory.

All the more so if one considers the figures given by Hahn for the Magdalenian site of Andernach: 32 burin spalls against 78 burin edges, i.e. more than twice as many burin edges.

5.2.9. *Combination tools (scraper/burin)* (total no.: 18; fig. 10: 60)

Of these tools 9 are made out of blades/flakes. The length varies from 16-39 mm, with an average value of 30 mm. Five specimens (of which 2 are complete) are combinations of a side scraper/burin, made out of core-preparation/renewal pieces, while the same basic material was used for making another four specimens (of which 3 are broken) with a combination of end scraper/ burin.

5.2.10. *Obliquely truncated blades* (total no.: 6; fig. 9: 56)

There are 4 complete specimens, and 2 fragmentary ones. Three oblique truncations are straight, 2 concavely retouched; in the case of the 6th specimen the truncation is almost perpendicular to the longitudinal axis of the artefact.

5.2.11. *Tools with notches* (total no.: 35)

This category includes 22 notched blades (including fragments). Distribution must be made between:
a. shallow notches, resulting from a small amount of fine retouch and
b. deep notches, resulting from a substantial application of retouch.

Notches of type a. occur on 2 complete and 9 broken blades (among the broken blades there is one particularly long specimen, 69 mm). On 6 of the blades traces of use can be seen on both sides of the notch. Of type b. 11 specimens are made out of blades. It is remarkable that for 4 examples the notch borders on to a transversal fracture. It is possible that these notches were made with the aim of breaking the blade on that spot. One specimen even has a notch on each side of a fracture. The presence of one typical notch remnant ("microburin",

Fig. 11. Photo of the profile wall of the Schipsloot with basin I. The culture level is indicated by the measuring-staff and is visible in the wall as a slightly paler band.

Fig. 12. Photo of basin I with the underlying sands.

fig. 9:57) in the material shows that this technique was already applied here (sporadically). On the remaining 7 specimens the notches do not border on to fractures.

There are 2 notched flakes. One example has a notch on a fracture surface, the other (very fragmentary) specimen has a notch next to the percussion bulb.

Seven core preparation pieces (of which 5 are broken) have a notch that has been crudely made, and finally 4 blocks have a notch too.

5.2.12. *Retouched blades, flakes, blocks* (total no.: 104; fig. 9:55)

These are all pieces with partial or marginal retouch (sometimes irregular), that is not restricted to any particular place.

6. CULTURAL RELATIONS

It is clear that we are not dealing with a site of the Ahrensburg tradition; typical tanged points

Fig. 13. Photo of the SW-NE profile of basin II. The white plastic squares indicate the location of the flint artefacts (see fig. 4 right).

are lacking, and the few B-points are of a different type (long) to that of the Ahrensburg tradition (short). Nor can we ascribe Een to the Hamburg tradition, despite the occurrence of a few atypical shouldered points and *Zinken.* Within the Hamburg tradition there are as a rule no backed blades or large quantities of Tjonger points.

The site must clearly be ascribed to the widely spread *Federmesser* tradition. As the name Tjonger tradition has been in general use for almost 40 years (Siebinga, 1944; Bohmers, 1947), in the Netherlands we mostly use this term (in France the term Azilian is used).

Various authors have subdivided the *Federmesser* tradition into groups on typological grounds, notably Schwabedissen (1954). Within his classification we can best ascribe Een to the so-called *Rissener Gruppe,* on account of: a. the limited size of the tools, b. the large group of short scrapers, and c. the fairly frequently occurring double scrapers. The *Rechteckmesser* that is present is however more a characteristic of his *Wehlener Gruppe* (cf. Schwabedissen, 1954: 17, fig. 5m) while Een also possesses characteristics of his *Tjonger Gruppe,* namely the presence of 3 Creswellian points, 3 Châtelperronian points and in addition a couple of tools that are reminiscent

of *Zinken.* The Creswellian points are small, however, and not very typical; and moreover few in number; the same applies equally to most of the *Federmesser* sites in Northern Germany (Bohnsack, 1956: 82 *"quantitativ wie qualitativ recht kümmerlich...").*

From the above it is clear that Een shows characteristics of all 3 groups mentioned by Schwabedissen. We therefore accede to the view of Paddayya (1971) that there is little sense in making any distinction between the *Tjonger, Rissener* and *Wehlener Gruppen.*

Another question concerns the relationship with the English Late Upper Palaeolithic, that Garrod (1925) called Creswellian. To this tradition we ascribe the Dutch sites of Haule V, Wijster, Siegerswoude II, Zeijen and Neer II, the Belgian sites of Lommel and Presle near Charleroi (Henegouwen), and the German site of Hohenholz near Steinhude (Lower Saxony). At these sites one finds greater numbers of Creswellian points, long B-points, Cheddar points, shouldered points and *Zinken* than at the normal Tjonger sites. For this tradition we prefer to continue using the term Creswellian (Campbell, 1977). We agree with Paddayya that the name Cheddarian (Bohmers, 1956) can better be dropped. Bohmers ascribed e.g. Siegerswoude II to the Cheddarian, although

Fig. 14. Photo of the NE-SW profile on basin III.

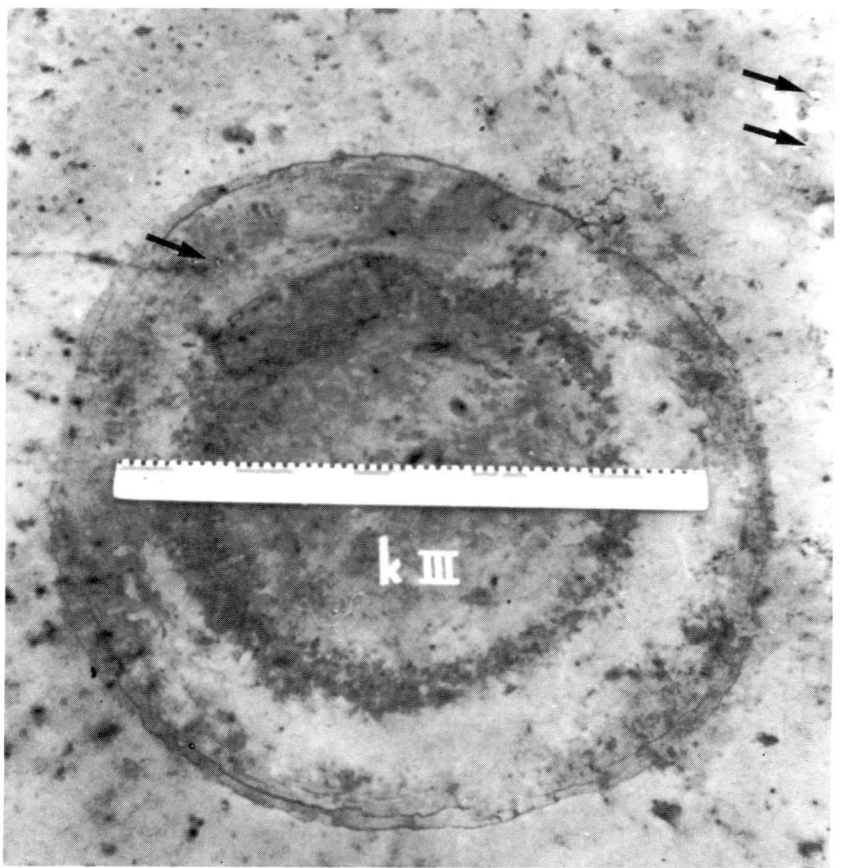

Fig. 15. Photo of basin III, as seen in a horizontal plane. The arrows indicate flint artefacts (see fig. 11 of Casparie and ter Wee).

not a single Cheddar point was found there. We find it hard to ascribe these sites to the *Federmesser* tradition, as Paddayya does. In Siegerswoude II and Zeijen, for example, there are no typical Tjonger points present, and at Siegerswoude II, Neer II, Zeijen and Hohenholz the characteristic backed blades are almost completely lacking. It is noteworthy that these do

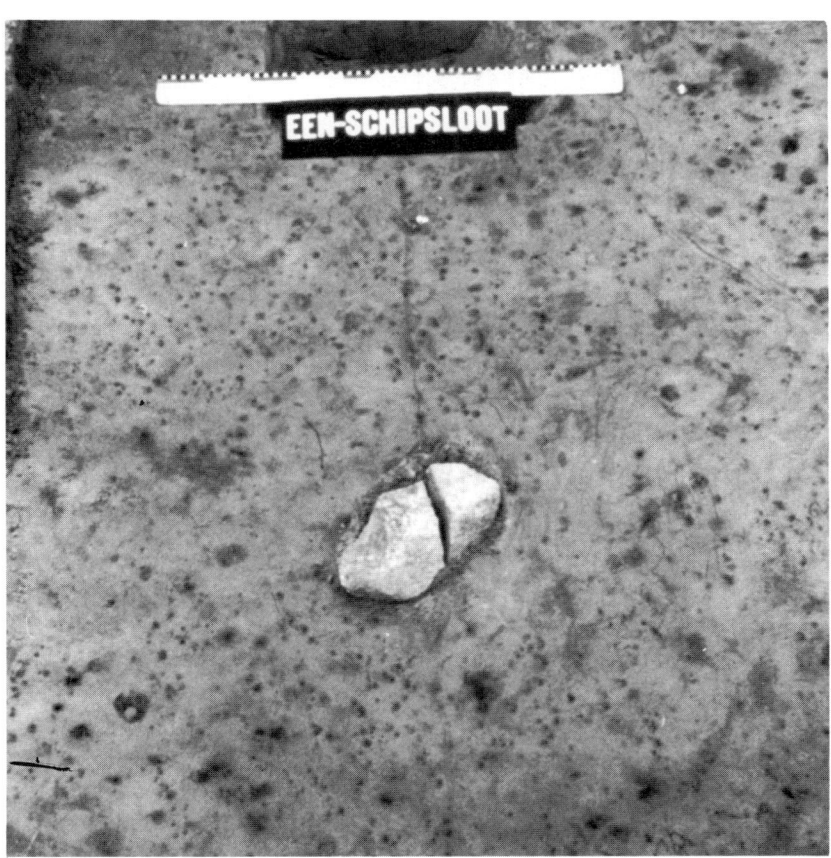

Fig. 16. Photo of a frost crack with a broken piece of granite.

occur in Lommel and Presle and at the English sites.

The first author does nog agree with Paddayya's view that the presence of B-points, Azilian points and tanged points in a *Federmesser* context are indicative of a younger phase (Paddayya, 1971: 266). So far there is no evidence for this. Long B-points also occur in the Hamburg tradition, and in a certain sense the Havelte points of this tradition can also be called tanged points.

About the shouldered points from Een-Schipsloot, that are represented by six specimens, including one complete artefact (plus another 6 fragments, which are possibly also broken shouldered points), can be said that they do not resemble the Hamburgian type very much. They are usually smaller, and in most cases it can be assumed that the retouch continued over one entire edge, without inter-ruption, as is also the general rule with Tjonger points. One could therefore also call them "Tjonger points with a shoulder". They are very similar to several types from Hengistbury Head (Mace, 1959: 246, fig. 4: 13-19).

The first author wonders whether the occurrence of points of this kind, that show affinity to the Hamburg type, is not perhaps indicative of rather a greater age.

As for the English Creswellian (within which shouldered points also occur), the precise age of this culture is still not known. From Aveline's Hole and Kent's Cavern we do know of biserial harpoons and these are also found in the French Magdalenian VI (Bohmers points this out already in 1947: 198; see also Lanting & Mook, 1977: 21-23).

Paddayya (1971: 262) points out that the Tjonger artefacts of Budel II and Waskemeer were found at those sites under the "layer of

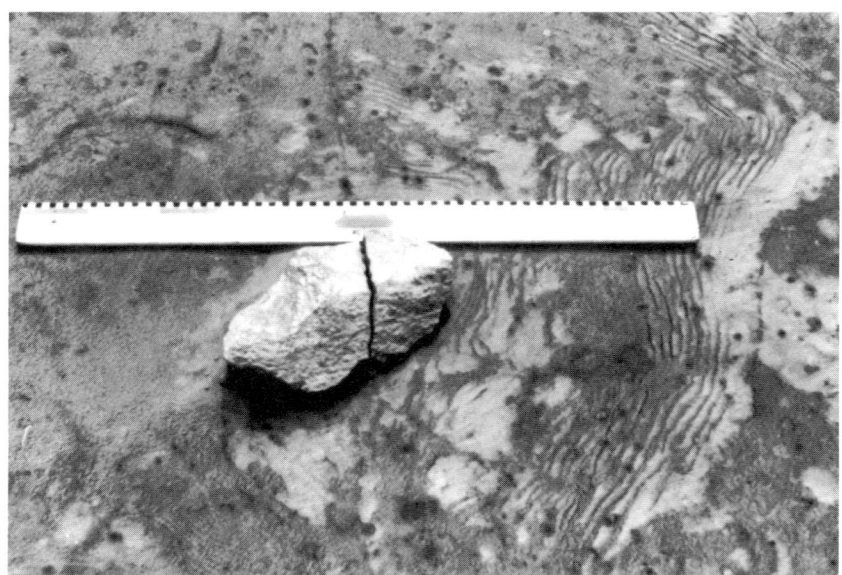

Fig. 17. Photo of the same stone as in fig. 16. Note the frost crack approximately perpendicularly above the middle of the staff. As a result of the wind blowing fine particles of sand across the excavated surface "fossil" wind ripples become visible once again in the culture level.

Usselo". The C 14-dating of Budel II is 9490 B.C. (GrN-1675). That of Waskemeer is 9200 B.C. (GrN-4871). Milheeze is dated to 8930 B.C. (GrN-2314), but the pollen analysis dates this site in the birch phase, i.e. the first part of the Allerød interstadial. In addition there is *Fundstelle* C of Westerkappeln, also a Tjonger site, with an age of 9850 B.C. (KI-271) (Günther, 1973; see for an overview of the C14-datings: Lanting & Mook, 1977).

The above-mentioned facts thus indicate that the Tjonger tradition is not restricted to the second part of the Allerød, but can also occur at an earlier stage (see also Paddayya, 1971: 262).

It is not impossible that the Hamburgian finds from Duurswoude II and IV actually date from the Bølling period (Bohmers & Houtsma, 1959). Between the end of the Bølling oscillation and the beginning of the Allerød there is a time difference of only a few hundred years. The similarity of the shouldered point, illustrated in fig. 8:37 (see 5.2.2), to those of the Havelte type from Duurswoude IV, is striking in this connection. These Havelte points also often have one or two small hafting notches opposite the shoulder. (These small notches also occur on shouldered points that are not of the Havelte type (Bohmers, 1956: 7). This raises the question whether the Hamburgian tradition in its last phase could not have been contemporary with the Tjonger tradition in an early phase.

The occurrence of backed blades type D and type C show furthermore, in our opinion, affinity with the Late Magdalenian of Central and Southern Germany. To name a few examples (Schwabedissen, 1954): Petersfels near Engen, Hegau (Tafel 90: 20-22); Ölknitz, Kr. Stadtroda, Thuringia (fig. 89; 2,4); and further north: Westerbeck, Kr. Gifhorn, Lower Saxony (fig. 44:30-31); Calbe-Kremkau, Kr. Salzwedel, D.D.R. (fig. 38:5).

To summarize it can therefore be said that the very variable point complex of Een-Schipsloot shows some affinity with both the English Creswellian and the Late Magdalenian of South and Central Germany. It could therefore be argued that the site of Een should be dated in an early phase of the Allerød interstadial. With regard to the C14-dating of the peat at the bottom of basin I, it can be remarked that this only means that the artefacts cannot be younger, while furthermore it is clear that most of the basins must have originated after the Upper Palaeolithic occupation.

The geological data available also make it clear, however, that part of the material in any case must date from the first centuries of the Late Dryas period, i.e. from after the Allerød interstadial (see 3: basin II, where the culture level continues undisturbed over the basin, while it is clear that the basin must have originated at the beginning of the Late Dryas) (see also stratigraphical table in Casparie & ter Wee, this volume). This clearly shows, that any view arrived at merely on the basis of typological evidence should be proffered at the very least with some caution. In this connection it must be remembered that the shouldered points from Een are atypical. The possibility remains, of course, that in Een we are dealing with artefacts of an early and a late phase, that now occur together in one level.

Usselo", that one would expect to find at this level, was not visible.

During the excavation ten basin-shaped depressions were found, that must have originated in a periglacial environment during the first part of the Late Dryas (see Casparie & ter Wee, this volume), as well as a number of frost cracks. It is clear that most of these basins originated (relatively soon) after the occupation. In the case of one basin, however, artefacts occurred both at the bottom of the basin and in the undisturbed culture level above the basin. This means that either this basin originated during the occupation period (which must in that case be dated in the first part of the Late Dryas), or that there were (at least) 2 occupation phases, of which in any case the latter must have the dating mentioned.

7. SUMMARY

In this article an Upper Palaeolithic site at Een (province of Drenthe, the Netherlands) is described. In the course of an excavation the first and third authors collected a total number of 8785 flint artefacts and several larger stones (including pieces of granite), that must have been brought to the spot by Palaeolithic hunters. An overview of the flint material can be found under 5.1 and 5.2.1. From a cultural point of view the site clearly belongs to the so-called *Federmesser* tradition, or (according to the currently accepted nomenclature in the Netherlands) the Tjonger tradition. Characteristic elements include: considerable numbers of Tjonger points among the points, backed blades, and large quantities of (usually short) scrapers (46.6 % of the tools). Some conspicuous components of the material include two Creswellian points, and six points resembling shouldered points, that are reminiscent of the Hamburgian tradition.

Stratigraphically the finds occur at the boundary between 2 layers of coversand, that are to be regarded as Younger Cover sand I and Younger Cover sand II, respectively. The "layer of

8. REFERENCES

ANDERSEN, S.N., 1972. Bro, en senglacial boplads på Fyn. *Kuml*, pp. 7-60.

BOHMERS, A., 1947. Jong-palaeolithicum en vroeg-mesolithicum. in: *Een kwart eeuw oudheidkundig bodemonderzoek in Nederland*. Gedenkboek A.E. van Giffen. Meppel, pp. 129-201.

BOHMERS, A. & Aq. WOUTERS, 1956. Statistics and graphs in the study of flint assemblages, I-III. *Palaeohistoria* 5, pp. 1-38.

BOHMERS, A., 1960. Statistiques et graphiques dans l'étude des industries lithiques préhistoriques, V. *Palaeohistoria* 8, pp. 15-37.

BOHMERS, A. & P. HOUTSMA, 1961. De prae-historie. In: *Boven-Boorngebied*. Rapport betr. het onderzoek van het Lânskip Genetysk Wurkforbân van de Fryske Akademy. Drachten, pp. 126-151.

BOHNSACK, D., 1956. Ein späteiszeitlicher Fund vom Hohenholz bei Steinhude. *Die Kunde* N.F. 7, pp. 67-84.

BOSINSKI, G. & J. HAHN, 1973. *Der Magdalénienfundplatz Andernach (Martinsberg)* (= Rheinische Ausgrabungen, Bd. 11). Bonn.

BUTTER, J. 1931. *Les silex de Budel.* Amsterdam.

CAMPBELL, J.B. 1977. *The Upper Palaeolithic of Britain.* Oxford.

CASPARIE, W.A. & M.W. TER WEE, 1981. Een Schipsloot-The geological-palynological investigation of a Tjonger site. *Palaeohistoria* 23, pp. 29-44.

CLARKE, D.L., 1968. *Analytical archaeology.* London.

DANTHINE, H., 1960. Fouilles dans un gisement préhistorique du Domaine de Presle (Hainaut). *Documents et rapports de la Société d'Archéologie et de Paléontologie de Charleroi* 50, pp. 1-38.

GARROD, D.A.E., 1926. *The Upper Palaeolithic Age in Britain.* Oxford.

GüNTHER, K. 1973. *Der Federmesser-Fundplatz von Westerkappeln, Kreis Tecklenburg* (= Bodenaltertümer Westfalens Bd. 13). Münster.

HEINZELIN DE BRAUCOURT, J. DE, 1960. *Principes de diagnose numérique en typologie* (= Mémoires de l'Académie Royale de Belgique). Bruxelles.

LANTING, J.N. & W.G. MOOK, 1977. *The pre- and protohistory of the Netherlands in terms of radiocarbon dates.* Groningen.

LEROI-GOURHAN, A. & M. BRéZILLON, 1972. *Fouilles de Pincevent; essai d'analyse ethnographique d'un habitat Magdalénien* (= Gallia Préhistoire, VIIe suppl.). Paris.

MACE, A., 1959. An Upper Palaeolithic opensite at Hengistbury Head, Christchurch, Hants. *Proceedings of the Prehistoric Society* 25, pp. 233-259.

MANBY, T.G., 1966. Creswellian site at Brigham, East Yorkshire. *The Antiquarian Journal* 46, pp. 211-229.

MORONEY, M.J., 1951. *Facts from figures.* London.

MOVIUS, H.L., N.C. DAVID, H.M. BRICKER & R.B. CLAY, 1968. *The analysis of a certain major classes of Upper Palaeolithic tools* (= American School of Prehistoric Research, Peabody Museum, Harvard Univ., Bull. 26). Cambridge.

MUSCH, J.E., 1974. Reconstructie van een jong-paleolithische wooneenheid in het Hoornseveld te Buinen, gem. Borger. *Nieuwe Drentse Volksalmanak* 91, pp. 139-160.

PADDAYYA, K., 1971. The Late Palaeolithic of the Netherlands. A review. *Helinium* 11, pp. 257-270.

PADDAYYA, K., 1973. A Federmesser site at Norgervaart, Prov. of Drenthe (Neth.). *Palaeohistoria* 15, pp. 167-213.

PETERS, E., 1930. *Die altsteinzeitliche Kulturstätte Petersfels.* Augsburg.

POPPING, H.J., 1931. Een Magdalénien-station op de Veluwe. *De Levende Natuur* 35, pp. 340-349.

POPPING, H.J., 1933a. Een en ander over de palaeolithische cultuur aan het riviertje „De Kuinder". *De Levende Natuur* 37, pp. 156-165.

POPPING, H.J., 1933b. Jong-palaeolithische werktuigvormen uit oppervlakte-vondsten. *Mensch en Maatschappij* 9, pp. 591-596.

POPPING, H.J., 1934. De jong-palaeolithische Kuinderculturen. Station Makkinga. *Mensch en Maatschappij* 10, pp. 378-400.

ROZOY. J.G., 1978. *Les derniers chasseurs; l'épipaléolithique en France et en Belgique: essai de synthèse.* Reims.

SACKETT, J.R., 1966. Quantitative analysis of Upper Palaeolithic stone tools. *American Anthropologist* 68, pp. 356-394.

SCHWABEDISSEN, H., 1954. *Die Federmesser-Gruppen des nordwesteuropäischen Flachlandes* (= Offa Bücher, Bd.9) Neumünster.

SCHWABEDISSEN, H., 1957. Das Alter der Federmesser-Zivilisation auf Grund neuer natur-wissenschaftlicher Untersuchungen. *Eiszeitalter und Gegenwart* 8, pp. 200-209.

SONNEVILLE-BORDES, D. DE & J. PERROT, 1954. Lexique typologique du paléolithique supérieur. I-II. *Bulletin de la Société Préhistorique Française* 51, pp. 327-355.

SONNEVILLE-BORDES, D. DE & J. PERROT, 1955, Lexique typologique du paléolithique supérieur. III. *Bulletin de la Société Préhistorique* 52, pp. 76-79.

SONNEVILLE-BORDES, D. DE & J. PERROT, 1956. Lexique typologique du paléolithique superieur IV-IX. *Bulletin de la Société Préhistorique Française* 53, pp. 408-412; 547-559.

TAUTE. W., 1963. Funde der spätpaläolithischen "Federmesser Gruppen" aus dem Raum zwischen mittlere Elbe und Weichsel. *Berliner Jahrbuch für Vor- und Frühgeschichte* 3, pp. 62-111.

TAUTE, W., 1968. *Die Stielspitzen-Gruppen im nördlichen Mitteleuropa* (= Fundamenta Reihe A, Bd. 5). Köln.

TROMNAU, G., 1975. *Neue Ausgrabungen im Ahrensburger Tunneltal* (= Offa Bücher, Bd. 33). Neumünster.

VERHEYLEWEGHEN, J., 1956. Le Paléolithique final culture périgordienne du gisement historique de Lommel (prov. de Limbourg-Belgique). Etudes géologique, stratigraphique et pétrographique par F. Gullentops. *Bulletin de la Société Royale Belge d'Anthropologie et de Préhistoire* 67, pp. 1-79.

SWIFTERBANT, OOST FLEVOLAND, NETHERLANDS:
EXCAVATIONS AT THE RIVER DUNE SITES, S21-S24, 1976
Final reports on Swifterbant III

T. Douglas Price*

CONTENTS

1. INTRODUCTION

2. GEOLOGY

3. PREVIOUS WORK AT PARCEL H46

4. THE 1976 INVESTIGATIONS
    4.1.   Mapping and borings
    4.2.   Excavations at S22
    4.3.   Excavation of S21, Grave XI
    4.4.   Excavation at S23
            4.4.1.   *Features*
            4.4.2.   *Artifacts*
    4.5.   Sondage at S24
    4.6.   Dating
            4.6.1.   *Isotopic*
            4.6.2.   *Artifactual*
    4.7.   Depositional Processes

5. SUMMARY AND CONCLUSIONS

6. ACKNOWLEDGEMENTS

7. BIBLIOGRAPHY

* Department of Anthropology, University of Wisconsin, Madison, U.S.A.

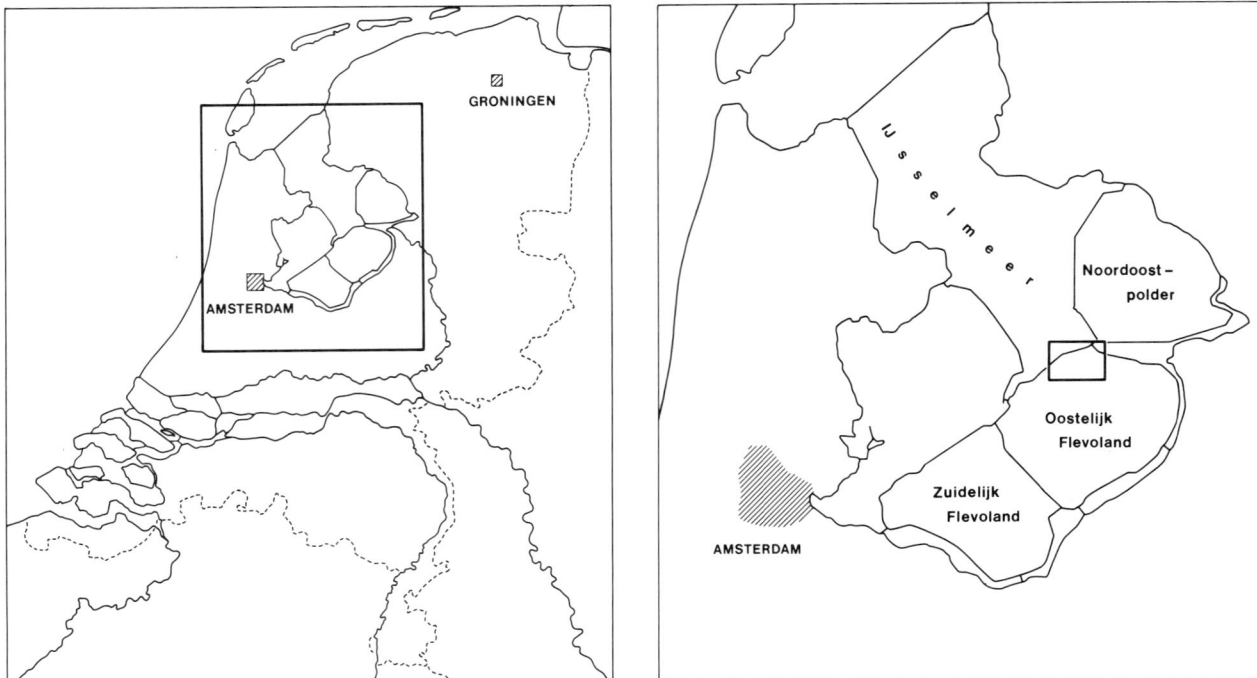

Fig. 1 a. The Netherlands at the present time, showing the location of the Swifterbant area.

1. INTRODUCTION

The region around the town of Swifterbant on the polder (reclaimed land) of Oost Flevoland in the Netherlands (fig. 1 a) offers rare opportunities to the archaeologist. Until roughly 3000 B.C.,* this area was part of a larger landscape that at one time extended across the dry bed of the English Channel. The gradual rise of sea level during the early Postglacial epoch transformed the region into a freshwater tidal delta at the mouth of the Old IJssel River, just prior to complete inundation. After 3000 B.C. this submerged surface was slowly covered with marine and freshwater deposits. Reclamation of the polder of Oost Flevoland, beginning with the closing of the dikes in A.D. 1956, once again has exposed the surface to the open air. Thus, beneath roughly one meter of sediments, at a depth of approximately five meters below modern sea level (Dutch N.A.P.: Nieuw

Amsterdams Peil), an intact land surface dating from prior to 3000 B.C. is preserved and accessible to the prehistorian.

The remains of Mesolithic and Neolithic materials were discovered on this buried surface initially in the 1960's in two distinct geomorphological situations: clay levees along the former stream channels and sand dunes along the former river banks. The investigation of the prehistoric human occupation of the area continued in the 1970's as a major research project under the overall direction of the Biologisch-Archaeologisch Instituut of the State University of Groningen.

This report details one portion of the larger research project, the excavations at one of the river bank dunes, located in the parcel designated as H46 on the polder of Oost Flevoland, to the north of the present town of Swifterbant (fig. 1 b). The excavation sites in this parcel are designated as S21, S22, S23, and S24 (fig. 2). These investigations were conducted by the University of Wisconsin-Madison in 1976. Geological investigations at the dune in

---

* In this paper, all radiocarbon dates cited are uncalibrated; therefore B.C. should be read as b.c.

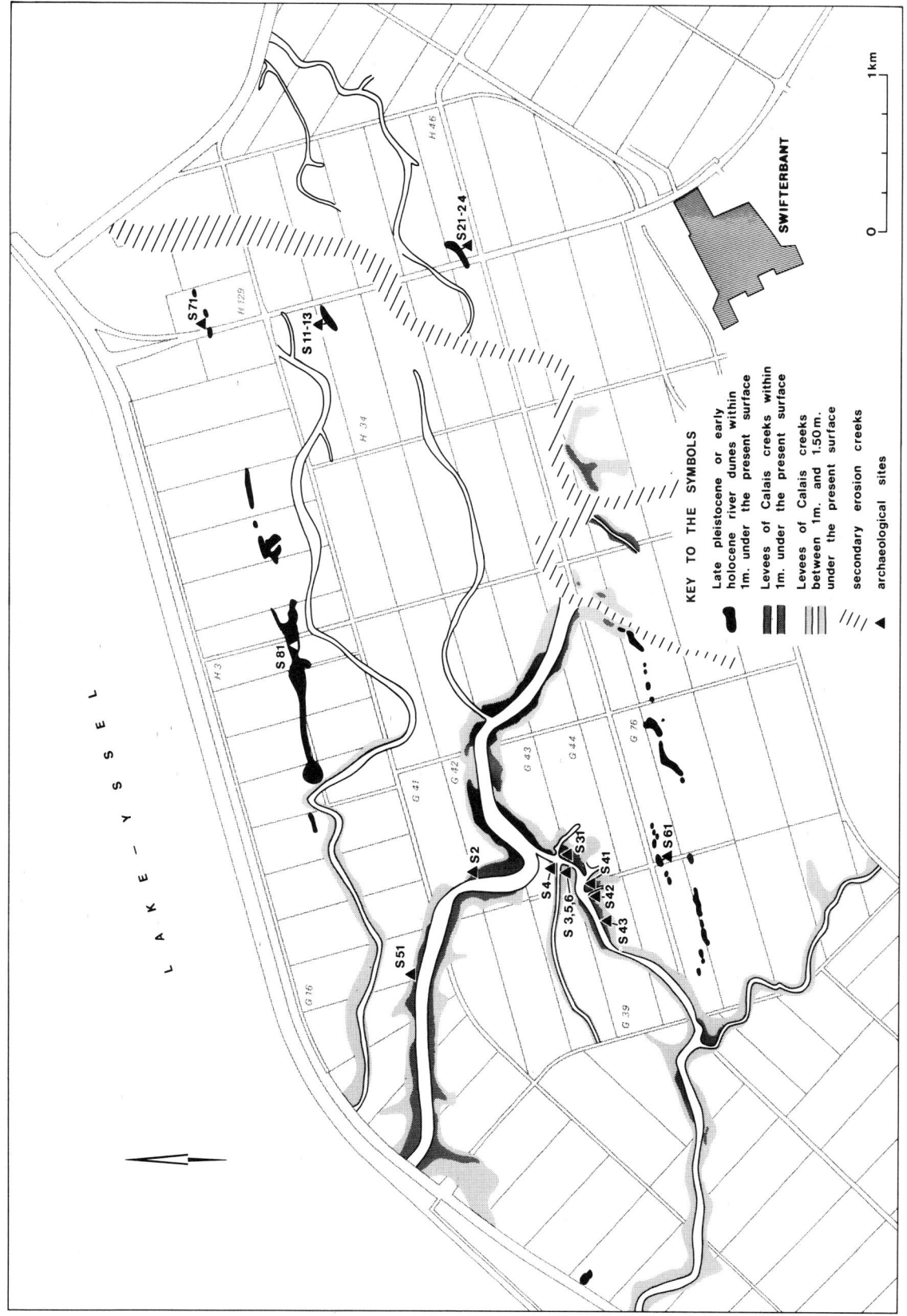

Fig. 1 b. Polder Oost-Flevoland, Swifterbant area. Location of sites on river dunes and on natural clay levees (Courtesy of the BAI).

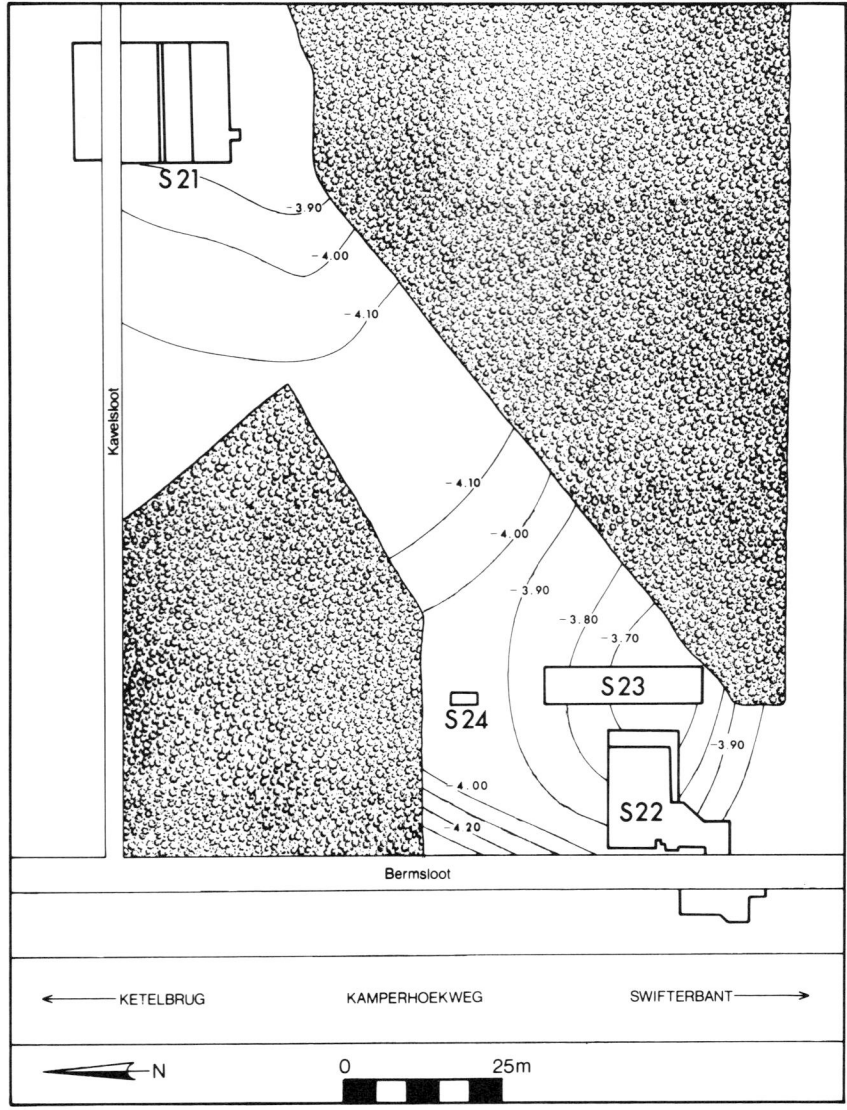

Fig. 2. Excavation units S21-S24 in parcel H46 at Swifterbant. Modern ground surface contours at 10 cm intervals. Areas of vegetation are woods and heavy undergrowth.

H46 are described by Ente (1971, 1976). A general discussion of the entire project appeared in Van der Waals and Waterbolk (1976). A list of all the publications of the Swifterbant project can be found in the bibliography.

2. GEOLOGY

Development of the land surface in the Swifterbant area in the Postglacial was primarily determined by the former Pleistocene land surface

and the presence of the Old IJssel River (Ente, 1976). Under the periglacial conditions prevalent at the close of the Pleistocene, broad areas of northwest Europe were subjected to eolian erosion and redeposition, creating the coversand topography that characterizes much of the northern Netherlands. During this same period, dunes were forming over vegetation along the banks of the Old IJssel River as it flowed through the Swifterbant area.

In the early Postglacial epoch, a relatively stable, wooded landscape developed along the channels and tributary streams of the river system. During the Atlantic period, after 5500 B.C., gradually rising sea levels transformed the area into a freshwater tidal delta, resulting in the deposition of clays and peats. After 3000 B.C. the region was completely submerged and the accumulation of sediments on top of the former land surface began.

Profiles from drainage canal cutting (Ente, 1976, fig. 7) and archaeological excavation (fig. 3) reveal portions of this depositional sequence in the stratigraphy. The profile of the dune can readily be seen in the section, marked by a slightly steeper slope to the south and a more gradual incline to the north toward the nearest stream channel. Soil development in the dune sand is seen in a weakly developed A and B horizon, more pronounced along the slopes of the dune. The grey A horizon grades quickly into a light-brown B horizon. On the higher portions of the dune, this soil profile is truncated with the disappearance of the A horizon and, on the highest parts of the dune, the disappearance of much of the B horizon. This problem will be considered in more detail below.

Directly above the surface of the dune, a layer of fen peat appears, formed during the higher water levels of the Atlantic period. The base of this peat, at a depth of -6.16 m N.A.P., has been radiocarbon-dated to 5610 ± 60 B.P. (GrN-5607) and correlated with the mean sea level at that time (Ente, 1976, p. 29). This peat layer is observed on both sides of the dune and may originally have covered the dune completely.

Analysis of the pollen in this peat horizon has indicated that during the second half of the Atlantic period the dune supported a mixed forest of oak, ash, and lime (Casparie et al., 1977, pp. 30; 33). Beneath the forest canopy, undergrowth was minimal with only small amounts of hazel growing. In the marshy areas surrounding the dunes, alder would have been the predominant species. Shortly after 3650 B.C., the influence of the sea and rising water levels were felt in this area and a fresh water tidal delta environment developed. Salt water

did not reach the dune, however, but open water stood in the nearby stream channels.

The final transgression of the sea over this area is noted clearly in the profiles by the presence of a thin layer of coarse, yellow, waterlaid sands, designated as the erosion layer. This lens, only a few centimeters in thickness, marks the last erosive activity of the sea. This wave erosion removed the top of the peat deposits and the crest of many of the river dunes. Above the erosion layer, several deposits are noted, designated as sediments of the Almere, Zuiderzee, and IJsselmeer phases of water levels in this area. These deposits are discussed in greater detail in Ente (1976) and will not be considered again here as they post-date the period of human occupation in the area.

The focus of archaeological interest lies in the sand deposits of the dune. Artifacts were found on the surface of the dune and throughout the A horizon where it appears. Cultural features, including hearths and graves, are found primarily in the B horizon along the crest of the dune where the A horizon has disappeared.

3. PREVIOUS WORK AT PARCEL H46

Shortly after the polder of Oost Flevoland was drained, archaeological remains in the form of stone tools, bone, and charcoal were discovered in the course of canal cutting and the subsequent geological inspection of these exposures (Ente, 1976). Because of its potential scientific value, a portion of parcel H46 was set aside as a reserve by the state, along with several other locales on the polder. The surface of the dune, as determined from preliminary borings, was planted in grass and the peat deposits along the margins of the dune were marked with shrubs and trees (fig. 2).

Preliminary archaeological excavations were undertaken in 1962 and 1966 at parcel H46 by the Research Division of the Polder Development Authority (= Rijksdienst voor de IJsselmeerpolders) (Van der Heide, 1966a; 1966b) to horizontally expose the areas originally ob-

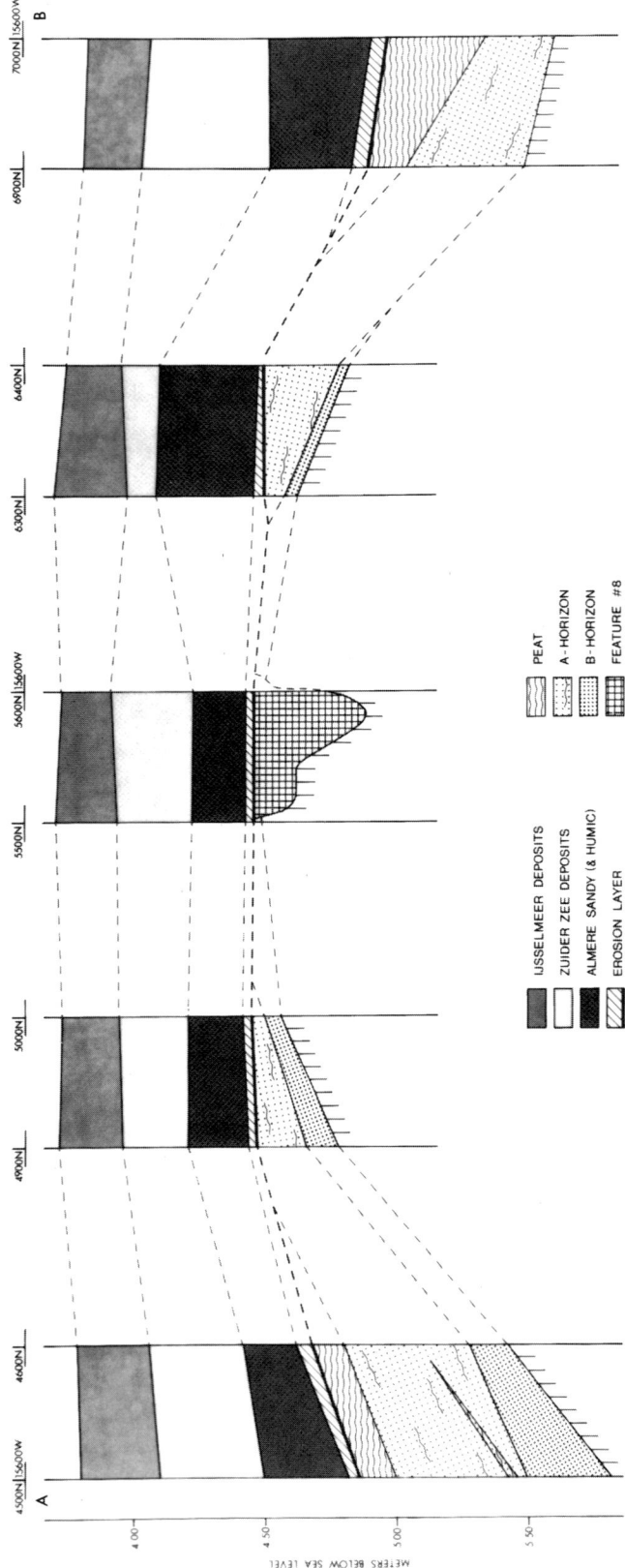

Fig. 3. Schematic profile of longitudinal section of west wall in excavation unit S23. Note that the vertical scale is exaggerated. The location of the section is indicated in figure 10.

served in the profiles of the drainage ditches. Two excavation units were opened at either end of the dune in parcel H46. These areas are now designated as S21: Working Floor 1 and S 22: Working Floors 4 and 5 (fig. 4). The excavated remains included pottery sherds, flint artifacts, a number of hearths, and several graves.

Investigation of the river dune as a part of the larger Swifterbant research project of the Institute in Groningen began in 1971 and continued in 1973 (Van der Waals & Waterbolk, 1976; de Roever, 1976). The preliminary excavations of Van der Heide were re-opened and expanded at both S21 as Working Floors 2 and 3 and at S22 as Working Floors 6a, 6b, and 6c (fig. 4). Additional artifacts were collected and a larger number of hearths and several more graves were revealed. A much clearer impression of the situation of the dune and the nature of soils was gained through these studies, as well as a more representative collection of archaeological remains. This fieldwork was not completed in two areas: one section in S22, designated as Working Floor 6c, and a burial in the south end of S21 (Grave XI).

4. THE 1976 INVESTIGATIONS

In 1974 the University of Michigan Museum of Anthropology began a series of investigations at a river dune in parcel H34 in the Swifterbant area (Whallon & Price, 1976). Excavation units in this parcel were designated as S11, S12, and S13 (fig. 1b). Borings and preliminary excavations uncovered a cultural horizon at S11 with both Mesolithic and Neolithic elements present. Radiocarbon dates of 4335 ± 45 B.C. (GrN-7214) and 4380 ± 45 B.C. (GrN-7215) argued that this site might also represent an occupation transitional between the Mesolithic and Neolithic periods. The 1974 project was under the direction of T. Douglas Price and Robert Whallon, Jr., and supported by a grant from the National Science Foundation (GS-42656).

One of the goals of the 1974 project had been to locate an intact occupation surface

dating from the Mesolithic period. Such information would help to document human adaptation in the Swifterbant area prior to the advent of farming cultures and provide additional data on Mesolithic occupation in the northern Netherlands — an extension of earlier work in the province of Drenthe (Price, Whallon & Chappell, 1974). Because of the absence of a demonstrably Mesolithic occupation at H34, Price, now at the University of Wisconsin-Madison, moved to a new locale in the Swifterbant area while Whallon continued work at the important site of S11.

The 1976 University of Wisconsin project was focused on another river dune located in parcel H46 (fig. 1b). Previous work at H46 had demonstrated that the dune contained both Mesolithic and Neolithic components. A radiocarbon date of 7775 ± 40 B.P. suggested that a pure Mesolithic occupation might be found at the site. Reasonably good bone preservation at H46 was indicated by the presence of several human skeletons and some animal bones from the earlier excavations at S21 and S22. These considerations directed our investigations to H46 as a potential site for the excavation of an intact Mesolithic settlement.

Beyond the recovery of a Mesolithic site, the 1976 project had additional goals. Some of the earlier excavations opened in 1973 by the Biologisch-Archaeologisch Instituut had not been completed and required termination. In addition, development of the stratigraphy of the dunes and the deposition of cultural materials remained problematical and we hoped to obtain more information on these questions.

Excavation units in the Swifterbant project are designated with the initial letter *S* (for Swifterbant) and then a sequential number appropriate to a new excavation. Thus, the original excavations at parcel H46, begun by Van der Heide and continued by the Biologisch-Archaeologisch Instituut, were designated as S21 and S22 (fig. 2). The new excavations opened in 1976 were designated as S23 and S24. These excavation unit abbreviations do not necessarily indicate separate sites or cultural components.

The 1976 field season began on 21 June and ended on 27 August. Several of the project goals were achieved but we were unable to locate an uncontaminated Mesolithic horizon and additional work at the site, planned for future seasons, was not undertaken. The project was supported by the National Science Foundation (grant BNS-11304).

4.1. Mapping and borings

A vertical datum for the site area in parcel H46 was established near the southwest corner of the field at an elevation of -3.94 m N.A.P. The horizontal grid for the area was set up incorporating the excavation units from the previous seasons in 1971 and 1973. These excavation units had been laid out along the network of drainage pipe ditches that cut through the parcel at roughly ten meter intervals on a north-south axis. An arbitrary point to the southeast of the reserve was set as the zero-zero datum point. Any spot in the area of investigation would then be designated by distance in centimeters to the north and west of this datum. Grid north for the excavations was set along the line of the drainage pipes in accordance with the earlier excavations. True north is approximately fourteen degrees east of grid north. A grid of ten meter squares was then set out over the entire field of the study area with the exception of the wooded margins. Vegetation was too dense in these areas for mapping or boring.

The field was mapped and ground surface elevations recorded using the established grid. Although the seabed of the former Zuiderzee was extremely flat, compaction of sediments subsequent to the drainage of this area has resulted in some slight relief. In the case of parcel H46, drying and compaction of the peat and lacustrine deposits surrounding the dune sand has left the top of the dune somewhat higher than the surrounding area. The map of the present ground surface thus reflects the contours of the buried dune (fig. 2).

Two distinct sections of the dune, or perhaps two distinct dunes, can be recognized at the east and west ends of the site area from the map of the ground surface contours. Although the longitudinal orientation of the two areas appears to be at right angles, the two ends of the field may simply represent two high portions of the same dune.

In order to verify the ground surface contour indications of the shape and extent of the dune and to map the sub-surface sediments, a series of borings were made following the site grid. In these borings, the depths below the surface of the peat and the dune sand were recorded and this information was used to construct a contour map of these major units of sediment beneath the more recent marine deposits. The results of the boring program confirmed the presence of two distinct areas of the dune, at the opposite ends of the site area. The borings also revealed a layer of peat in the middle of the site area between the two ends. On the basis of the ground surface contour map and the results of the borings, it was clear that the dune curved sharply through the wooded area to the south and that the vegetation, intended to mark the course of the dune, was partially misplaced (fig. 2).

4.2. Excavations at S22

The eastern portion of unit S22, designated as 6c (fig. 4), had been opened in 1971 but not completed. In this eleven by three meter area, the overburden of marine deposits had been removed and approximately fifteen centimeters of the top of the dune had been excavated. Some artifacts, a few hearths, and a portion of one grave were uncovered but time did not permit the remainder of the artifact horizon or the hearths to be excavated. The burial (number VI) was removed from the grave and the entire area was covered with heavy plastic and buried underneath twenty centimeters of backdirt.

In 1976, this backdirt and the plastic were

Fig. 4. Excavation units S21 and S22: floor plans (cf., de Roever, 1976, Figs. 2, 3 & 5).

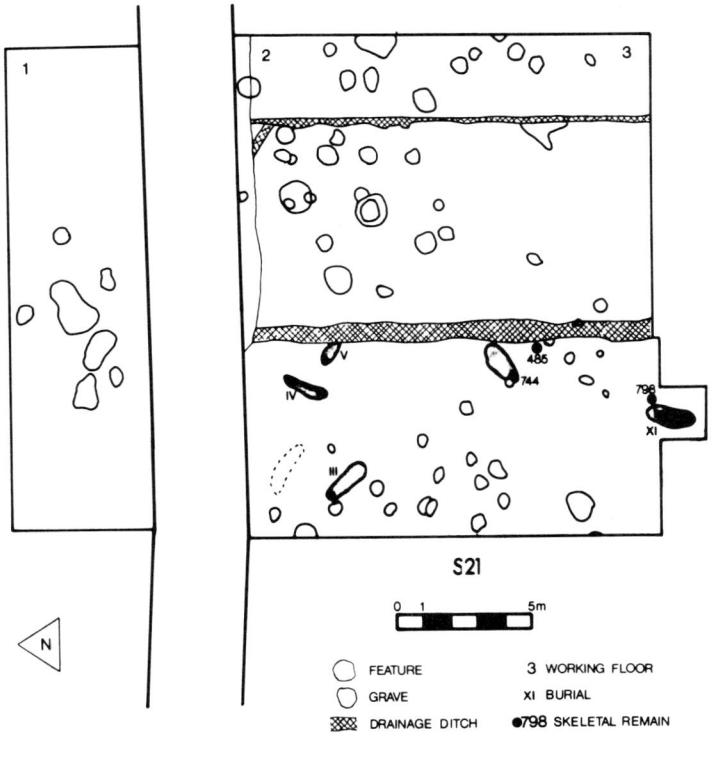

S21

0 1   5m

FEATURE   3 WORKING FLOOR

GRAVE   XI BURIAL

DRAINAGE DITCH   ●798 SKELETAL REMAIN

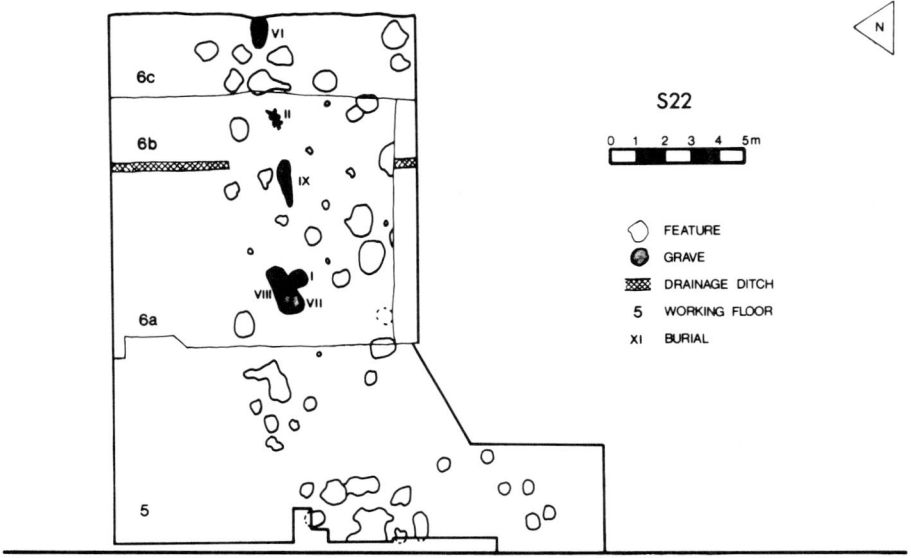

S22

0 1 2 3 4 5m

FEATURE

GRAVE

DRAINAGE DITCH

5 WORKING FLOOR

XI BURIAL

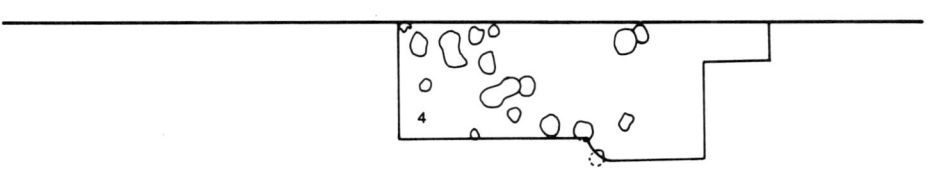

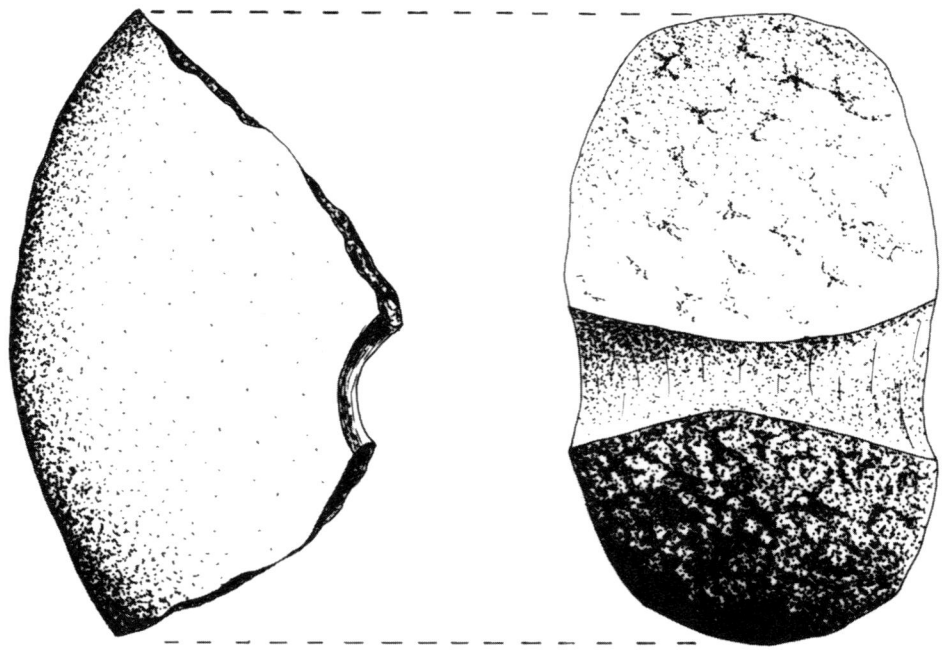

Fig. 5. "Macehead" fragment from S22: Feature 4. Full size.

removed and the excavation floor once again exposed. The profile of the east wall of the excavation unit indicated that the A horizon of the soil had originally extended across most of the dune at this point, truncated only in the north-central portion of the excavation. The 1971 excavations in Working Floor 6c had removed much of the A horizon.

On the floor of this excavation unit the B horizon was exposed with the exception of two areas. At the north end of the unit, a compact, light-grey A horizon was intact where the dune began to slope down to the north. In the south-central portion of the unit, a second pocket of grey sand was observed with remnants of the B horizon on either side. This pocket of grey sand was unusual in several respects. It was clearly intrusive through the B horizon in this area of the excavation; the base of the grey sand pocket rested directly on the sterile dune sands without an intervening B horizon. This base of the grey sand pocket was irregular but occurred at a maximum depth of 30 cm below the A horizon. Many of the artifacts from S22 and a hearth

were found within this pocket. Numerous charcoal flecks were observed throughout the fill of this depression. This pocket clearly represents a former depression in the surface of the dune, natural or man-made, that filled with grey sand. The cause of the depression and the means by which it was filled are unclear.

The majority of the features in S22:6c were hearths and appeared clearly as black concentrations in the lighter surrounding sands of the B horizon. Charcoal and burned artifacts were encountered in the fill of these features. The location and outline of these features are shown in figure 4.

During the course of the 1976 work at S22, exact provenience coordinates were recorded for a total of 492 items, including 428 flaked stone artifacts (86.3%), six small pebbles (1.2%), five pieces of rock (1.0%), and fifty-four large pieces of charcoal (10.5%). Virtually all of the artifacts were found within the grey sand of the A horizon or in the central depression. Only a very few items were encountered in the B horizon.

All but one of the rocks from the excavation were small fire-cracked fragments. The exception was a segment of a "macehead" found in the bottom of a hearth (Feature 4). The macehead fragment is not quite one-quarter of the complete object, broken vertically through the shaft hole (fig. 5). This hole is an hourglass perforation with an estimated complete diameter of three centimeters. The object is made of metamorphic sandstone and the entire surface is pecked and ground to a smooth finish. The shape of the macehead would have been almost spherical with a flattened base. The weight of the fragment was 415 grams. This object is similar to those reported by Hulst and Verlinde (1976).

Flint artifacts were by far the most common remains encountered in the excavations at S22 in 1976. Table 1 presents the counts and percentages of the various production stages observed among the 428 lithic artifacts. Of the identifiable flint materials, fifty-six (13.1%) showed evidence of deliberate retouch or utilization. The counts and percentages for the various categories of retouched and utilized pieces are given in table 2.

4.3. Excavation of S21, Grave XI

Attention was turned next to the grave that had originally been encountered in 1973 in unit S21. A portion of this grave had been observed in plan and profile at the south end of the excavation unit (de Roever, 1976, Figs 2 & 4). Roughly half of the grave was excavated to the level of the burial. The remainder of the grave (and the burial) continued under the profile and could not be recovered during the 1973 season. The exposed portion of the grave and skeleton were covered with heavy plastic and partially re-buried.

Knowing the exact location of the grave, it was possible in 1976 to remove the overlying deposits directly above the remainder of the grave. A 1.5 by 1.5 meter extension was added at the south end of S21 (fig. 4) to uncover the remainder of the grave. The southern half of the grave was clearly visible in the A horizon

present on this portion of the dune at a depth of -4.46 m N.A.P. At this level the outline of the grave was an elongated oval, approximately 200 cm in length and 80 cm in width (fig. 6). The fill was light grey sand. This grey sand was distinct in the A horizon, and continued through the B horizon and into the sterile dune sands. The grave had been dug after the soil had formed on the dune. In all probability, the fill of the grave is simply the result of the mixing of the soil and sterile sand removed during the original excavation of the basin-shaped pit. A number of artifacts were encountered during the excavation of the grave but none were in direct association with the burial. These artifacts were apparently simply incorporated in the fill.

The grave extended to a depth of -5.08 m N.A.P., some 62 cm below the present surface of the dune. A single extended burial was found in the grave, oriented north-south along the long axis of the dune (fig. 6). This burial was of an adult female and only the major bones were preserved. The burial consists of a skull with the apex of the vault and the right side of the face missing, together with the partial post-cranial skeleton. The burial was removed to the Instituut voor Anthropobiologie of the State University Utrecht for further cleaning and analysis. This material is reported along with the remainder of the human skeletal material from the Swifterbant project in Meiklejohn and Constandse-Westermann (1978) and Constandse-Westermann and Meiklejohn (1979).

4.4. Excavations at S23

One of the major goals of the 1976 season was to open a new area for investigation completely across the top of the dune and along its slopes. Earlier work at H46 had focused on the crest of the dune where erosional activities had removed much of the soil horizon and the artifacts. It was possible that *in situ* occupation floors would be preserved on the slopes of the dune, buried beneath peat and other deposits. On the basis of the results of the mapping and borings, a large trench was established in the western section of the dune. This 25 x 6 meter unit, designated as S 23, crossed the longitudinal axis of the dune at a right angle and included both the north and the south slopes of the dune (fig. 2).

The overburden of deposits was removed mechanically above the erosion layer. The remaining sediments, directly above the erosion layer, were removed by hand. We began by cleaning an area three meters in width, running the length of the excavation unit. The remaining three meter wide area of the trench was removed as time permitted. The north and south ends of this latter area were not excavated during the field season (fig. 7). A total of 126 square meters were excavated at S 23.

The erosion layer of coarse yellow, waterlain sand is three to five centimeters thick across the dune and contains a number of artifacts. These materials are in secondary position, having been removed from the surface of the dune by wave erosion. The erosion layer was shovel-skimmed and screened for artifacts and other materials. Counts of the lithic artifacts from the erosion layer at S 23 are presented in tables 1 and 2. No ceramic fragments were encountered in this layer and these were likely destroyed by the erosive activity of the sea. A number of pieces of bone, primarily of fish, and a few scales were found in the screening of the erosion layer. These almost certainly date from the period of erosion and are not associated with the occupation of the dune. A few acorn shells were also found in the erosion layer and may belong to the period of prehistoric occupation. No nutshell was found during the excavation of the dune at S 23 but similar materials have appeared in the excavations at S 11 in parcel H 34 (Whallon & Price, 1976).

Excavations on the dune itself, beneath the erosion layer, were done by shovel-skimming and trowelling in areas of higher artifact density. All artifacts were plotted in three dimensions to provide exact provenience data for the location of these materials. Cultural features — non-portable human remains such

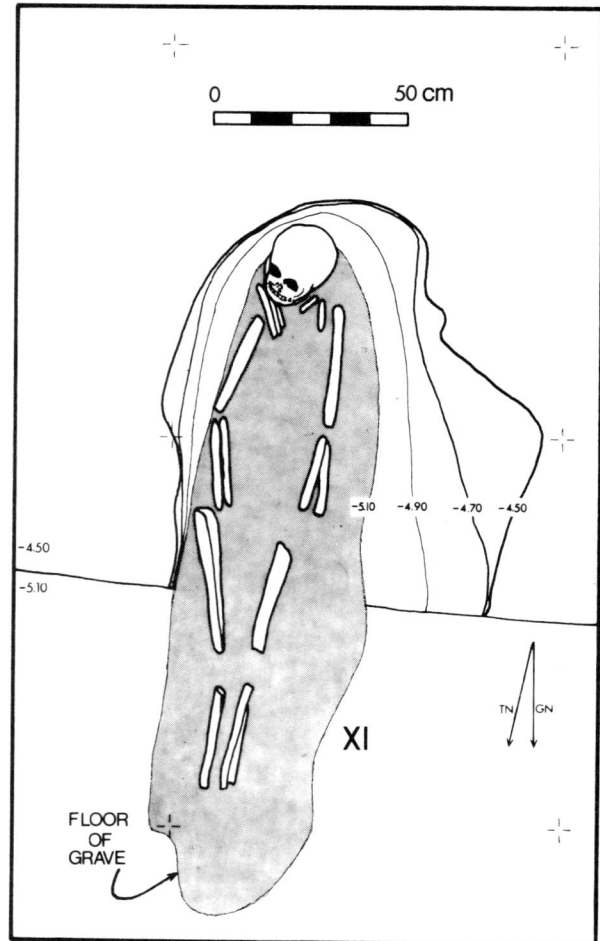

Fig. 6 a. S 21: Grave XI. Plan view. East-west line in the drawing marks the south wall of S 21 from 1973.

as hearths and pits — were mapped, photographed, and then removed in quarter sections. Two cross-sections of each feature were drawn when possible. Fill from the features was saved for flotation and charcoal was collected when available.

4.4.1. Features

In the central portion of S 23, all of the A horizon and most of the B horizon was absent due to the erosion of the dune crest. Very few artifacts were present in this area but the majority of the features were found here

Fig. 6 b-c. S 21: Grave XI. (b) Photograph of the grave in section and plan. (c) Photograph of the burial in the bottom of the grave.

(fig. 7). The features were observed as black, or dark grey, sandy concentrations in the dune sand.

A total of thirty-nine cultural features were recorded at S 23. Summary information on these features is presented in table 3. The large majority of these features are hearths. Hearths were distinguished from pits by the relative absence of charcoal in the pits. Three features (numbers 14, 15, and 22) appeared as dark stains on the surface of the dune. Cross-sections of these stains, however, revealed them to be the result of animal burrowing in the dune — probably foxes or moles. Two of the features (20 and 24) were not excavated because such small portions were intact within the excavation. Feature 36 may be a grave; it is a linear feature and a single unidentifiable bone was found in the feature fill. However, because the feature continued under the profile it was impossible to determine its exact nature.

Many of the features were relatively shallow and it should be noted that these features,

Table 1. Counts and percentages of production stages for lithic artifacts from the 1976 excavations, S22, S23. EL = erosion layer. DS = dune sand.

Stage	S22:DS		S23:EL		S23:DS	
	Count	Percent	Count	Percent	Count	Percent
Nodule	2	0.4	9	0.3	3	0.1
Decortication flake	23	5.4	142	4.9	107	4.5
Decortication flake	11	2.6	91	3.1	76	3.2
Block	15	3.5	87	3.0	86	3.6
Core rejuvenation flake	8	1.9	58	2.0	30	1.3
Flake core	4	0.9	3	0.1	3	0.1
Blade core	20	4.7	62	2.1	43	1.8
Flake	91	21.3	421	14.5	262	11.1
Blade	72	16.8	477	16.5	391	16.5
Microburin	0	0	0	0.0	1	0.1
Sharpening removal	0	0	0	0.0	3	0.1
Shatter	182	42.5	1545	53.4	1362	57.6
	428	100.0	2895	99.9	2367	100.0

like the dune crest, have been truncated by the erosive activity of the sea. The features observed along the crest of the dune in S23 are in effect only the remnants of the deeper pits that at one time reached the higher surface of the dune. The features are noticeably deeper the further they are located from the highest part of the dune and the major zone of erosion.

There was some suggestion of slight vertical separation among the features, although it was difficult to document because of the slope of the dune. A few features were not observed immediately beneath the erosion layer but only after some of the soil horizon had been removed. Feature 16, for example, on the south slope of the dune was recorded initially at a depth of -4.78 m N.A.P. The surface of the dune above this feature was located at approximately -4.65 m N.A.P.

Feature 6 at S23 is a typical example of the features of this site (fig. 8). This roughly circular, basin-shaped pit is located at approximately 5400N and 15300W on the site grid. The feature has an outer diameter of 80 centimeters. The compact, dark core of the feature is surrounded by a lighter grey sandy zone that is less compact. Large quantities of charcoal were found in the feature, primarily in the dark central core. A "log" of charcoal was observed at the surface of the feature (indicated by the black object in fig. 8). This feature is almost certainly a hearth. The lighter grey zone surrounding the core of the feature is the result of the mixing of the actual hearth fill (the central core) with the surrounding dune sand.

One definite grave (XII) was also recorded as S23 (fig. 9). The grave was directly underlain by a hearth (Feature 27). The outline of the grave was indistinct and roughly elliptical. The maximum dimensions of the grave at a depth of -4.49 m N.A.P. were approximately 1.30 m in length and 0.65 m in width. The

Table 2. Counts and percentages of retouched and utilized artifacts from the 1976 excavations, S22, S23. EL = erosion layer. DS = dune sand.

Type	S22:DS		S23:EL		S23:DS	
	Count	Percent	Count	Percent	Count	Percent
A point			1	0.4		
B point			1	0.4		
C point			1	0.4		
Needle-shaped point			1	0.4		
Lancette point	1	1.8				
Double point			1	0.4		
Surface Retouched point			1	0.4		
Broken point	1	1.8	2	0.9	10	4.0
Isoceles triangle			1	0.4		
Short Scalene triangle	1	1.8			3	1.2
Broad symmetric trapeze			3	1.3		
Right-angle trapeze			1	0.4		
Broken trapeze			2	0.9		
Backed Blade			20	8.8	13	5.1
Triangular backed blade			3	1.3	1	0.4
Short Blade borer			1	0.4	2	0.8
Long Blade borer			3	1.3	2	0.8
Short flake borer	1	1.8	5	2.2	7	2.8
Burin					2	0.8
Simple flake scraper	6	10.7	50	22.1	10	4.0
Short convex end scraper					1	0.4
Short straight end scraper					2	0.8
Long convex end scraper					1	0.4
Convex Side scraper	1	1.8	1	0.4		
Blade knife					1	0.4
Retouched flake	7	12.5	34	15.0	80	31.6
Notched flake	2	3.6	13	5.7	5	2.0
Retouched blade	4	7.1	11	4.9	29	11.5
Notched blade	1	1.8	1	0.4		
Truncated blade	4	7.1	4	1.8	3	1.2
Piece Esquillee					4	1.6
Utilized flake	14	25.0	49	21.7	33	13.0
Utilized blade	13	23.2	16	7.1	44	17.4
Total Tools	56	100.0	226	99.4	253	100.2
Total Tools / Total Lithics	$\frac{56}{428}$ = 13.1%		$\frac{226}{2895}$ = 7.8%		$\frac{253}{2369}$ = 10.7%	

Table 3. S23: Feature attributes.

Feature number	Coordinates			Shape	Dimensions (cm)			Number of artifacts	Interpretation
	North	West	Surface		Max.	Min.	Depth		
1	6000	15300	-4.59 NAP	Irregular oval	89	75	26	12	Hearth
2	5900	15400	-4.59	Circular	97	94	23	36	Hearth
3	6000	15400	-4.59	Teardrop	32	23	10	2	Small hearth
4	5800	15500	-4.59	Circular?	128	65	22	4	Hearth
5	5900	15200	-4.49	Circular	69	60	9	6	Hearth / Pit
6	5700	15300	-4.48	Circular	82	79	17	0	Hearth
7	5700	15500	-4.50	Irregular	76	65	12	0	Hearth·
8	5500	15500	-4.49	Oval	99	113	43	16	Hearth
8x	5400	15400	-4.49	Circular	65	59	37	0	Hearth
8z	5200	15400	-4.49	Circular	32	23	20	6	Hearth / Pit
9	5500	15200	-4.45	Irregular	176	151	16	9	3 Hearths / Pits
10	5300	15200	-4.46	Irregular	134	120	39	9	2 Hearths
11	5100	15100	-4.51	Oval	98	85	30	56	Hearth
12	6000	15100	-4.49	Circular	72	68	17	9	Hearth
13	6000	15000	-4.49	Oval	58	47	25	14	Hearth
14	Animal burrow							0	
15	Animal burrow							0	
16	4800	15500	-4.78	Circular	79	76	16	12	Pit?
17	5000	15300	-4.55	Circular	80	75	18.	5	Hearth
18	5800	15100	-4.49	Oval	109	68	23	19	Hearth
19	5500	15100	-4.48	Circular	100	91	34	3	Hearth
20	Not excavated							0	
21	5500	15000	-4.50	Teardrop	77	61	10	3	Pit
22	Animal burrow							0	
23	5900	15000	-4.48	Circular	68	62	24	8	Hearth
24	Not excavated							0	
25	5800	15000	-4.50	Oval	63	55	12	6	Hearth
26	5800	15200	-4.49	Teardrop	65	62	22	4	Hearth
27	5200	15400	-4.49	Teardrop	135	94	34	31	Hearth
28	5200	15200	-4.48	Circular	65	60	30	59	Hearth
29	5100	15300	-4.51	Circular	92	87	37	41	Hearth
30	5400	15100	-4.53	Circular	61	58	13	4	Hearth
31	5300	15100	-4.51	Circular	53	53	24	1	Hearth
32	5200	15100	-4.51	Oval	92	55	26	19	Hearth
33	4900	15400	-4.61	Circular	56	55	12	3	Hearth
34	5000	15400	-4.55	Oval	101	70	14	12	Hearth
35	5000	15000	-4.68	Circular	77	74	19	3	Hearth
36	5200	15000	-4.51	Elongated oval	115	55	31	60	Pit / Grave?
37	5100	15100	-4.51	Circular	55	30?	21	2	Hearth
38	4900	15100	-4.70	Oval	111	77	32	31	Hearth
39	4800	15000	-4.83	Teardrop	101	103	60	25	Hearth
40	5300	15200	-4.48	Circular	48	41	20	37	Hearth

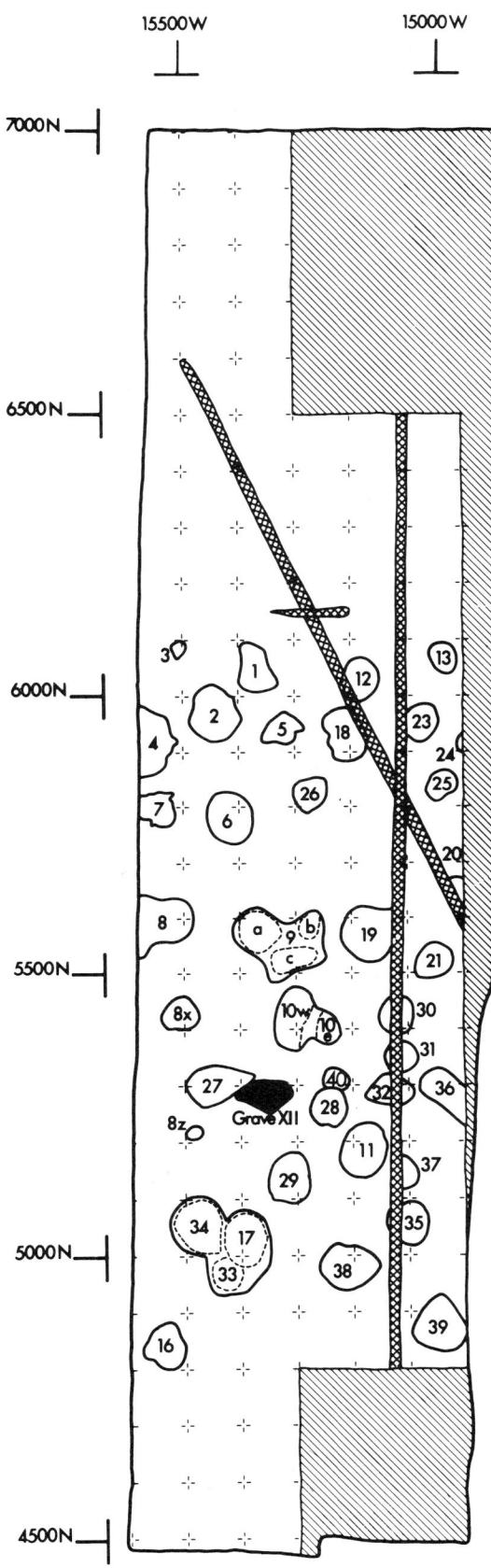

15500W 15000W
7000N
6500N
6000N
5500N
5000N
4500N

grave had clearly been truncated by the erosion of the top of the dune and only the lower portion of the pit was intact. The orientation of the long axis of the grave was east-west, approximately eighty-eight degrees east of magnetic north. This orientation parallels the main axis of the dune. The fill of the grave was light grey sand, the result of the mixing of the soil horizon and the sterile dune sand when the grave was originally excavated and filled. A number of flint artifacts were incorporated in the fill of the grave but were not in direct association with the burial. The portion of the grave that had not disappeared due to erosion was relatively shallow, only 35 cm in depth.

The burial was very fragmentary; only a few bones were encountered and these were badly preserved. Some of these pieces of bone appeared to be calcined. The only identifiable fragments were a segment of long bone diaphysis and a radial diaphysis. The size of the remains suggests either an older adolescent or an adult (Meiklejohn & Constandse-Westermann, 1978).

Disturbances in the excavation were also mapped (fig. 7). The narrow linear disturbance running north-south on the east side of excavation unit S23 is a drainage pipe ditch. These ditches cross parcel H46 at ten meter intervals. The long, diagonal disturbance and the smaller one that intersects it are tracks from anchors or boat keels on the former sea bed. These were not immediately identifiable as evidence of shipping activities. It is difficult to remember, while standing in the field, that at one time five meters of sea water stood above the excavations.

4.4.2. *Artifacts*

Four major classes of materials were recorded in the excavation of the dune at S23: flint, stone, pottery and bone. The counts and per-

Fig. 7. S23: Distribution of features and disturbances. Diagonal hatching indicates unexcavated areas. Cross-hatching indicates disturbances.

centages of these classes are presented in table 4. These data do not include the material from the erosion layer.

The horizontal distribution of these materials at S23 points out the effects of erosion on the dune crest (fig. 10). Very few artifacts occur *in situ* in the central area of S23 where the highest remaining portion of the dune is located. Most artifacts are found along the north and south slopes of the dune. Artifact density per square meter (fig. 11) indicates those areas of the dune that have not been eroded and where at least a portion of the A horizon is still intact. The erosion of the dune removed the surface from roughly 5000 north to 6400 north on the grid plan.

These classes of material are relatively equally distributed on both sides of the dune but some differences can be noted. Pottery fragments are more abundant on the north slope while rock is somewhat more common on the south. These differences may be due to the nature of the mechanical processes of erosion on the dune, the larger area of intact A horizon to the north, or actual differences in the location of prehistoric occupations.

The vertical distribution of artifacts in the dune sand was variable, depending upon location. Artifacts further from the crest of the dune, and further down the slope, tended to show greater vertical dispersion. At a maximum, artifacts were distributed through a horizon over 70 cm in thickness. The vertical dispersal of materials at the site will be considered in more detail below.

A total of 2369 flint artifacts were recorded in the excavations of the dune sand at S23. Production stages represented in these materials are presented in table 1. The percentages of these stages are very similar to the materials from the erosion layer. Shatter, created by heat, frost, or flaking, is by far the most common category. Blades and flakes comprise the next larger categories of production stages. The remaining categories are represented in relatively small number.

The counts and percentages of retouched and utilized pieces of flint are given in table 2.

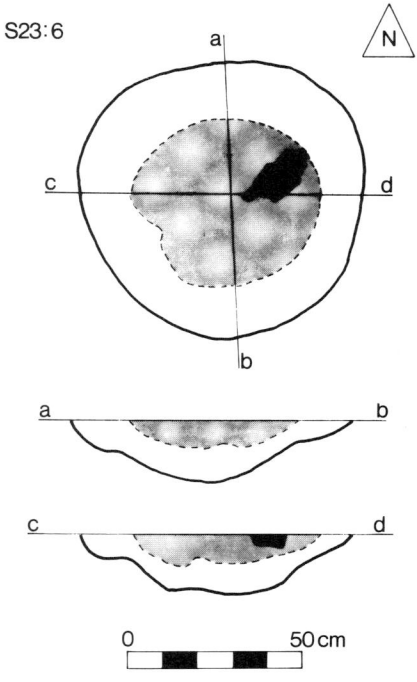

Fig. 8 a. S23: Feature 6. Plan view.

Some differences can be noted between the erosion layer and the dune sand for these types. Scrapers and utilized flakes are more common in the erosion layer, a total of 44.2%, while in the dune sand these two categories comprise only 18.6% of the total retouched and utilized pieces. Retouched flakes are more common in the dune excavations. Intriguing

Table 4. Counts and percentages of major material classes at S23. Erosion layer not included.

Material	Count	%
Flint	2369	96.5%
Rock	39	1.6
Bone	6	0.2
Pottery	41	1.7

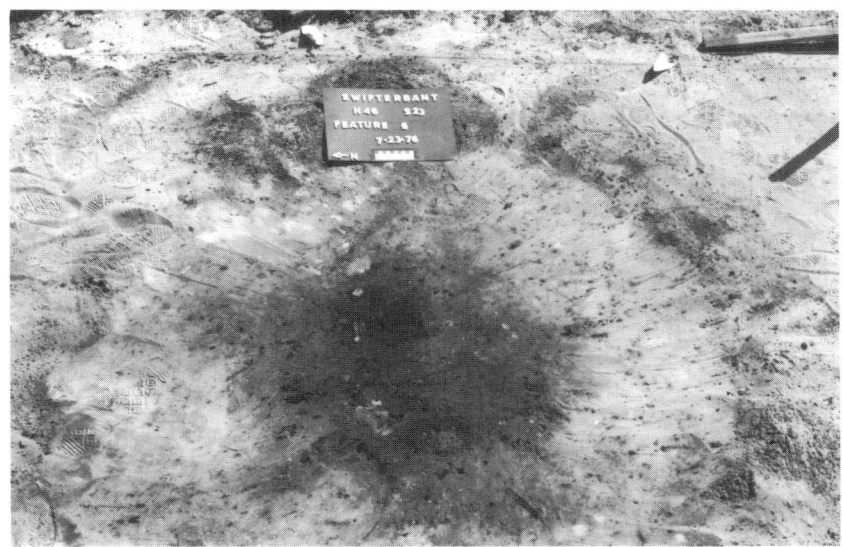

Fig. 8 b-c. S23: Feature 6. (b) Photograph of plan view. (c) Photograph of profile.

also are the absence of trapezes, the smaller number of backed blades, and the abundance of utilized blades in the dune sand. No clear patterning in these differences could be discerned.

Ceramic materials in the dune sand comprised only 1.7% of the major material classes. Forty-one pieces of pottery were recovered and analyzed by J.P. de Roever (1979). The following is a summary of her description of this material. The majority of the sherds are quite small (less than 3 cm in diameter) and the surfaces are often heavily weathered or absent. There is only one large fragment of a flaring, concave neck of a vessel – the upper section of an *S*-shaped profile. Five of the sherds are slightly convex and the remainder are too small to be indicative of vessel shape.

Only twenty of the sherds are well preserved and none show any decoration. The sherds are brown to grey/black in color and are tempered with sand, grit, and some fine organic

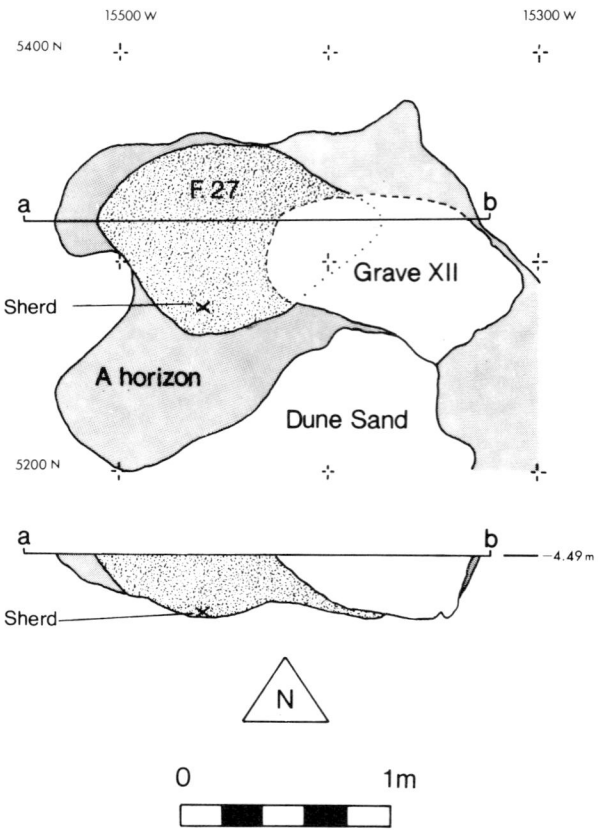

Fig. 9. S23: Feature 27 and Grave XII. Plan view and cross-section.

Grave XII. All of the bones were small, heavily eroded, and unidentifiable.

4.5. Sondage at S 24

Following the completion of the work at S23, a final excavation unit was opened in an area fifteen meters to the north of S23 at 15600 west and 8500 north on the grid plan (fig. 2). This unit was designated as S24 and intended to expose the north slope of the dune at a deeper level. Several questions were under investigation: (1) how far to the north did the slope of the dune extend, (2) did the occupation debris continue along the slope of the dune or was it confined to the area around the crest, (3) would the occupational layer at some point be found within the peat horizon, and (4) would the conditions of organic preservation improve in the lower areas of the dune where peat deposits were deeper?

A sondage of approximately four by two meters was excavated by machine through the deposits of peat and clay to the surface of the dune at a depth of -8.40 m N.A.P. The lower peat and the dune sand were brought up in block and spread on the surface of the meadow. This material was handpicked and a number of lithic artifacts were found in the sand. No organic remains were observed. Clearly the occupation of the dune extended for some distance along the slope and away from the crest. No cultural material was observed in the peat deposits above the dune sand in S24.

4.6. Dating

4.6.1. *Isotopic*

Following the formation of the river dunes at the close of the Pleistocene, the Swifterbant area was inhabited by Mesolithic and, later, Neolithic occupants. Materials recovered in the pre-1976 excavations documented the presence of both time periods at H46 in the form of Mesolithic projectile points and "Neolithic" sherds. Three radiocarbon dates were obtained from charcoal found in hearths which cor-

material. The pottery is coil-made and has heavily-smoothed surfaces. Wall thickness of the sherds ranges between 6 and 11 mm with a mean of 8 mm. The neck fragment and eight other sherds have cooking residues present.

Of the thirty-nine rocks encountered in the dune sand at S23, the majority were small fragments without indications of use. Only six showed any sign of modification (table 5). Of these six, one was a hammerstone, one a fragment of a "pick", and four had smoothed or polished surfaces. These rocks are of local origin available from ground moraine which outcrops at the ridges of Urk and Schokland, a few kilometers to the north and east of the site respectively.

Six pieces of bone were recovered in the excavations, not including the material from roborated this picture (de Roever, 1976, pp. 2 & 7):

GrN-6709   7745±40 B.P.   S21 Hearth
GrN-6710   6875±45 B.P.   S22 Hearth
GrN-6708   6670±35 B.P.   S21 Hearth

These determinations should date Mesolithic features at the dune.

An additional radiocarbon date has been obtained from S23:

GrN-8248   6240±50 B.P.   S23 Hearth

The charcoal sample for this determination was taken from a hearth (Feature 27) in which a potsherd was found. Although the possibility of intrusion cannot be ruled out, there was no visible disturbance in the hearth. The sherd was found at the bottom of the hearth at a depth of approximately -4.80 m N.A.P. The hearth itself directly underlies a grave (XII) in profile (fig. 9). The radiocarbon date thus provides a minimum date for the sherd as well as a *terminus post quem* for the grave.

The date is earlier than expected. Neolithic occupation on the clay levees in parcel G43 at Swifterbant is dated to around 5300 B.P.

(Van der Waals, 1977). The date from Feature 27 corresponds more closely to the dates from another dune site, S11, in parcel H34. Two dates from two separate features at S11 (GrN-7214 6285±45 B.P. and GrN-7215 6330±45 B.P.; Whallon & Price, 1976, p. 226) are approximately 1000 years earlier than the dates from the Neolithic levee sites and only slightly younger than Bandkeramik dates from Limburg in the southern Netherlands. These dates from S11 and the date from S23 may belong to the very end of the Mesolithic or to a period transitional between the Mesolithic and the Neolithic. This question will be discussed in more detail in the conclusion of this report.

4.6.2. *Artifactual*

The artifacts from S23, and from S21-22, are also suggestive of both Mesolithic and early "Neolithic" occupations. Projectile points are generally the best temporal markers for the early Postglacial in northwest Europe. Traditional Mesolithic point types, following the established typology (Deckers, 1979; Price, 1975; Newell & Vroomans, 1972; Bohmers, 1963; Bohmers & Wouters, 1956), are

Table 5. Modified rock at S23.

S23 #	Maximum dimension (cm)	Weight (g)	Description
38	5.3	22.3	Flake with one smoothed surface
42	4.6	15.8	Flat pebble with one smoothed surface
1189	6.2	96.8	Ovoid hammerstone with two pitted ends and one smoothed side
1350	9.1	257.9	Flat, pear-shaped cobble with bifacially flaked edge along one side ("chopping tool") and one polished side
1372	12.2	493.9	"Pick" fragment, ground to a point at one end, broken transversely, no indication of hole. Slightly convex in longitudinal cross-section.
1498	4.6	10.9	Small fragment with one smoothed side

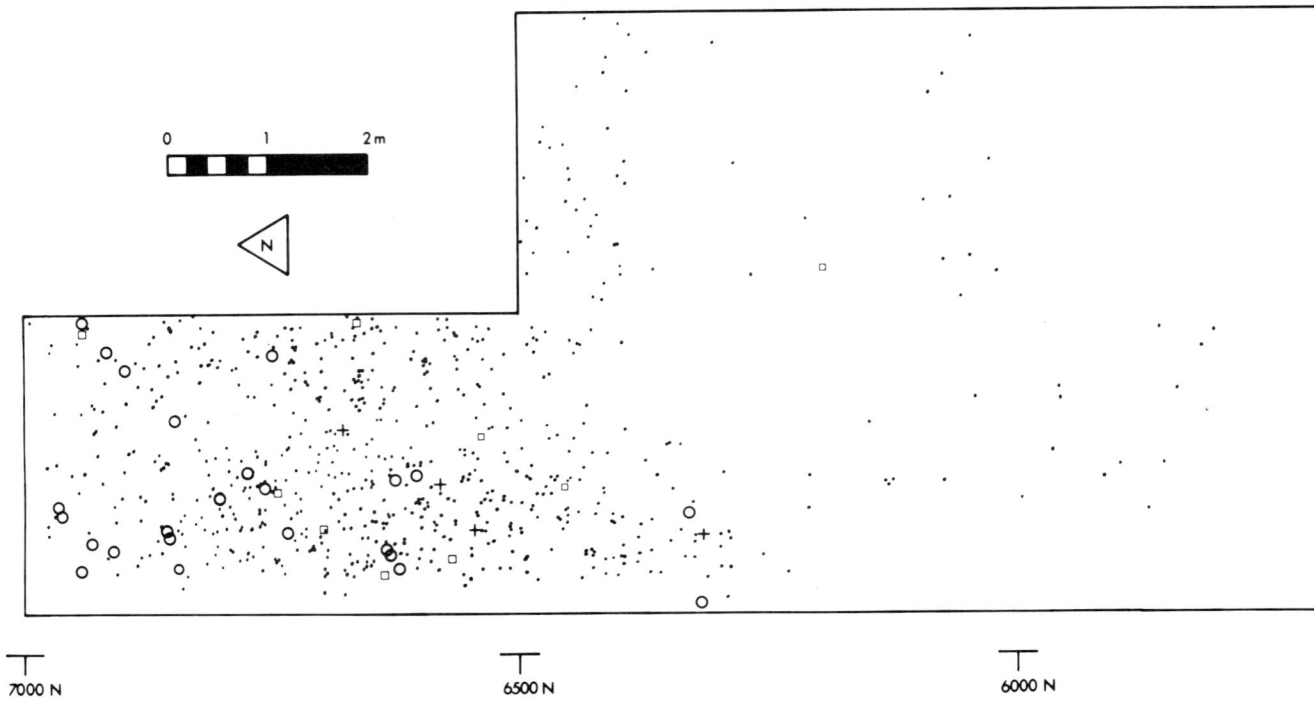

Fig. 10. S 23: Artifact distribution.

present at the dune in parcel H46 (fig. 12). A, B, and C points, needle-shaped points, double points, trapezes and triangles, all appear in the Mesolithic by 5500 B.C. (Newell, 1973, Graph 3) and correlate with the earlier radiocarbon dates from S 21 and S 22. Other point types, present at H46, are not characteristic of the Mesolithic and are indicative of a somewhat younger occupation. The surface-retouched point found in the erosion layer at S 23 (fig. 12) is not typically Mesolithic. Two transverse points were also recovered from Grave VIII at S 21 where they were found among the bones of the burial (de Roever, 1976). The *pièce esquillée* is unusual in a Mesolithic context in the Netherlands. Other artifact types such as the scrapers and borers, as well as the "macehead" and the "pick" fragments, are more ubiquitous and less diagnostic.

Interestingly, the lithic artifacts from S 23 are remarkably similar to those excavated in 1974 at S 11 (Whallon & Price, 1976, p. 227).

Counts and percentages for the major type groups at these two sites are presented in table 6. With the exception of a few more trapezes and knives and fewer retouched flakes at S 11, the differences between the two sites are insignificant.

Ceramics from H46 are generally thick with S-shaped profiles, flaring rims, occasional decoration and pointed bases. These sherds do not resemble the TRB materials from the northern Netherlands and are more closely related to the pottery of the Ertebølle-Ellerbek cultures of northern Germany and southern Scandinavia (de Roever, 1979).

4.7. Depositional processes

It is not possible at the present time to assign accurate dates to the occupations at S 23 on the basis of geological events. We can only establish an approximate beginning and end for the period of potential habitation. The formation of the dune at the close of the Pleisto-

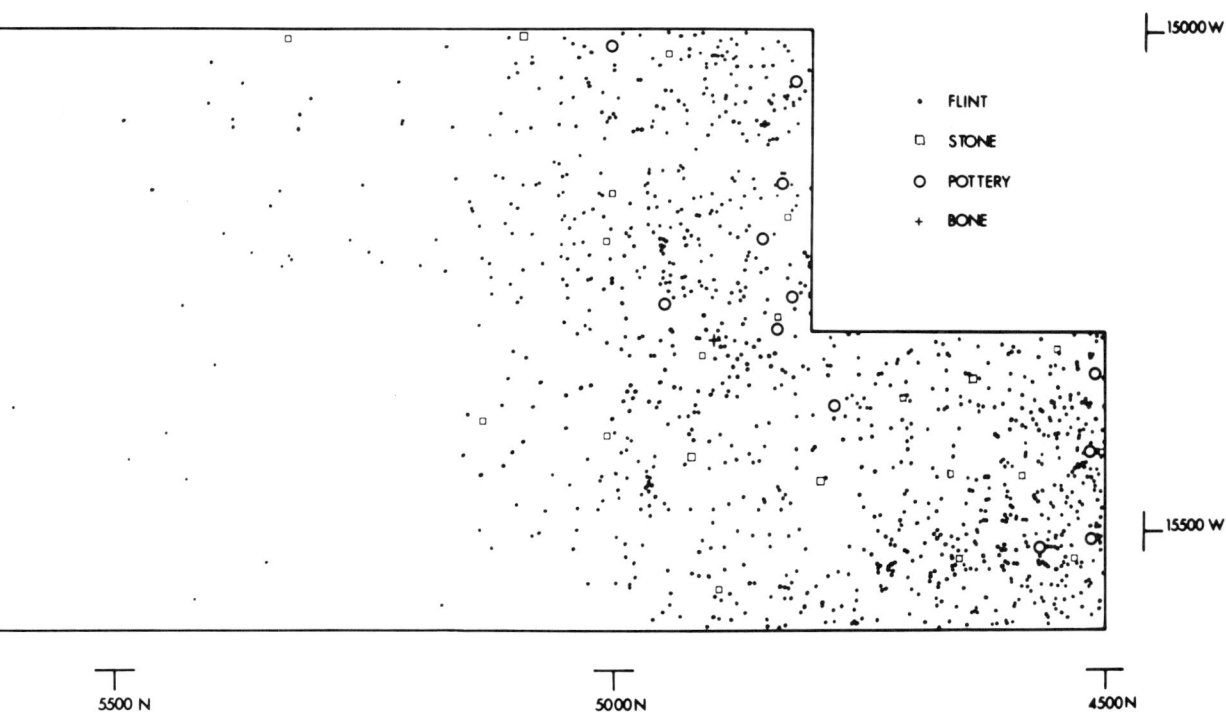

FLINT

STONE

POTTERY

BONE

15000 W

15500 W

5500 N   5000N   4500N

cene around 8000-7000 B.C., marks one end of this sequence. Submergence of the dune by rising water levels and peat growth by 3000 B.C. or shortly thereafter (Jelgersma, 1961; de Roever, 1976, p. 218) concludes the occupational history of the dune. A date of 3660 B.C. (GrN-5067 5610±60 B.P.; Ente, 1976) from peat just at the surface of the dune at -6.15 m N.A.P. supports this picture.

The processes of erosion, deposition, and soil formation occurring within the roughly 5000-year period of potential occupation are not well understood but some insight may be gained from the distribution and character of archaeological materials at the dune. Because of the deep vertical distribution of artifacts along the slopes of the dune, stratigraphic separation of the various archaeological components initially would appear as a means of establishing both chronological and depositional contexts for the material.

Figure 13 shows the south wall of the excavation unit at S23 (location, see fig. 10) with

a backplot of the artifacts found within one meter of this wall. The vertical dispersal of the artifacts is evident on this backplot. Within this one meter zone at the south wall of the excavation the slope of the dune is less than 25 cm and thus should not contribute significantly to the deep vertical dispersal of the artifacts. The band of erosion sand within the profile (layer 2, fig. 13) is almost certainly a later intrusion, perhaps along channels left by decayed tree roots.

A histogram of the number of artifacts by depth (fig. 14) in this profile does indicate differential distribution of the artifacts through the deposit. Two modes are seen in the histogram: one at approximately -5.05 m N.A.P. and a second at -5.45 m N.A.P. These modes are suggestive of at least two stages in the deposition of artifacts. The relatively rapid rise of the first mode at -5.05 m on the histogram may correspond to a living floor directly on the present surface of the dune. This surface would have been covered by peat growth some-

Table 6. S 11 (1974) and S 23 (1976): Retouched tools, counts and percentages from both erosion layer and dune sand.

Type	S23		S11	
	#	%	#	%
Points	18	5.8	8	6.5
Trapezes	6	1.9	7	5.6
Triangles	4	1.3	2	1.6
Backed Blades	37	11.9	17	13.7
Borers	20	6.4	10	8.1
Burins	2	0.6	4	3.2
Scrapers	65	20.9	25	20.2
Knives	1	0.3	5	4.0
Retouched Flakes	114	36.7	31	25.0
Retouched Blades	40	12.9	14	11.3
Pièce Esquillée	4	1.3	1	0.8
n =	311	100.0	124	100.0

time around 3000 B.C. according to Jelgersma's curve for the rise of sea level (Jelgersma, 1961; Ente, 1976). As we observed in S24, artifacts continue on the surface of the dune, beneath the peat, at a depth of -8.40 m N.A.P. Thus, portions of the upper occupation surface of the dune, particularly those lying along the deeper slopes, should pre-date 3660 B.C., the radiocarbon date from peat just above the surface of the dune at a depth of -6.15 m N.A.P.

The second mode in the vertical distribution of artifacts centers on a dark humic band in the profile (fig. 10) that likely represents a buried soil and former surface of the dune. The sands above this lower surface do not appear to be water-lain. No bedding is observable in the profile and there is no sorting of artifacts by weight as might be expected with water activity. Figure 15 is a scatterplot of flint artifact weight by depth in this profile. Flint material weighing between 0.01 and 3.0 grams is plotted on this graph. For the sake of clarity, pieces larger than 3 grams were not plotted, but these few larger pieces show a

similar homogeneous distribution through the profile. The absence of sorting suggests that human activity, eolian erosion, and/or slope wash may be responsible for the build up of sand between the lower surface and the top of the dune. The vertical dispersal of artifacts through these two major levels is likely due to natural activities in the soil such as root growth, animal burrowing, and other processes of bioturbation.

Although no definitively diagnostic artifacts are present in this south profile at S23, several artifact categories are suggestive of a date for the development of the dune above the lower surface. Backed blades, typical of the Mesolithic period, are found throughout the vertical spread of material. A *pièce esquillée*, rare in a Mesolithic context, occurs at -5.48 m N.A.P. within the lower level of the distribution.

A major question concerns the vertical distribution of pottery at the dune sites. Four pieces of pottery were found close to the south profile at S23. All of these pieces are within or

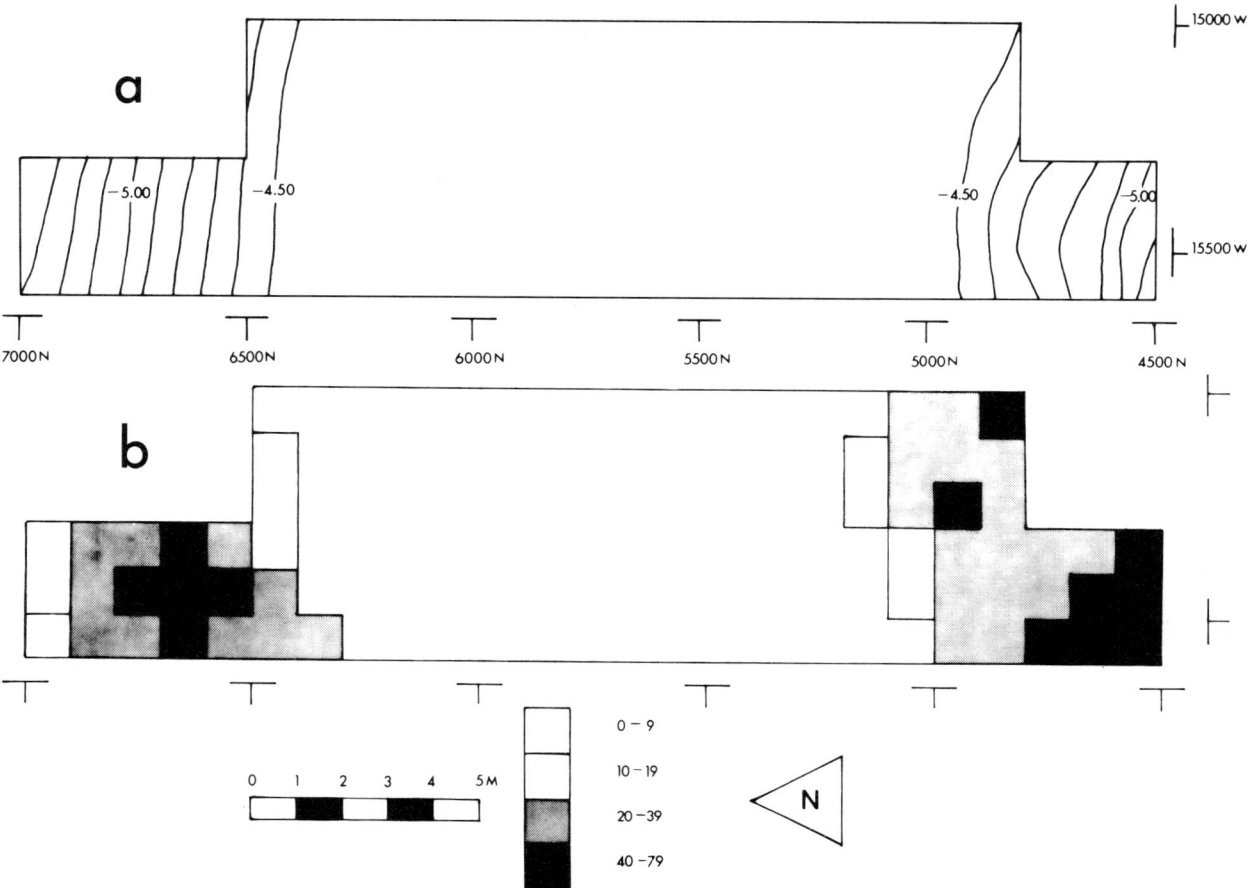

Fig. 11 a-b. S23: (a) Contours of the dune surface in the excavation unit. Contour interval is 10 cm. (b) Artifact density per square meter in the excavation unit.

closer to the upper level in the profile (fig. 13). At S22, ceramics were found at or near the surface of the dune (de Roever, 1976, p. 215) and thus may also be restricted to the upper portion of the deposits. At S11 there is also a tendency for ceramic materials to occur higher in the vertical distribution of materials (Whallon & Price, 1976, p. 226) although the sherds are distributed over 30 cm vertically in some areas of the site. If pottery is confined to the upper levels in the dune sites as it appears, then some stratigraphic separation of purely Mesolithic occupations from a higher, later ceramic horizon is indicated.

The dates for the ceramic horizon can only

be approximated. The radiocarbon determination of 3600 B.C. on peat at the surface of the dune should provide a *terminus post quem* for this upper horizon. At present only a single date of 4300 B.C. can be directly associated with the pottery at S23. This is the date from Feature 27, a hearth which contained a potsherd in the fill. Other dates from H46 of approximately 5800 B.C. and 4800 B.C. are attributable to purely aceramic, Mesolithic occupations and may belong with the lower surface of the deposits. I would argue that at some point, perhaps shortly after 4800 B.C., there was a period of major dune disturbance with deposition of sand along the slopes of

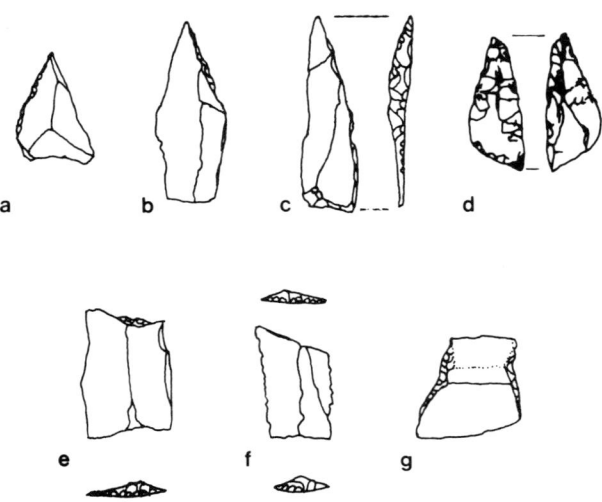

Fig. 12 a-g. Projectile points from S23: (a) A point, (b) B point, (c) C point, (d) surface-retouched point, (e) trapeze, (f) right-angle trapeze, (g) trapeze. Full size.

the dune, followed by a new occupation around 4300 B.C. by a pottery-using population.

Some information is also available on the occupations from the features at S22 and S23, in spite of the wave erosion of the peat and sand deposits at the top of the dune. The features found within this zone of erosion (fig. 7) are generally not vertically separated but appear at the same level in the dune. There is some slight vertical separation of a few features in areas on the margin of the erosion zone where the A horizon is still intact. Most features appeared as black ovals in the A horizon while others did not appear until much of the A horizon had been removed (e.g., Feature 16). Again, two possible occupation horizons are suggested.

Finally, the fill of the graves was distinguishable in the A horizon (cf., fig. 6) and consisted of a homogeneous, light-grey sand. This grave fill is almost certainly the result of the mixing of the A and B horizons and the sterile dune sand when the graves were originally excavated and filled. The graves thus appear to have been dug following the formation of the soil on the dune and very

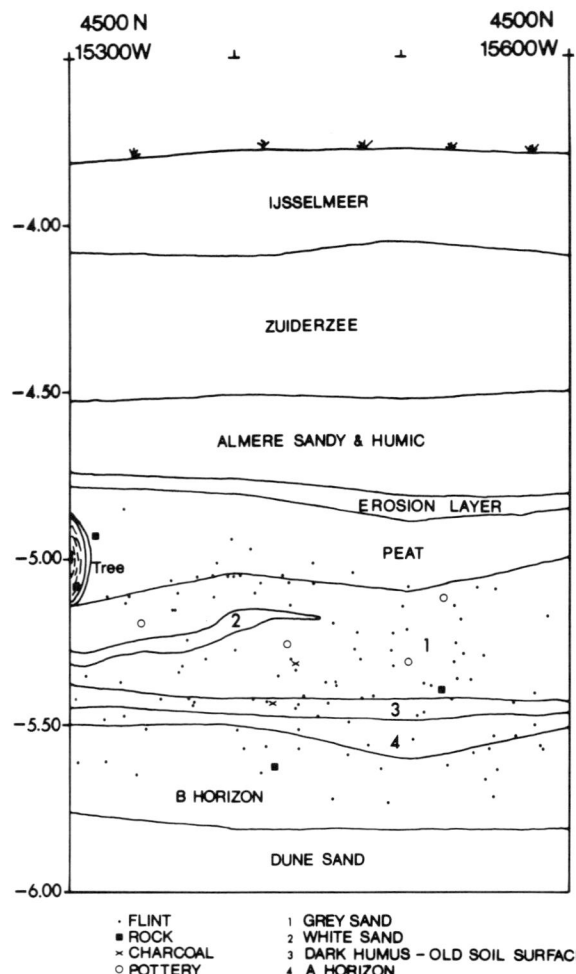

Fig. 13. S23: Profile of the south wall with vertical artifact distribution. Location of the section is shown in figure 10.

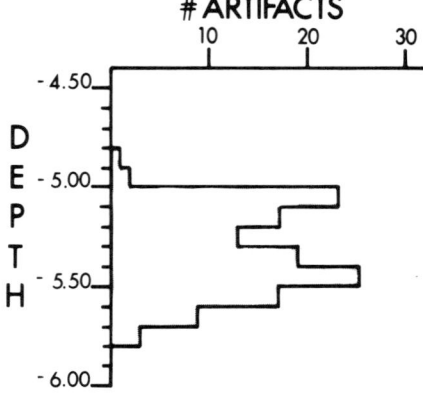

Fig. 14. Histogram of vertical artifact distribution in the three south squares of S23.

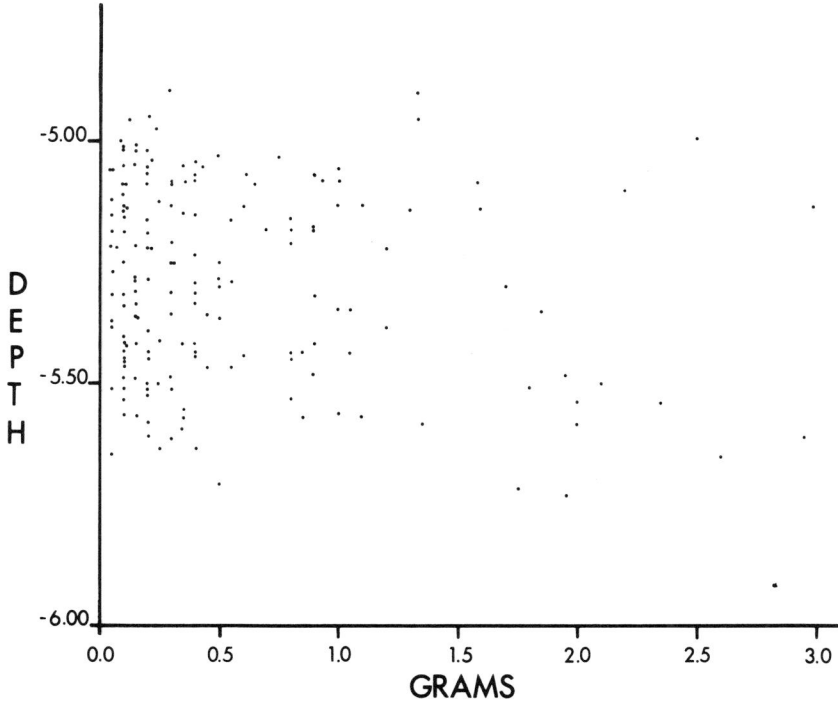

Fig. 15. Scatterplot of flint artifact weight by depth in the three south squares of S23.

5. SUMMARY AND CONCLUSIONS

probably after the occupation of the upper horizon.

Continuing investigations in the Swifterbant area over the last twenty years have resulted in the recovery of the remains of human occupation dating from both the Mesolithic and the Neolithic periods. The somewhat higher elevations of the river dunes appear to have been favored locales for habitation.

During the Boreal period of the early Post-glacial, this area would have been a fairly typical coversand environment in a riverine regime. The landscape would have been dominated by a relatively stable Boreal forest. As conditions became moister and sea level rose after 5500 B.C. in the Atlantic period, the area was transformed into a bog and creek system in a fresh water tidal delta environment. A marsh forest would have been present with much alder, oak, and pine. Lime trees would

have been common on the river dunes (Caspa-rie *et al.*, 1977). Finally the area would have been inundated by rising water levels some-time after 3000 B.C.

Although much of the occupation horizon on the river dune at H46 has been truncated by erosion, some of the materials collected from the site appear to document a *Ceramic Mesolithic* occupation. Ceramic Mesolithic sites are also recorded along the Atlantic and Baltic coasts of Europe. In northern Germany and southern Scandinavia pottery is found with the Ertebølle-Ellerbek cultures of the Mesolithic. Although organic materials are not preserved and macrolithic tools such as flint axes are missing in the river dune sites at Swifterbant, the date of 4300 B.C. at S23 falls within the range of the Ertebølle of Denmark which begins around 4600 B.C. (Brinch Petersen, 1973). Ceramics in the Danish Ertebølle, similar in size and shape to the Swifterbant material (de Roever, 1979), appear after 3800 B.C. in southern Scandinavia. In France, ceramics in association with Mesolithic remains have been reported along the Atlantic coast at Roucadour (Roussot-Laroque, 1977). This material is dated to approximately 4000 B.C. and the ceramics are similar to the Swifterbant material with pointed bases and heavy-walled construction. The evidence from these areas supports the earlier hypothesis of Schwabedissen (1966) that ceramics may have diffused into northern Europe along the coast prior to or simultaneously with the appearance of inland pottery-using farming cultures.

An earlier utilization of the dune is indicated as well. Radiocarbon dates of roughly 5800 B.C. and 4800 B.C., along with certain lithic artifacts, provide evidence for earlier Mesolithic occupations, corresponding to the Boreal and Late Mesolithic periods of Newell (1973). The graves on the dune, intrusive through the developed soil horizon, must postdate 4300 B.C. These graves are likely from the end of the period of potential occupation on the dune. The facts that the graves are intrusive through the soil and that the bones of the burials are partially preserved argue for interment at a time when water levels were higher in the area. These graves may well be contemporaneous with the Neolithic settlements and burials from the clay levee sites in the Swifterbant area. The levee burials are dated to 3590 B.C. by a radiocarbon date on bone collagen from a grave in parcel G42 (GrN-5606 5540±65 B.P.; Van der Waals & Waterbolk, 1976, p. 7).

The graves at H46 show a distinct orientation (figs. 4 & 7). With only one or two exceptions, the longitudinal axes of the graves follow the top of the dune and are located near the former crest.

From the angle of the north and south slopes of the dune (fig. 3), the crest of the dune must have stood at least 75 cm higher than its present surface, or at approximately -3.75 m N.A.P. The orientation and placement of the graves thus also argues for the original location of the graves in the highest and driest areas available and at a late date in the period of potential occupation.

Although one of the major goals of the 1976 project, the location and recovery of an intact Mesolithic occupation surface was not achieved, the season was successful. Earlier excavations at the dune were completed and an additional burial removed before it was destroyed by the drying of the dune. The new excavation unit, S23, completely transected the dune and provided valuable information on the occupational history of the site.

In essence we have learned that the higher areas of the dune have been badly disturbed. In areas along the slopes of the dunes, two probable cultural horizons are present and vertically separate, albeit with some overlap. Artifactual material, vertical separation, and radiocarbon dates argue for a possible Ceramic Mesolithic in the Swifterbant area. However, because of the disturbance of the dune and the overlap of the cultural horizons, successful recovery of an intact surface from this period may be possible only in deeper excavations – where the water table will make discovery difficult – or on those lower dunes (e.g., S11) that were not truncated by marine erosion. Clearly, more investigation is needed with better stratigraphic separation to adequately document the settlements of the pre-Neolithic inhabitants of the Swifterbant area.

6. ACKNOWLEDGEMENTS

The Swifterbant project is the result of the efforts of many individuals and institutions, too numerous to name here. Special mention must go, however, to the Biologisch-Archaeologisch Instituut of Groningen and its personnel for their interest, hospitality, and cooperation. Peter Deckers and J.D. van der Waals made useful comments on the content of this report. Paulien de Roever very kindly made available the results of the earlier excavations at H46 and undertook the analysis of the ceramic materials from S23. Her comments on the manuscript were also most helpful. Thanks are due also to Dr. P.J. Ente of the Research Division of the Polder Development Authority for his advice and involvement. The support of the National Science Foundation is gratefully acknowledged. Finally, the crew members from the University of Wisconsin and from Jugoslavia — Peter Gendel, Zilka Kujundžić, Michael Malpass, Janet McPhail, and David Rose — deserve my special thanks for an enjoyable and productive summer.

A final note concerns an earlier publication. Due to an unfortunate set of circumstances, certain information mistakenly appeared in my article "Mesolithic settlement systems in the Netherlands" in 1976 in *The Early Postglacial Settlement of Northern Europe* edited by Dr. P.A. Mellars. Dr. R.R. Newell of the Biologisch-Archaeologisch Instituut had requested that data from several Mesolithic sites in the Netherlands not be published in that article and I was unable to have the information omitted. My apologies to Dr. Newell for any inconvenience.

7. BIBLIOGRAPHY

BOHMERS, A., 1963. A statistical analysis of flint artifacts. In: D. Brothwell & E. Higgs (eds.), *Science in Archaeology*. London, pp. 469-481.

BOHMERS, A. & A. WOUTERS, 1956. Statistics and graphs in the study of flint assemblages. *Palaeohistoria* 5, pp. 27-28.

BRINCH PETERSEN, E., 1973. A survey of the Late Paleolithic and the Mesolithic of Denmark. In: S.K. Kozłowski (ed.), *The Mesolithic in Europe*. Warsaw, pp. 77-127.

BRINKHUIZEN, D.C., 1976. De visresten van Swifterbant. *Westerheem* 25, pp. 246-252.

CASPARIE, W.A., B. MOOK-KAMPS, R.M. PALFENIER-VEGTER, R.C. STRUIJK & W. VAN ZEIST, 1977. The palaeobotany of Swifterbant — a preliminary report. Swifterbant Contribution 7. *Helinium* 17, pp. 28-55.

CLASON, A.T., 1978. Worked bone, antler and teeth — a preliminary report. Swifterbant Contribution 9. *Helinium* 18, pp. 83-86.

CLASON, A.T. & D.C. BRINKHUIZEN, 1978. Swifterbant. Mammals, birds, fishes. A preliminary report. Swifterbant Contribution 8. *Helinium* 18, pp. 69-82.

CONSTANDSE-WESTERMANN, T.S. & C. MEIKLEJOHN, 1979. The human remains from Swifterbant. Swifterbant Contribution 12. *Helinium* 19, pp. 237-266.

DECKERS, P.H., 1979. The flint material from Swifterbant, Earlier Neolithic of the northern Netherlands. I. Sites S-2, S-4, and S-51. Final Reports on Swifterbant II. *Palaeohistoria* 21, pp. 143-180.

ENTE, P.J., 1971. Sedimentary geology of the Holocene in the Lake Ijssel region. *Geologie en Mijnbouw* 50, pp. 373-382.

ENTE, P.J., 1976. The geology of the northern part of Flevoland in relation to the human occupation in the Atlantic

time. Swifterbant Contribution 2. *Helinium* 16, pp. 15-35.

HACQUEBORD, L., 1974. De geologie van de noordwesthoek van Oostlijk Flevoland. *Berichten Fysisch Geografische Afdeling der Rijksuniversiteit Utrecht* 8, pp. 43-51.

HACQUEBORD, L., 1976. Holocene geology and palaeogeography of the environment of the levee sites near Swifterbant (polder Oost Flevoland, section G36-41). Swifterbant Contribution 3. *Helinium* 16, pp. 36-42.

HEIDE, G.D. VAN DER, 1966a. Enkele aantekeningen betreffende prehistorische bewoning van het oostelijk deel van het Zuiderzeegebied. *Kamper Almanak*, pp. 200-214.

HEIDE, G.D. VAN DER, 1966b. Opgraving bij Swifterbant. *Fibula* 7, pp. 86-89.

HULST, R.S. & A.D. VERLINDE, 1976. Geröllkeulen aus Overijssel und Gelderland. *Berichten van de Rijksdienst voor het Oudheidkundig Bodemonderzoek* 26, pp. 93-126.

JELGERSMA, S., 1961. *Holocene sea level changes in the Netherlands*. Maastricht.

MEIKLEJOHN, C. & T.S. CONSTANDSE-WESTERMANN, 1978. The human skeletal material from Swifterbant, Early Neolithic of the Netherlands — inventory and demography. Final Reports on Swifterbant 1. *Palaeohistoria* 20, pp. 39-89.

NEWELL, R.R., 1973. The Postglacial adaptations of the indigenous population of the Northwest European Plain. In: S.K. Kozłowski (ed.), *The Mesolithic in Europe*. Warsaw, pp. 399-440.

NEWELL, R.R. & A. VROOMANS, 1972. *Automatic Artifact Registration and Systems for Archaeological Analysis with the Philips P1100 Computer: A Mesolithic Test Case*. Oosterhout.

PRICE, T.D., 1975. Mesolithic settlement systems in the Netherlands. Unpublished Ph.D. dissertation. University of Michigan, Ann Arbor.

PRICE, T.D., R. WHALLON, JR. & S. CHAPPELL, 1974. Mesolithic sites near Havelte, province of Drenthe (Netherlands). *Palaeohistoria* 16, pp. 7-61.

ROEVER, J.P. DE, 1976. Excavations at the river dune sites, S21-S22. Swifterbant Contribution 4. *Helinium* 16, pp. 209-221.

ROEVER, J.P. DE, 1979. The pottery from Swifterbant — Dutch Ertebølle? Swifterbant Contribution 11. *Helinium* 19, pp. 13-36.

ROEVER-BONNET, H. DE, A.C. RIJPSTRA, M.A. VAN RENESSE & C.H. PEEN, 1979. Helminth eggs and gregarines from coprolites from the excavations at Swifterbant. Swifterbant Contribution 10. *Helinium* 19, pp. 7-12.

ROUSSOT-LAROQUE, J., 1977. Néolithisation et Néolithique ancien d'Aquitaine. *Bulletin de la Société Préhistorique Française* 74, pp. 559-582.

SCHWABEDISSEN, H., 1966. Ein horizontierter "Breitkeil" aus Satrup und die mannigfachen Kulturverbindungen des beginnenden Neolithikums im Norden und Nordwesten. *Palaeohistoria* 12, pp. 409-468.

WAALS, J.D. VAN DER, 1972. Die durchlochten Rössener Keile und das frühe Neolithikum in Belgien und den Niederlanden. In: H. Schwabedissen (ed.), *Die Anfänge des Neolithikums vom Orient bis Nordeuropa*. Teil V A. Köln, pp. 153-184.

WAALS, J.D. VAN DER, 1977. Excavations at the natural levee sites, S2, S3/5, and S4. Swifterbant Contribution 6. *Helinium* 17, pp. 3-27.

WAALS, J.D. VAN DER & H.T. WATERBOLK, 1976. Excavations at Swifterbant — discovery, progress, aims, and methods. Swifterbant Contribution 1. *Helinium* 16, pp. 4-14.

WATERBOLK, H.T., 1976. Oude bewoning in het waddengebied. In: *Waddenzee, natuurgebied van Nederland, Duitsland en Denemarken.* Harlingen/'s-Gravenhage, pp. 211-221.

WHALLON, R., JR. & T.D. PRICE, 1976. Excavations at the river dune sites S11-S13. Swifterbant Contribution 5. *Helinium* 16, pp. 223-229.

SEEDS AND FRUITS FROM THE SWIFTERBANT S3 SITE
Final reports on Swifterbant IV

Willem van Zeist and Rita M. Palfenier-Vegter

CONTENTS

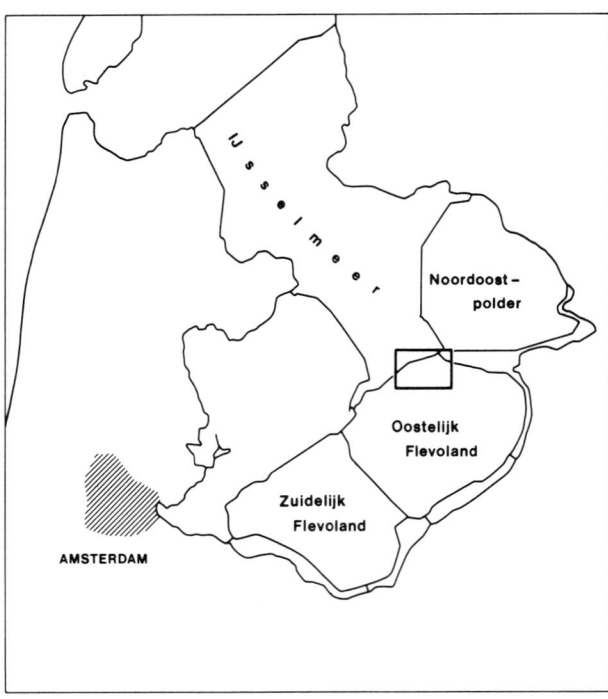

Fig. 1. Location of Swifterbant.

1. INTRODUCTION

This paper constitutes the final report on the examination of seeds, fruits and other macroscopic plant remains, except charcoal and non-carbonized wood, from the Swifterbant S3 settlement site. Prior to the discussion of the palaeobotanical results a brief survey will be presented of the palaeogeography of the area and of the prehistoric habitation. The main objectives of this review are the following: 1) to focus attention on the unusual environmental conditions of the S3 site and other similar habitation sites in the area; 2) to provide information on the physical milieu of the area indispensable for the interpretation of the botanical data; and 3) to point at the particular conditions with respect to the preservation of plant remains at S3 and other sites. For more detailed information the reader is referred to the publications cited below.

1.1. Prehistoric geography and habitation of the Swifterbant area

In the northern part of the Polder Eastern Flevoland (Oost Flevoland), near Swifterbant, municipality of Dronten (fig. 1), a prehistoric tidal estuary was discovered in the subsoil during geological investigations by the Research Division of the Polder Development Authority in 1961 and following years (Ente, 1976). Later on, systematic borings carried out by Hacquebord (1976) have for some particular areas refined the picture of the buried tidal delta landscape. A reconstruction of the landscape to the west and the north of the present town of Swifterbant around 3300 B.C., the period of relevance to the subject of this paper, is shown in fig. 2. The area was dissected by larger and smaller creeks bordered by natural levees. The levees were best developed along the large streams. Behind the levees were extensive, low-lying back-swamps. It was a fresh-water tidal estuary, in which the differences between high and low tide would usually have been no more than 10-20 cm (Ente, 1976). Ranges of late Pleistocene or early

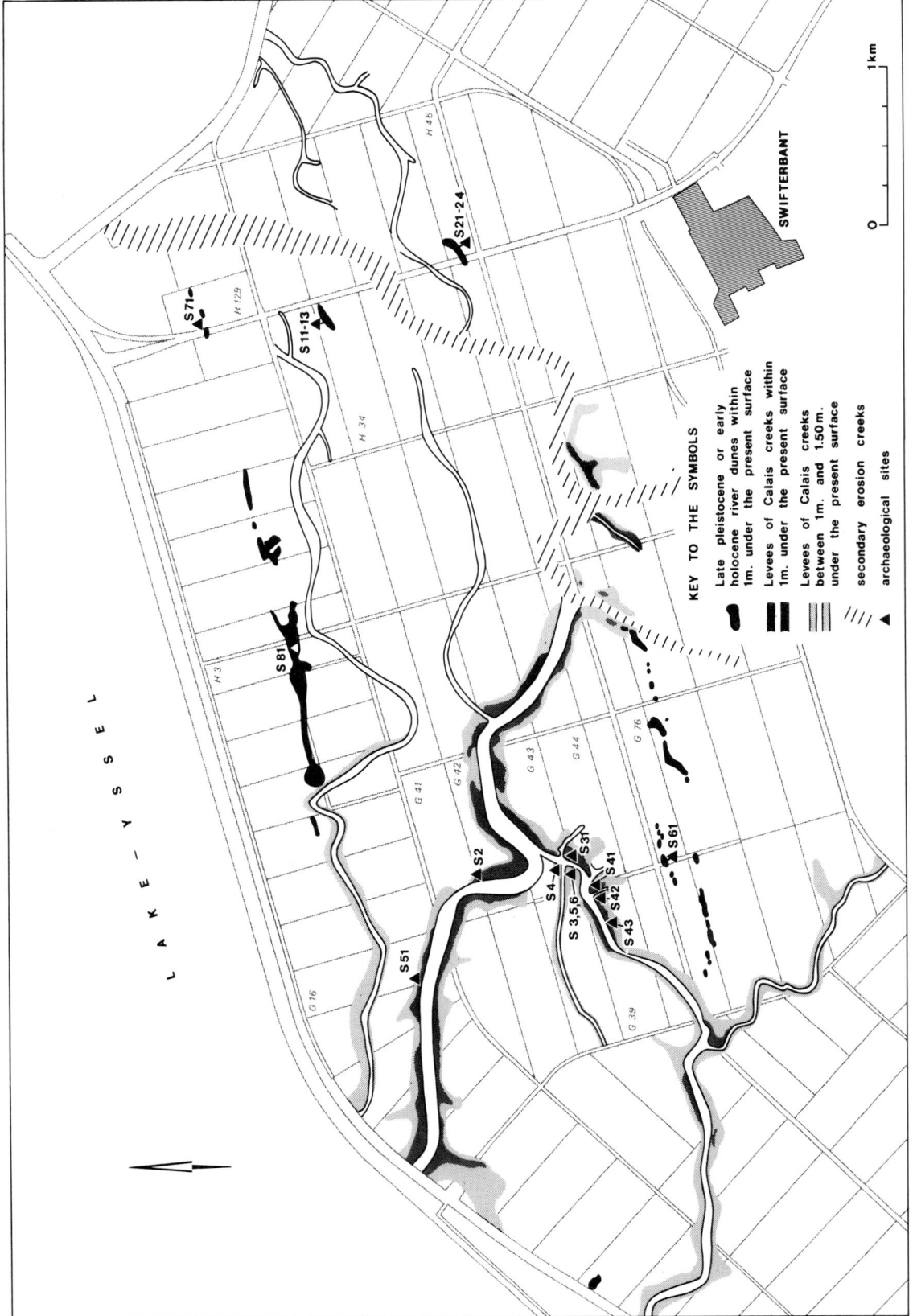

Fig. 2. Swifterbant area. Situation of creeks, levees and river-bank dunes and location of archaeological sites (after Deckers, 1979, fig. 1).

Holocene river-bank dunes delimited the valley in which the creek system developed to the north and the south.

In the course of the geological survey by the Research Division of the Polder Development Authority indications of human occupation on the highest parts of the natural levees along the creeks were established. Traces of human habitation were also found on the river-bank dunes (cf. Van der Waals & Waterbolk, 1976). Close-distance borings brought some more settlement sites to light (H. Fokkens, unpublished report). The river-bank dunes must have attracted prehistoric man over a long period. Radiocarbon dates and archaeological evidence point to an intermittent use of the dunes from c. 5800 B.C. to c. 3300 B.C. (Deckers et al., 1980). On the dunes, traces of human occupation of various periods are found in the same level. The habitation on the natural levees embraced a much shorter period, viz. c. 3500-3200 B.C. The location of the prehistoric settlement sites in the Swifterbant area is indicated in fig. 2.

About 3000 B.C. the area became inundated, as a result of which marine clay, up to 30 cm thick, was deposited. This clay layer seals off the fresh-water tidal landscape and consequently also the occupation remains on the natural levees. Only the highest parts of the river-bank dunes still stuck out. These were eventually covered by peat formed in the extensive marshes which developed in the area after the deposition of the clay. The greater part of the peat was eroded by the rapidly extending Almere, the medieval predecessor of the Zuiderzee (cf. Deckers et al., 1980).

Systematic investigations of the Swifterbant sites were started by the archaeological department of the Research Division of the Polder Development Authority. In 1962-1967, under the direction of G.D. van der Heide, excavations were carried out at the river-bank dune sites S21-22 and at the natural levee site S2 (cf. Van der Waals & Waterbolk, 1976). In 1971, the excavations were resumed, now by the Biologisch-Archaeologisch Instituut. In addition, teams from the Museum of Anthropology of the University of Michigan (R. Whallon) and from the Department of Anthropology of the University of Wisconsin (T.D. Price) worked on Swifterbant dune sites (cf. Price, this volume).

1.2. The choice of the S3 site for botanical examination

In the sites on the river-bank dunes only charred plant remains are preserved. In the sites on the natural levees, conditions for the preservation of vegetable material are much better. As a matter of fact, the habitation layers consist here of dark clay rich in organic material. In contrast to the rather dry conditions on the dunes at the time of the habitation, the banks bordering the creeks constituted a moist environment. The levee sites were flooded regularly or occasionally, particularly in the winter season.

So far, eight settlement sites on natural levees have been localised: two (S2 and S51) alongside a major gully, the others on the banks of a smaller creek (S3, S5 and S6 are taken here as one site). On four of them excavations have been carried out. These sites were also tested for their suitability for palaeobotanical research. To this purpose, one litre samples of soil were systematically examined for numbers of seeds and fruits (for a description of the method see 2.1.). The samples from S51, S2 and S4 were poor or barren in seeds. On the other hand, at least a part of the samples from S3 yielded somewhat greater numbers of uncarbonized seeds. It must be left undecided here whether the scarcity of seeds and fruits in S2, S4 and S51 is of primary or of secondary nature; in other words, whether only few seeds were incorporated in the settlement layers or whether due to less favourable conditions most of the seeds originally present decayed in the course of time. As only the S3 site offered good prospects for palaeobotanical research, it was decided to confine a more detailed examination to this site.

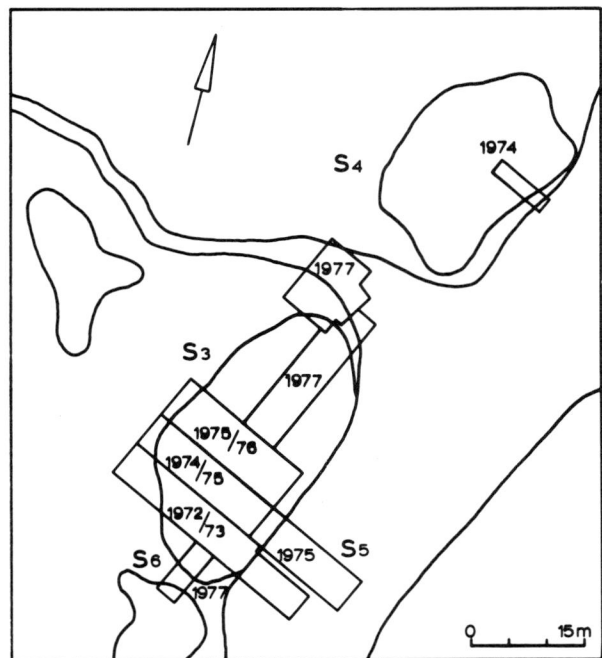

Fig. 3. Trenches excavated in successive years in site S 3 and adjacent sites (after Deckers *et al.*, 1980, fig. 19).

1.3. The S 3 site

The gully along which the S 3 site was situated was initially a tributary creek which subsequently made contact with the southwestern creek, so that a break-through channel was formed (Hacquebord, 1976). It is likely that the natural levees along the break-through channel did not silt up as high as those along the major creek on which S 2 and S 51 were founded. The moister environment at the locality of the S 3 site may have favoured the preservation of organic remains, thus contributing to the accumulation of the habitation deposit, up to 75 cm thick. In the field, layers of reed stems, scattered wood chips, bark and twigs were frequently observed.

The S 3 site, on the western levee of the break-through channel, is more or less oval in outline. The length is c. 35 m and the width does not exceed 20 m. Excavations of the site were carried out in 1972-1977. In fig. 3 the areas excavated in the successive years are indicated. A deep trench through the creek

alongside S3 is designated as S5, although the finds from this cutting must have originated from S 3. Numerous pottery sherds, flints, worked and unworked bone fragments, and, more rarely, beads and pendants occurred dispersed in the habitation layers. Of the structural elements, hearths and posts are mentioned. The remains of fire-places consisted of clay which was partly burnt and of layers of white ash. Remains of about 750 stakes and posts were found, but it was not possible to arrive at configurations of huts or other structures (Van der Waals, 1977; Deckers *et al.*, 1980).

Along the creek, settlement layers had been washed away at an early stage of the habitation. The erosion was soon followed by rapid sedimentation in the same zone, and thereupon the settlement expanded again towards the creek (Deckers *et al.*, 1980).

2. FIELD AND LABORATORY PROCEDURES

2.1. Field work

The excavations of the S 3 site and of the other natural levee sites were performed in square metre parcels and per layer of c. 10 cm. The grid system over the S 3 settlement is indicated in figs. 6-33. Thus, each square is designated by a Roman and an Arabic figure. The successive 10 cm layers are indicated by letters F up to and including K (figs. 6-15).

Systematic sampling for palaeobotanical examination started in 1973. During the 1973 campaign, all soil dug away per layer of c. 10 cm and per square metre was dissolved with water in a concrete mixer and subsequently washed through two sieves, with meshes of 2 and 1 mm respectively, placed one on top of the other. This method was applied in order to recover small finds which had escaped notice during the excavation. The organic fraction retrieved on both sieves, or part of it, was set apart for botanical examination in the laboratory in Groningen. This method appeared to be less satisfactory. It soon became clear that for practical reasons only a minor part of the great numbers of samples brought to the laboratory could be examined, and what is worse, that the choice of the samples to be examined had to be completely arbitrary. Some of the samples were fairly rich in seeds, whereas from others only small numbers could be recovered. Moreover, small seeds could have been washed through the 1 mm sieve, thus effecting the relative frequencies of the seed and fruit types.

The experience with the samples of the 1973 campaign and a modification in the field processing of the excavated soil induced us to apply another strategy with respect to the sampling for palaeobotanical research. The washing through the 1 mm sieve had turned out to be very time-consuming and cumbersome. For that reason it was decided to wash only through a 2 mm sieve from 1974 on. It became clear that most of the seeds were washed through the 2 mm sieve, so that the residues left on the sieve were not suitable for our purposes. From then on, unprocessed soil samples had to be taken to the laboratory. It would be most efficient if soil samples sufficiently rich in seeds and fruits could be selected for possible examination in the laboratory. For determining in the field the seed concentration of the excavated soil a kind of crash procedure was set up by W. A. Casparie and R. M. Palfenier-Vegter. This method, which turned out to be very satisfactory, is described below.

1. From the soil excavated per square metre and per 10 cm layer a one litre sample was soaked in water to which a detergent had been added for 24 hours.
2. The next day the sample was washed through three sieves, with meshes of 2.0, 0.5 and 0.2 mm respectively.
3. Thereupon, the fractions were examined under a binocular stereo-microscope for uncarbonized seeds and fruits and carbonized cereal grains. No identifications were made, but only the numbers of seeds and fruits were recorded. Moreover, an estima-

tion of wood and charcoal fragments was made. On an average, half an hour was spent per sample.
4. Based upon the numbers of seeds and fruits four categories of samples are distinguished:
   — 20 and more (rich in seeds)
   — 10-19 (moderately rich in seeds)
   — 1-9 (rather poor in seeds)
   — no seeds.
   It should be emphasized that the terms rich and moderately rich in seeds are relative indications. Compared to the Iron Age coastal settlement sites, for instance, the Swifterbant material is poor in seeds and fruits. The seed frequencies per one litre form the basis of figs. 11-15 and 16-24 (section 6).
5. Of the soil from square metre layers which had turned out to be sufficient rich in seeds, a three litre sample was set apart for possible examination in Groningen. It is self-evident that this was unprocessed soil.

The palaeobotanical field work was carried out by R. M. Palfenier-Vegter (1974, 1975, 1976, 1977) and Y. M. Koster (1976). The one-litre samples from S51 were examined in Groningen, those from S2 in part. In washing the soil for retrieving small archaeological objects, also carbonized grains, fragments of hazelnut shells and other large botanical remains were recovered from the residues on the 2 mm sieve. These plant remains are included in the discussion (table 3, figs. 6-10). The samples concerned are referred to in this paper as "sieve residues".

2.2. Choice of samples to be examined

In the laboratory the samples were processed and subsequently examined for seeds, fruits and other plants remains according to the methods usual in the palaeobotany department of the Biologisch-Archaeologisch Instituut. For a rather detailed description of the laboratory treatment of similar samples the reader is referred to Van Zeist (1974, section 2.2.).

In spite of the greatly reduced numbers of samples brought to the laboratory, only a minor part of them could be examined, this because of the very time-consuming procedure. In the selecting of the samples it was attempted to comply with two conditions:

1. Of each of the layers F-K a representative number of samples should be analysed.
2. Per layer the samples should not all be derived from one and the same area, but they should be distributed as well as possible over the site.

The latter condition was curtailed by the fact that concentrations of metre squares with relatively great numbers of seeds occurred (see 6.1.1.), implying that the samples brought to the laboratory were not more or less evenly distributed over the site. The origin of the samples examined in the laboratory is indicated in figs. 25-33. It should be taken into account that the 10 cm levels of figs. 25-33 do not correspond to the layers F-K employed in the field. There are five "field" layers against nine 10 cm levels (see 6.1.2.). In spite of the fact that the spatial distribution as well as the numbers of the samples examined may be criticized, we believe that the study presented in this paper provides a fair picture of the plant species represented in the site.

2.3. Presentation of the results

The results of the analyses are presented in table 1. No numbers of seeds and fruits are shown but percentages. In this connection the following should be remarked. Only occasionally were all three fractions of one sample (2.0, 0.5 and 0.2 mm sieve) wholly analysed. Of the middle and fine fractions usually only a part was examined. Before adding the results of the three fractions together, the numbers of seeds and fruits found in a subsample of a fraction were converted to the total numbers for that fraction. The frequencies of the seed and fruit types are expressed as percentages of the total numbers of seeds and fruits per sample. A plus sign is given in the case of remains other than seeds and fruits or if only one or a few seed fragments were found. The mosses are shown separately in table 2. The moss remains have been identified by Dr. H.J. During (Utrecht). As the volume of soil of all samples was approximately the same, viz. about 3 litres, the total numbers of seeds and fruits shown at the bottom of table 1 are directly comparable with one another.

2.4. Aims of the investigation

The aims of the palaeobotanical investigation of the Swifterbant S3 site are rather straightforward, namely:

- to assess the exploitation of the wild flora by the inhabitants of the site;
- to ascertain which plants were cultivated;
- to attempt a reconstruction of the vegetation;
- to look for possible indications of seasonal habitation.

No groundplans of huts or houses were recognized in the field. For that reason, in the selection of samples for palaeobotanical examination it was not possible to take account of possible activity areas, such as work- and living-floors. Nevertheless, in working out the results it turned out that certain correlations between seed content of the soil and the distribution of archaeological objects seem to exist (6.1.3., 6.2.1.).

The analyses of the samples were carried out by R.M. Palfenier-Vegter. Both authors are responsible for the identifications. W. van Zeist takes the sole responsibility for the interpretation of the results and for the text.

2.5. The preliminary report

In the preliminary report on the palaeobotany of Swifterbant (Casparie *et al.*, 1977) two pollen diagrams, one of them prepared for the upper layers of the fill of the creek along which the S3 site was situated, are presented. Since then no other pollen diagrams have been prepared for the area. In the same report the species identifications of wood and charcoal systematically collected from 39 metre squares

Table 1. Seeds and fruits from Swifterbant S3. No numbers are shown but percentages. At the bottom of the table the actual numbers counted and the (estimated) numbers of seeds per 3 litres of soil are presented. A plus-sign indicates a few seed fragments or vegetable remains other than seeds. cf. = identification uncertain. Samples 13 and 19 are not in the correct order due to a misreading of the labels.

Sample number	1	2	3	4	5	6	7	8	9	10	11	12	13	14	15	16	17	18	19	20	21	22	23
Sample designation	I-17-i	I-18-i	II-16-H	III-14-G	III-20-F	III-20-G	IV-13-F	IV-17-G	IV-20-H	IV-24-F	IV-24-G	V-13-K	II-18-G	V-18-K	V-21-G	V-23-i	VI-25-K	VII-22-K	III-17-F	VIII-17-F	VIII-23-F	VIII-23-H	VIII-24-H
Agrostis spec																0.4						0.6	
Alisma plantago-aquatica										2.7		0.2											
Alnus glutinosa			0.1							2.7								0.3					
Alopecurus cf. geniculatus																							
Anthriscus sylvestris			0.1	2.1																			
Arctium cf. lappa	0.1		0.1		0.2			0.2						0.1			0.7		0.1			1.4	0.2
Artemisia vulgaris				2.1																			
Aster tripolium						1.2	0.3															0.6	0.1
Atriplex hastata/patula	6.9	8.3	9.1	6.2	55.6	30.9	1.0		9.1		26.5	3.6	36.3	24.8	5.1	16.8	9.4	2.5	13.1	3.6	4.4	12.9	10.6
Betula spec.			0.1																				
Bromus mollis/secalinus											19.3												
Caltha palustris														0.3									
Capsella bursa-pastoris														0.3									
Carduus crispus												0.8	0.5				0.1						
Carex disticha																							
Carex nigra-type			0.1																				
Carex paniculata			0.1			0.5																	
Carex pseudocyperus		0.1																					
Carex riparia			0.2																				
Carex rostrata/vesicaria			0.1		0.3				2.7					0.3									
Carex serotina-type																					2.2		
Carex spec.			0.5																				
Ceratophyllum submersum														0.1									
Chenopodium album	0.1	0.1	0.4		3.8	3.0	1.7	1.8		+	0.8	2.5	2.6	0.5	31.5	0.4			0.2	36.0	25.1	13.3	0.6
Chen. rubrum/glaucum																						0.6	
Cirsium arvense			0.2					0.2								0.7							
Cladium mariscus	0.1		0.1			0.5	1.0	0.5		2.7								0.2		8.7	2.2	2.2	
Conium maculatum	0.1					0.3														2.2	2.2		
Corylus avellana	+		+			0.1														+		+	
Crataegus monogyna	0.1			+							0.4										26.6		
Eleocharis palustris																						0.6	
Galeopsis tetrahit/speciosa		0.1																					
Galium aparine				1.8				0.3								0.9							
Galium palustre																							
Glyceria fluitans																							
Gramineae indet.								0.2															
Hordeum vulgare nudum	0.1		0.1	0.7	0.3	1.4			3.3		1.8			0.1			1.8	0.9	13.1	3.6	6.6	1.4	0.1
Hydrocotyle vulgaris																							
Juncus gerardii																							
Lapsana communis			0.1						8.3														
Lycopus europaeus													0.3	0.3								0.6	0.3
Lychnis flos-cuculi																							
Mentha aquatica	0.1		0.1				0.3			2.7		0.2											
Menyanthes trifoliata															0.5								
Moehringia trinervia																					3.6		
Nymphaea alba	0.8	0.4	0.8	0.7		6.4	0.7	0.9	0.1		1.3			0.5	2.6		0.2	0.6	0.5			1.1	0.3
Oenanthe aquatica																							
Phragmites australis	0.6	0.8	8.8	1.5	5.9	13.7	9.7	80.6	0.3			2.6	4.5	14.4		15.1	2.3	2.7			8.9	29.5	51.3
Plantago major				0.3										0.6			0.4	0.6	0.5		4.4		
Poa pratensis/trivialis			0.1	0.3									0.3	0.6			0.4			8.7			
Polygonum aviculare	6.8	5.4	72.8	2.6	4.2	16.1		0.5	3.1	2.7	41.1		1.0	9.3		33.6	12.1	18.0				29.1	20.9
Polygonum lapathifolium	69.5	67.3	3.3	3.4	5.2	14.2	0.2		1.4	75.8	2.7	3.1		3.3	36.5		3.7	52.2	63.5	10.9	3.6	4.0	0.5
Polygonum persicaria	3.5	12.1	0.4	0.2	13.6	4.6		0.3	6.9		3.1		0.7	2.8		3.4	1.0	3.0				1.2	0.8
Potamogeton crispus																							0.3
Potamogeton spec.																							
Pyrus malus			0.1	0.2	+			0.1	0.1										0.1			0.1	
Ranunculus acer											0.4												
Ranunculus sceleratus										2.7										10.8			
Rosa canina/rubiginosa	0.1																					0.2	
Rosa spec.					+																		
Rubus fruticosus																							
Rumex conglomeratus						0.5																	

24	25	26	27	28	29	30	31	32	33	34	35	36	37	38	39	40	41	42	43	44	45	46	Sample number
VIII-26-H	VIII-26-K	IX-14-G	IX-15-F	IX-19-F	IX-25-H	IX-29-F	X-15-G	X-22-F	XI-16-F	XI-23-K	XII-15-G	XIV-17-i	XIV-23-G	XV-19-K	XV-24-G	XV-27-G	XVI-25-G	XVIII-19-i	XX-19-i	XXIV-19-i	XXIX-18-F	XXXIII-18-F	Sample designation
.	.	.	.	.	.	.	.	.	.	.	.	.	.	.	.	.	.	.	.	.	.	.	Agrostis spec
.	.	.	.	.	.	.	.	.	.	.	.	.	0.8	.	.	.	.	.	.	.	.	.	Alisma plantago-aquatica
.	.	.	.	2.8	.	.	.	7.5	.	.	.	.	.	.	.	.	.	.	.	.	.	.	Alnus glutinosa
.	.	.	.	.	.	.	.	.	.	.	.	.	.	.	.	.	.	.	.	0.1	.	.	Alopecurus cf. geniculatus
.	.	.	.	.	.	.	.	.	.	.	.	.	.	.	.	.	.	.	.	1.4	.	.	Anthriscus sylvestris
.	.	.	.	.	.	.	.	.	.	.	.	.	.	.	.	.	.	.	.	.	.	.	Arctium cf. lappa
.	.	.	.	.	.	.	.	.	.	.	.	.	.	.	.	.	.	.	.	0.2	.	.	Artemisia vulgaris
.	.	.	.	.	0.2	.	.	.	.	.	.	.	.	.	.	.	.	.	.	.	.	.	Aster tripolium
29.9	3.2	.	1.7	8.4	14.1	12.6	0.2	.	1.2	9.7	0.5	2.8	.	7.7	.	9.3	7.0	20.3	10.5	8.4	.	1.0	Atriplex hastata/patula
.	.	.	.	.	.	.	.	.	.	.	.	.	.	.	.	.	.	.	.	.	.	.	Betula spec.
.	.	.	0.6	.	.	.	.	.	.	.	.	.	.	.	.	.	.	.	.	.	.	.	Bromus mollis/secalinus
.	.	.	.	.	.	.	.	.	.	.	.	.	.	.	.	.	.	.	.	.	.	.	Caltha palustris
.	.	.	.	.	.	.	.	.	.	.	.	.	.	0.1	.	.	.	.	.	.	.	.	Capsella bursa-pastoris
.	.	.	.	.	.	.	.	.	.	.	.	.	.	.	.	.	.	.	.	0.1	.	.	Carduus crispus
.	.	.	.	.	0.4	.	.	.	.	.	.	.	.	.	.	.	.	.	0.1	.	.	.	Carex disticha
.	.	.	.	.	.	.	.	.	.	.	.	.	.	.	.	.	.	0.1	.	.	.	.	Carex nigra-type
.	.	.	.	.	.	.	.	.	.	.	.	.	.	.	.	.	1.1	.	.	.	.	.	Carex paniculata
.	.	.	.	.	.	.	.	2.5	.	.	.	.	.	.	.	.	.	.	.	.	.	.	Carex pseudocyperus
.	.	.	.	.	.	.	.	.	.	.	.	.	.	.	.	.	.	.	.	.	0.6	.	Carex riparia
.	.	.	.	.	.	.	.	.	.	.	.	.	.	.	.	.	.	.	.	.	0.6	.	Carex rostrata/vesicaria
.	.	.	.	.	.	.	.	.	.	.	.	.	.	.	.	.	.	.	.	.	.	.	Carex serotina-type
.	.	.	.	.	.	.	.	.	.	.	.	.	.	.	.	.	.	.	.	.	.	.	Carex spec.
.	.	.	.	.	.	.	.	.	.	.	.	.	.	.	.	.	.	.	.	.	.	.	Ceratophyllum submersum
.	.	31.4	36.0	49.4	.	2.4	38.1	17.5	39.0	0.2	13.4	42.5	40.2	.	32.1	1.3	4.7	4.5	10.5	1.0	44.5	29.0	Chenopodium album
.	.	.	.	.	.	.	.	.	.	.	.	.	.	.	1.0	0.6	1.0	.	.	.	.	.	Chen. rubrum/glaucum
.	.	.	.	.	.	.	.	.	.	.	.	.	.	.	.	.	.	.	.	.	.	.	Cirsium arvense
.	.	.	1.4	.	.	.	.	2.5	.	.	.	.	2.2	.	.	.	.	.	.	.	8.8	.	Cladium mariscus
.	.	.	0.3	.	.	.	1.1	.	0.2	.	1.0	0.4	.	.	.	.	.	.	.	0.1	1.7	1.6	Conium maculatum
.	.	.	.	.	.	.	.	+	.	.	.	.	+	.	.	.	.	.	.	.	+	.	Corylus avellana
.	.	0.1	.	.	.	.	.	0.1	.	0.1	.	.	.	.	.	.	.	.	.	.	.	0.1	Crataegus monogyna
.	.	.	.	.	.	.	0.2	.	.	.	.	.	0.8	.	.	.	.	.	.	.	.	.	Eleocharis palustris
.	.	.	.	.	.	.	.	.	.	.	.	.	.	.	.	.	.	.	.	0.1	.	.	Galeopsis tetrahit/speciosa
.	.	.	.	0.6	.	.	.	.	0.1	.	.	.	.	.	.	.	.	.	.	.	0.1	.	Galium aparine
.	.	.	.	.	.	.	.	.	.	.	.	.	0.8	.	.	.	.	.	.	.	.	.	Galium palustre
.	.	.	.	.	.	.	.	.	.	.	.	.	.	.	.	0.6	.	.	.	.	.	.	Glyceria fluitans
.	.	.	.	.	.	.	0.5	.	.	0.2	.	.	1.7	.	3.4	.	.	.	.	.	.	.	Gramineae indet.
.	.	0.1	0.5	5.6	0.5	0.9	0.2	5.0	0.8	.	0.4	.	1.1	.	1.4	.	0.2	.	.	.	0.4	0.7	Hordeum vulgare nudum
0.4	.	.	.	.	.	.	.	.	.	.	.	.	.	.	.	.	1.1	.	.	.	.	.	Hydrocotyle vulgaris
.	.	.	.	.	.	.	.	.	.	.	.	.	.	.	.	.	2.3	.	.	.	.	.	Juncus gerardii
.	.	.	.	.	.	.	.	.	.	.	.	.	.	.	.	.	.	.	.	.	.	.	Lapsana communis
.	.	.	.	.	0.2	.	.	2.5	.	.	.	.	.	.	.	.	.	.	.	.	.	.	Lycopus europaeus
.	.	.	.	.	.	.	.	.	.	.	.	.	.	.	.	.	.	.	.	.	0.6	.	Lychnis flos-cuculi
.	0.4	.	.	.	.	.	.	0.2	.	.	.	.	.	0.1	1.0	.	.	0.1	.	0.1	0.6	.	Mentha aquatica
.	.	.	.	.	.	.	0.1	.	.	.	.	.	.	.	.	.	.	.	.	.	.	.	Menyanthes trifoliata
.	.	.	.	.	.	.	.	.	.	.	.	.	.	.	.	.	.	.	.	.	.	.	Moehringia trinervia
.	.	.	.	.	0.8	.	.	.	.	1.2	.	.	.	.	.	.	.	0.1	+	.	.	.	Nymphaea alba
.	.	.	.	0.3	.	.	.	.	.	.	.	0.4	.	.	.	.	.	.	.	.	.	.	Oenanthe aquatica
5.7	0.1	.	.	.	27.0	.	0.3	.	.	0.4	3.9	.	2.8	0.1	2.2	4.7	3.5	0.4	0.5	2.1	.	.	Phragmites australis
.	0.4	.	.	.	.	.	.	.	.	.	.	.	.	.	.	.	.	.	0.1	.	.	.	Plantago major
0.2	.	.	.	.	0.4	.	.	.	.	.	.	.	.	.	.	.	.	.	.	0.5	13.0	0.3	Poa pratensis/trivialis
49.3	2.3	.	.	.	37.1	.	0.2	.	.	34.7	.	.	0.8	2.7	.	.	.	6.2	9.8	0.6	.	.	Polygonum aviculare
.	87.6	.	.	2.8	2.2	.	.	.	23.0	0.3	.	.	.	.	.	.	.	1.9	2.2	.	.	.	Polygonum lapathifolium
.	1.9	.	0.6	.	0.2	.	.	.	2.1	.	.	.	.	.	.	.	.	.	.	.	.	.	Polygonum persicaria
.	.	.	.	.	.	.	.	.	.	.	.	.	.	.	1.0	.	.	.	.	.	.	.	Potamogeton crispus
.	.	.	.	.	.	.	.	.	.	.	.	.	.	.	.	.	.	.	.	.	.	.	Potamogeton spec.
.	.	.	.	.	.	.	.	.	.	.	.	.	0.3	.	.	.	.	.	.	.	.	.	Pyrus malus
.	.	.	.	.	.	.	.	11.3	.	.	.	.	.	.	.	.	.	.	.	.	.	.	Ranunculus acer
.	.	.	.	.	.	.	.	2.5	.	.	.	.	0.8	.	3.0	.	.	0.1	.	.	.	.	Ranunculus sceleratus
.	.	.	.	.	.	.	.	.	.	.	.	.	.	.	.	.	.	.	.	.	0.1	.	Rosa canina/rubiginosa
.	.	.	.	.	.	.	.	.	.	.	.	.	.	.	.	.	.	.	.	.	.	.	Rosa spec.
.	.	.	.	0.3	.	.	.	.	0.4	.	0.2	.	.	.	.	.	.	.	.	.	.	.	Rubus fruticosus
.	.	.	.	.	.	.	.	.	.	.	.	.	.	.	.	.	.	.	.	.	.	.	Rumex conglomeratus

Table 1 (continued).

Sample number	1	2	3	4	5	6	7	8	9	10	11	12	13	14	15	16	17	18	19	20	21	22	23
Sample designation	I-17-i	I-18-i	II-16-H	III-14-G	III-20-F	III-20-G	IV-13-F	IV-17-G	IV-20-H	IV-24-F	IV-24-G	V-13-K	II-18-G	V-18-K	V-21-G	V-23-i	VI-25-K	VII-22-K	III-17-F	VIII-17-F	VIII-23-F	VIII-23-H	VIII-24-H
Rumex hydrolapathum	.	.	0.1	.	.	.	.	.	.	.	.	.	.	.	.	.	.	.	.	.	.	.	.
Rumex maritimus	.	.	.	.	.	.	.	.	.	.	.	.	.	.	.	.	.	.	.	.	.	0.6	.
Rumex spec.	.	0.1	.	.	.	.	0.3	.	.	.	0.4	.	.	.	.	.	.	.	.	.	.	.	.
Salicornia europaea	0.1	.	.	.	.	.	.	.	.	.	.	.	.	.	.	.	.	.	.	.	.	.	.
Scirpus lacustris ssp.lacustris	0.1	.	.	1.0	.	.	.	.	.	.	.	.	.	.	.	.	.	.	.	.	.	.	.
Scirpus maritimus	.	.	.	0.5	0.3	.	.	0.5	.	.	0.4	0.2	0.3	.	.	.	.	.	.	0.2	4.4	0.6	0.3
Scirpus tabernaemontani	0.1	.	0.1	0.5	3.1	1.5	0.7	1.1	.	8.3	2.2	0.5	3.7	0.3	.	.	.	.	.	.	2.2	0.9	1.4
Sium erectum	.	.	.	.	.	.	.	.	.	.	.	.	.	.	.	0.4	.	.	.	.	.	.	.
Sium latifolium	.	.	.	0.5	.	.	.	.	.	.	.	.	.	.	.	.	.	.	.	.	.	.	.
Solanum dulcamara	0.1	.	.	.	.	.	0.3	.	.	.	0.4	.	0.5	.	.	.	.	.	.	7.2	.	.	.
Solanum nigrum	0.2	0.9	0.5	1.0	2.1	1.0	.	0.9	0.3	4.6	1.3	0.5	3.1	0.3	2.3	0.7	0.8	0.3	.	3.6	4.4	0.8	.
Sonchus asper	.	.	0.1	.	.	.	.	.	.	.	.	.	0.3	.	.	.	.	0.2	.	.	.	.	.
Sonchus palustris	.	.	.	.	.	.	.	.	.	.	.	.	.	.	.	.	.	.	.	.	.	.	.
Stellaria media	0.1	.	0.1	49.0	.	.	34.5	.	0.2	2.7	4.9	26.6	32.3	0.6	3.7	0.4	.	0.2	.	3.6	.	1.2	0.1
Trifolium repens	.	.	.	.	+	.	.	.	.	.	.	.	.	.	.	.	.	.	.	.	.	.	.
Triticum dicoccum	.	.	.	0.2	.	.	.	.	.	.	.	.	.	.	.	.	.	.	.	.	.	.	.
Typha latifolia/angustifolia	.	.	.	.	.	.	.	.	.	.	.	.	.	.	.	.	0.3	.	.	.	.	.	.
Urtica dioica	10.6	4.3	1.8	26.8	2.4	3.9	48.9	11.5	0.3	33.0	13.3	62.3	9.6	5.2	54.2	22.1	19.9	7.1	.	35.9	22.2	11.2	12.8
Zannichellia palustris	.	.	.	.	.	.	.	.	.	.	.	.	.	.	.	.	.	.	.	.	.	.	.
Numbers of species	22	12	28	22	17	18	14	12	14	16	15	11	18	19	8	16	13	17	9	10	12	22	15
Numbers of seeds counted	1418	747	1137	198	286	210	221	214	623	36	225	585	390	338	134	237	487	629	23	28	29	177	272
Total numbers of seeds per 3 litres of soil	9141	10402	2719	388	286	1014	597	8703	1822	109	225	2384	2321	1944	216	535	2489	5102	23	28	45	1289	2152

24	25	26	27	28	29	30	31	32	33	34	35	36	37	38	39	40	41	42	43	44	45	46	Sample number
VIII-26-H	VIII-26-K	IX-14-G	IX-15-F	IX-19-F	IX-25-H	IX-29-F	X-15-G	X-22-F	XI-16-F	XI-23-K	XII-15-G	XIV-17-i	XIV-23-G	XV-19-K	XV-24-G	XV-27-G	XVI-25-G	XVIII-19-i	XX-19-i	XXIV-19-i	XXIX-18-F	XXXIII-18-F	Sample designation
.	.	.	.	.	.	.	.	.	.	.	.	.	.	.	.	.	.	.	.	.	.	.	Rumex hydrolapathum
.	.	0.6	.	.	.	.	.	.	.	.	.	.	.	.	.	.	.	.	.	.	.	.	Rumex maritimus
.	.	.	.	.	.	.	.	.	.	.	.	.	.	.	.	.	.	0.1	.	.	.	.	Rumex spec.
.	.	.	.	.	.	.	.	.	.	.	.	.	.	.	.	.	.	.	.	.	.	.	Salicornia europaea
.	.	.	.	.	.	.	.	.	.	.	.	.	0.8	.	.	.	.	.	.	.	.	.	Scirpus lacustris ssp.lacustris
.	.	.	.	.	.	.	0.2	.	.	.	.	.	0.8	0.1	.	.	0.5	0.8	0.1	.	.	.	Scirpus maritimus
.	.	0.4	0.6	1.4	1.6	0.7	.	.	0.8	.	.	2.8	34.4	0.5	0.6	.	.	1.0	0.8	0.2	3.2	.	Scirpus tabernaemontani
.	.	.	.	.	.	.	.	.	.	.	.	0.4	.	.	.	.	.	.	.	.	.	0.3	Sium erectum
.	.	.	.	.	.	.	.	.	.	.	.	.	.	.	.	.	.	.	.	.	.	0.5	Sium latifolium
.	.	0.4	.	1.4	.	.	0.2	.	.	.	.	1.0	.	.	.	.	.	.	.	1.1	.	0.5	Solanum dulcamara
0.7	.	3.6	7.0	.	.	.	1.1	11.3	0.8	0.2	4.4	2.4	5.9	0.8	0.6	4.7	2.8	1.8	1.2	.	1.7	1.2	Solanum nigrum
.	.	.	.	.	.	.	.	.	.	.	.	.	.	.	.	.	.	.	.	.	.	.	Sonchus asper
.	.	.	.	.	.	0.6	.	.	.	.	.	.	.	.	.	.	.	.	.	0.1	.	.	Sonchus palustris
0.4	0.3	10.6	11.6	2.8	.	5.9	9.5	2.5	7.3	0.5	6.9	4.0	1.7	82.4	.	18.7	14.0	40.7	35.0	32.0	0.6	4.5	Stellaria media
.	.	.	.	.	.	.	.	.	.	.	.	.	.	.	.	.	.	.	.	.	.	.	Trifolium repens
.	.	.	.	.	.	.	.	.	.	.	.	.	.	.	.	.	.	.	.	+	.	+	Triticum dicoccum
.	.	.	.	.	.	.	.	.	.	0.4	.	.	.	.	.	.	.	.	.	.	.	.	Typha latifolia/angustifolia
3.5	4.4	52.8	40.7	22.5	14.9	76.7	48.1	35.0	49.1	24.0	72.9	40.5	6.7	5.4	53.4	60.0	61.8	21.8	28.6	39.3	35.7	60.8	Urtica dioica
.	.	.	.	.	.	.	.	.	.	.	.	.	.	.	.	.	.	.	.	.	0.6	.	Zannichellia palustris
7	8	9	11	14	14	8	14	12	12	12	9	11	17	10	11	8	12	15	14	17	16	12	Numbers of species
256	684	270	190	92	237	179	521	40	252	432	392	241	120	1543	98	122	97	1143	1559	851	174	303	Numbers of seeds counted
790	9615	557	10317	712	993	2526	1591	160	1042	3906	1168	499	358	6247	492	150	430	7032	6889	3551	683	1570	Total numbers of seeds per 3 litres of soil

during the 1975 campaign and of c. 250 posts are discussed. No more wood identifications have been carried out for the S3 site. In this paper the wood identifications will come up for discussion in the section on the reconstruction of the former vegetation (section 4; table 12).

A few plant names enumerated in table 5 (pp. 48-49) of the preliminary report need to be corrected. Thus, on closer inspection the seeds of *Veronica anagallis-aquatica* appeared to be of recent origin (modern contamination) and should be discarded. *Rumex obtusifolius* should read *Rumex* spec. The fruit initially identified as *Sonchus arvensis* is of *Sonchus palustris*.

3. DISCUSSION OF THE PLANT REMAINS

In this section some remarks will be made on the remains of wild plants recovered from Swifterbant except wood. The cereal grains will be treated in section 5. Most of the seeds and fruits found at Swifterbant have already been discussed in an earlier study (Van Zeist, 1974). For those types no descriptions will be given here, but the discussion will remain confined to presenting the measurements and to some remarks on the ecology of the species. For syntaxonomic units which are occasionally mentioned in this section but which form the basis of the reconstruction of the former vegetation (section 4) the reader is referred to Westhoff & Den Held (1969).

3.1. Musci

Moss remains are not particularly abundant at Swifterbant. Only 21 of the 46 samples processed in the laboratory yielded moss remains, generally in very small quantities (table 2). The scarcity and the incompleteness of the remains handicapped the identification of the mosses which has been carried out by Dr. H.J. During (Institute for Plant Taxonomy, State University of Utrecht). Many mosses represented at Swifterbant are epiphytic. Unlike our presen-

tation of data for the higher plants (Spermatophyta), for the mosses no families are given here. The mosses are discussed in alphabetical order of the genera. The data on the habitats of the moss species are after Landwehr (1966).

Anomodon viticulosus (Hedw.) Hook et Tayl.
On old tree trunks, particularly on elm but also on ash and lime. In 4 samples.

Antitrichia curtipendula (Hedw.) Brid.
On very old oak and beech trunks. In 1 sample.

Barbula convoluta Hedw.
This moss species occurs particularly in disturbed habitats; also in frequently trodden places. In 1 sample.

Brachythecium cf. *populeum* (Hedw.) Schimp.
Particularly on ash trunks, occasionally on those of hornbeam (not relevant for Swifterbant) and elm. In 1 sample.

Brachythecium cf. *rutabulum* (Hedw.) Schimp.
This species is found in a great variety of habitats, from dry to wet and from sunny to shaded. On mineral soils as well as on rotten tree stumps and on the roots and at the foot of deciduous trees. In 3 samples.

Brachythecium cf. *velutinum* (Hedw.) Schimp.
Particularly on rotten tree stumps, on the roots and at the foot of deciduous trees. Also on moist, shaded walls and roofs. In 1 sample.

Calliergonella cuspidata (Hedw.) Loeske
In various moist habitats. In 3 samples.

Drepanocladus cf. *aduncus* (Hedw.) Warnst.
In eutrophic to mesotrophic marsh vegetations. In 1 sample.

Eurhynchium striatum (Hedw.) Schimp.
In rather dry to moist deciduous forests. A characteristic species of the Fagetalia sylvaticae. In 1 sample.

Homalothecium sericeum (Hedw.) Schimp.

On tree trunks, particularly of elm, willow and ash. In 5 samples.

Hypnum cupressiforme Hedw.
A very variable species which is found in a great variety of habitats. Different varieties occur on tree trunks. In 4 samples.

Isothecium myosuroides Brid.
On trunks of oak and beech. In 3 samples.

Leucodon sciuroides (Hedw.) Schwaegr.
On old trees, particularly on elm and willow. In 3 samples.

Neckera complanata (Hedw.) Hueb.
On tree trunks, particularly on beech, ash and elm. This species is relatively well represented at Swifterbant: in 7 samples.

Neckera crispa Hedw.
On shaded limestone rocks and on old trees. In 1 sample.

Philonotis cf. *fontana* (Hedw.) Brid.
In and along running water (clear-water brooklets, springs). The presence of this species at Swifterbant (in one sample) is surprising, and in none of the vegetation units postulated for the Swifterbant area can this moss species be placed.

Sphagnum imbricatum Hornsch.
In raised bogs. In 7 samples.

Sphagnum palustre L.
In bogs and in acid alder, birch and willow carr vegetations. In 2 samples.

Sphagnum papillosum Lindb.
In bogs. This species was found in 6 samples.

It is somewhat surprising that the typical raised bog species *Sphagnum imbricatum* and *papillosum* are comparatively well represented at Swifterbant. Remains of other species from raised bogs, such as *Eriophorum, Calluna* and *Erica*, have not been recovered. Moreover, the assumption of a fresh-water tidal flat area makes the presence of raised bogs rather unlikely. One wonders whether the *Sphagnum* remains could have been of secondary origin, viz. from peat deposits which had been cut into by the creeks.

3.2. Alismataceae

Alisma plantago-aquatica
Of this species one damaged fruit and two "inner fruits" (the fruit wall of which had not been preserved) were recovered. Apparently, this species from fresh-water marsh vegetations is never well represented in settlement sites.

3.3. Betulaceae

Alnus glutinosa
Although *Alnus glutinosa* must have been a common tree in the Swifterbant area (see 4.2.), fruits of this species are not particularly numerous. Only five samples yielded one or a few fruits. Two well-developed, undamaged fruits measure: 2.7 x 2.4 and 2.3 x 2.2 mm. In addition, a small number of fruiting-cone remains (portions of the central axis with the basal parts of the scales) was recovered (table 3).

Betula spec.
Betula is represented by only one fruit. Because the wings have not been preserved, this fruit cannot be identified to the species level.

Corylus avellana
Carbonized and non-carbonized nutshell remains of *Corylus avellana* were found regularly. Particularly from the sieve residues fairly great numbers of shell fragments were recovered. One may assume that hazel-nuts were collected intensively by the inhabitants of the site.

3.4. Caryophyllaceae

Two caryophyllaceous species are represented by only one seed, viz. *Lychnis flos-cuculi* (0.8 x 0.7 mm; sample XXIX-18-F) and *Moehrin-*

Table 2. Moss remains from Swifterbant. Epiphytic mosses (growing on tree trunks) are indicated by an asterisk.

Sample number	1	3	5	6	8	10	12	13	14	16	18	19	21	22	23	28	30	32	39	41	45
Sample designation	I-17-i	II-16-H	III-20-F	III-20-G	IV-17-G	IV-24-F	V-13-K	II-18-G	V-18-K	V-23-i	VII-22-K	III-17-F	VIII-23-F	VIII-23-H	VIII-24-H	IX-19-F	IX-29-F	X-22-F	XV-24-G	XVI-25-G	XXIX-18-F
*Anomodon viticulosus	+	+	+	.	.	.	.	.	+	.	.	.	.	.	.	.	.	.	.	.	.
*Antitrichia curtipendula	.	.	.	.	.	.	.	.	+	.	.	.	.	.	.	.	.	.	.	.	.
Barbula convoluta	.	.	.	.	.	.	.	.	.	.	.	.	+	.	.	.	.	.	.	.	.
Brachythecium cf. populeum	.	.	.	.	.	c	.	.	.	.	.	+	.	.	.	.	.	.	.	.	.
Brachythecium cf. rutabulum	.	.	.	.	.	.	.	.	.	+	.	.	+	.	.	+	.	.	.	.	.
*Brachythecium cf. velutinum	.	.	.	.	.	.	.	.	.	.	.	+	.	.	.	.	.	.	.	.	.
Calliergonella cuspidata	.	.	.	.	.	.	.	.	.	.	.	+	+	+	.	.	.	.	.	.	.
Drepanocladus cf. aduncus	.	.	.	.	.	.	.	.	.	.	.	.	.	.	.	.	.	+	.	.	.
Drepanocladus spec.	.	.	.	.	.	.	.	.	.	.	.	.	.	.	+	.	.	.	.	.	.
Eurhynchium striatum	+	.	.	.	.	.	.	.	cf	.	.	.	.	.	.	.	.	.	.	.	.
*Homalothecium sericeum	+	+	+	.	+	.	.	.	.	.	+	.	.	.	.	.	.	.	.	.	.
Hypnum cupressiforme	.	.	.	+	.	+	.	.	+	.	+	.	.	.	.	.	.	.	.	.	.
*Isothecium myosuroides	.	+	.	+	.	.	.	.	+	.	.	.	.	.	.	.	.	.	.	.	.
*Leucodon sciuroides	+	+	.	+	.	.	.	.	.	.	.	.	.	.	.	.	.	.	.	.	.
*Neckera complanata	.	+	.	+	+	.	cf	.	+	+	+	.	.	.	.	.	.	.	.	.	.
*Neckera crispa	.	+	.	.	.	.	.	.	.	.	.	.	.	.	.	.	.	.	.	.	.
Philonotis cf. fontana	.	.	.	.	.	.	.	.	.	.	.	.	.	.	.	.	.	+	.	.	.
Sphagnum imbricatum	.	.	.	.	+	+	.	.	.	.	.	.	.	.	+	.	+	.	+	+	+
Sphagnum palustre	.	+	.	.	.	.	.	.	.	.	.	.	.	.	.	.	.	.	.	.	+
Sphagnum papillosum	.	.	.	.	.	+	.	.	.	.	.	.	.	.	+	.	+	.	+	+	+

Table 3. Total numbers of seeds, fruits and other plant remains recovered on 2 mm sieve (washing of soil for retrieving archaeological objects).

Hordeum vulgare var. nudum	1788
Triticum dicoccum	71
Triticum cf. aestivum	1
Corylus avellana	c. 30 + many fragments
Crataegus monogyna	24,5
Pyrus malus, pips	3
Pyrus malus, carbonized fruits	3
Cornus sanguinea	1
Phragmites australis, stem fragments	22
Alnus glutinosa, remains of fruiting-cones	16
Menyanthes trifoliata	3
Ceratophyllum submersum	1
Nymphaea alba	4
Galeopsis tetrahit/speciosa	1
Polygonum aviculare	1
Polygonum persicaria	1
Vicia spec.	1
Claviceps purpurea	1
Cenococcum geophilum	1

gia trinervia (1.1 x 0.8 mm; sample VIII-17-F). *Lychnis flos-cuculi* is found in wet grasslands. *Moehringia trinervia* grows in deciduous forests and in willow carr vegetations.

Stellaria media

The seeds of a third species of the Caryophyllaceae, viz. of *Stellaria media*, are quite frequent. This seed type was found in most of the samples, not infrequently in large numbers (up to 82.4%), suggesting that *Stellaria media*, a plant from fields, waste places and roadsides, was very common in the vicinity of the site.

The greatest diameter of *Stellaria* seeds in four samples from different layers does not show significant differences (table 4). On the other hand, the *Stellaria* seeds from Tritsum and Feddersen Wierde are larger, and those from Elisenhof distinctly smaller than the Swifterbant seeds. No obvious explanation can be presented for the size differences between the sites. Differences between Swifterbant and the other sites listed in table 4 could possibly be attributed to differences in environment: predominantly fresh water at Swifterbant and brackish conditions at the other sites. However, this leaves one with the question why the Elisenhof *Stellaria* seeds are nearly half the size of those from Tritsum.

3.5. Ceratophyllaceae

Ceratophyllum submersum

Fruits bi-convex, oval in outline (fig. 4:9). One carbonized fruit (in a sieve sample) and one non-carbonized specimen were recovered. The middle of the rounded apex is slightly pointed, indicating the place of attachment of the short apical spine. Dimensions: 4.4 x 2.5 mm (non-carbonized) and 3.6 x 2.6 mm (carbonized fruit).

Ceratophyllum demersum and *submersum* occur both in shallow basins with stagnant fresh or slightly saline water. At present *C. submersum* is a rare species in the Netherlands and largely confined to the coastal area, whereas *C. demersum* is rather common. It is possible that the late-Atlantic climate, with higher summer

temperatures and milder winters, favoured the growth of *C. submersum*.

3.6. Chenopodiaceae

Atriplex

No attempt has been made to distinguish between the seeds of *Atriplex hastata* and *A. patula*, and it is doubtful if this is at all possible. The seeds of *A. hastata/patula* were found in most of the samples, sometimes in large numbers. The greatest diameter of *Atriplex* seeds in 3 samples from different layers are shown in table 5. The Swifterbant seeds are, on an average, slightly larger than those from Tzummarum and Tritsum, with mean greatest diameter of 1.36 and 1.48 mm, respectively (Van Zeist, 1974, p. 268).

Körber-Grohne (1967, p. 268) claims that the seeds of *A. hastata* (1.0-1.9(1.3) mm, N = 50) can be separated from those of *A. patula* (1.3-1.9(1.5) mm, N = 50). For Elisenhof, Behre (1976, p. 95-96) distinguishes equally between *A. patula*-type (diameter: 1.2-2.1 (1.75) mm for 50 seeds) and *A. hastata/littoralis*-type (diameter: 1.1-2.3(1.46) mm for 50 seeds). However, we do not feel able to make a satisfactory distinction between both seed types. *A. hastata* and *A. patula* have about the same ecological requirements. They are both species from fields, waste places and other disturbed habitats.

One large *Atriplex* seed (diameter 2.8 mm) was met with (V-18-K). This seed is not necessarily of the halophytic taxa *A. littoralis* or *A. hastata* var. *salina*, but it can also have originated from an inland variety of *A. hastata* (Körber-Grohne, 1967, p. 268).

Chenopodium

Chenopodium album is represented in most of the samples examined, but this seed type is less numerous than the *Atriplex hastata/patula*-type. The greatest diameter of the seeds in four samples shows no significant differences (table 5). The *Chenopodium album* seeds from Swifterbant are, on an average, slightly smaller than those from Tritsum (1.55 mm) and Paddepoel

Table 4. Greatest diameter of *Stellaria media* from Swifterbant and other sites.

	min.	aver.	max.
Swifterbant IV-13-F (N = 50)	0.9	1.05	1.2
II-18-G (N = 50)	0.9	1.06	1.2
XVIII-19-i (N = 50)	0.9	1.07	1.2
XV-19-K (N = 50)	1.0	1.07	1.2
Tritsum (N = 100)	1.0	1.35	1.6
Elisenhof (N = 50)	0.6	0.76	1.0
Feddersen Wierde (N = 28)	1.0	1.2	1.4

Table 5. Greatest diameter of Chenopodiaceae from Swifterbant.

	min.	aver.	max.
Chenopodium album			
XI-16-F (N = 23)	1.3	1.43	1.6
X-15-G (N = 41)	1.3	1.43	1.6
XIV-17-i (N = 29)	1.3	1.45	1.8
XX-19-i (N = 50)	1.3	1.49	1.8
Atriplex patula/hastata			
III-20-F (N = 50)	1.2	1.59	2.1
XVIII-19-i (N = 50)	1.4	1.66	1.9
XV-19-K (N = 11)	1.4	1.69	2.0

(1.53 mm), but somewhat larger than those from Elisenhof (1.31 mm) (Van Zeist, 1974, p. 269; Behre, 1976, p. 96).

Chenopodium rubrum/glaucum seeds were found in only four samples. The seeds of these *Chenopodium* species which are both found in nitrate-rich habitats cannot be separated satisfactorily. Dimensions for 7 seeds: 0.9-1.1 mm (aver. 0.97 mm).

Attention should be drawn here to the conspicuous absence of *Chenopodium ficifolium* seeds at Swifterbant. This *Chenopodium* species is represented by great numbers of seeds in various coastal settlement sites (Behre, 1976, p. 97; Körber-Grohne, 1967, p. 271; Van Zeist, 1974, p. 269). The absence of *Chenopodium ficifolium* cannot be ascribed to unfavourable climatic or ecological conditions. The late-Atlantic climate must have been more favourable for this thermophilous species than that of early-Subatlantic times, and nitrate-rich habitats must have been common in and near the settlement. One may wonder whether in the second half of the fourth millennium B.C. *Chenopodium ficifolium* was already present in western Europe. One should consider the possibility that this species was introduced by prehistoric farmers, be it unintentionally. *Chenopodium ficifolium* is not reported for Bandkeramik settlement sites in the Rhineland area and South Limburg (Bakels, 1979; Knörzer, 1972, 1974, 1977, 1980). It is equally not represented in the Rössen site near Langweiler (Knörzer, 1971).

Salicornia europaea

One U-shaped seed of *Salicornia europaea* was counted. One should consider the possibility that this species, which is characteristic of salt-marsh vegetations, occurred in the Swifterbant area, be it rarely.

3.7. Compositae

Arctium cf. lappa

On the basis of the shape and the size, the *Arctium* achenes from Swifterbant have, with some reserve, been attributed to *A. lappa*.

Nine achenes from various samples measure 6.3(5.8-6.8) x 3.0(2.6-3.3) mm. *Arctium lappa* as well as other *Arctium* species are found in ruderal habitats.

Artemisia vulgaris

From two samples a few achenes of *Artemisia vulgaris* were recovered. The achenes are oblong in outline, tapering towards both ends (fig. 5:7). The surface is longitudinally striate. The longitudinal ribs and the annular ring at the apex are not preserved in the subfossil specimens. Dimensions: 1.7 x 0.7 and 1.6 x 0.6 mm. *Artemisia vulgaris* is a species from roadsides, fields and other disturbed habitats.

Aster tripolium

A few fruits of *Aster tripolium*, a species which is characteristic of halophytic vegetations, were found. As usual in subfossil, non-carbonized achenes of *Aster tripolium*, the outer wall and the annular ring at the apex have disappeared. Dimensions of three inner fruits: 3.0 x 0.9, 2.5 x 0.9 and 3.0 x 0.8 mm. The size of the *Aster tripolium* achenes from Swifterbant corresponds with those obtained for specimens from pre-Roman Iron Age Tritsum and Roman Iron Age Sneek, situated in a salt-marsh environment: 2.84 x 0.89 and 2.94 x 0.81 mm, respectively (Van Zeist, 1974, pp. 270-71).

Carduus crispus

Carduus crispus, the achenes of which distinguish themselves from those of *Cirsium* species by a fine transverse wrinkling of the fruit wall, is represented in a small number of samples. Dimensions for 3 achenes: 2.8 x 1.6, 3.0 x 1.1 and 3.0 x 1.2 mm. *Carduus crispus* is a species from disturbed habitats rich in nitrates.

Cirsium arvense

A few *Cirsium* achenes recovered from the Swifterbant samples are attributed to *C. arvense*. They differ from those of *C. vulgare* by the size and the more slender shape and from those of *C. palustre* by the absence of narrow longitudinal ribs. Dimensions for 4

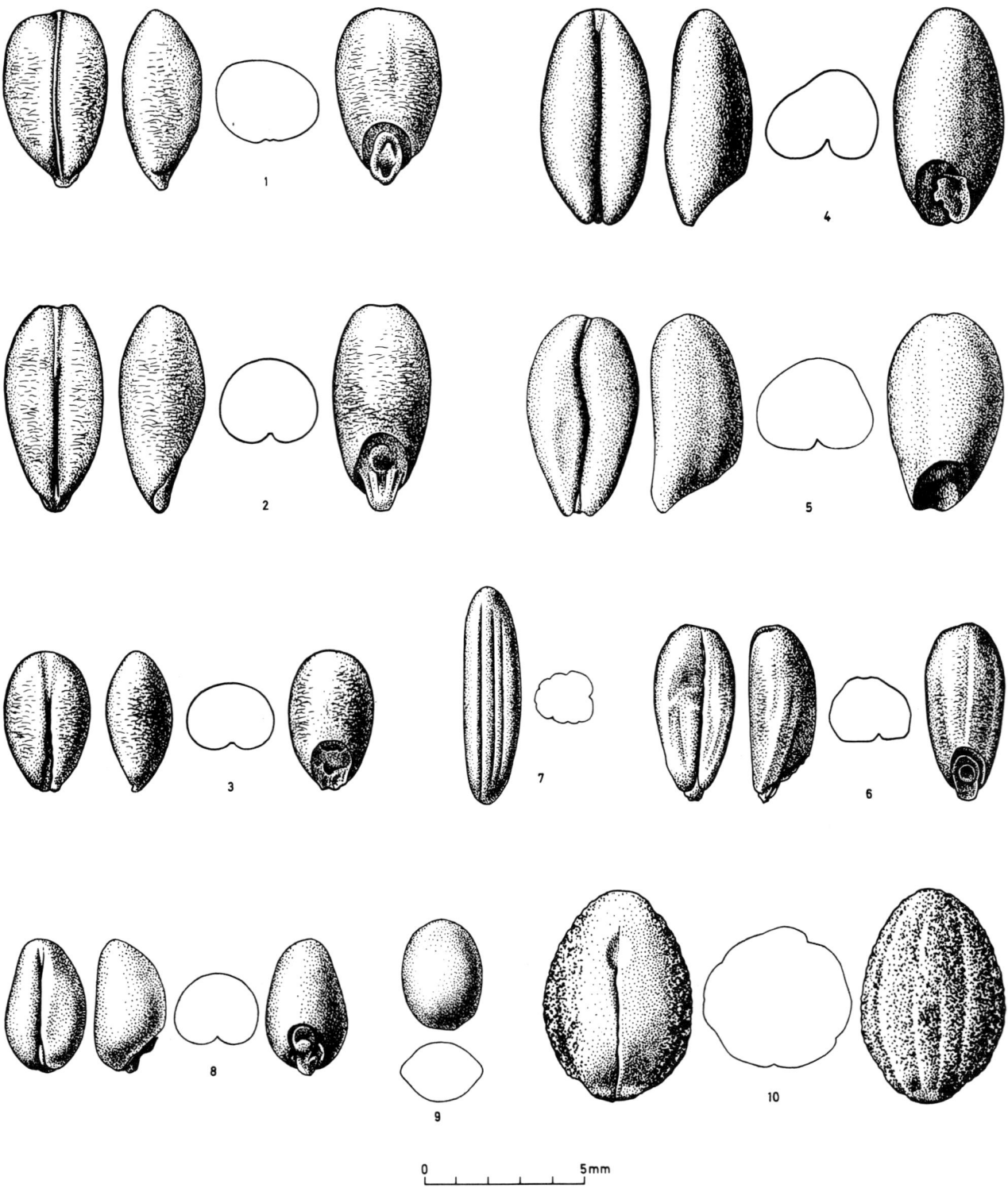

Fig. 4. 1-3: *Hordeum vulgare* var. *nudum* (VIII-21-i); 4: *Triticum dicoccum* (XII-19-F); 5: *Triticum dicoccum* (XII-22-F); 6: *Triticum dicoccum* (XVI-20-F); 7: *Claviceps purpurea* (XIII-19-i); 8: *Triticum* cf. *aestivum* (XII-15-i); 9: *Ceratophyllum submersum* (XV-20-F); 10: *Crataegus monogyna* (XVI-22-F).

specimens: 3.0 x 1.3, 3.4 x 1.4, 2.7 x 1.1 and 2.8 x 1.0 mm. *Cirsium arvense* is a species from fields and other disturbed habitats.

Lapsana communis

Lapsana communis, a species from fields and forest edges, is represented in two samples. The achenes are lanceolate, slightly curved, with a large number of longitudinal ribs. Dimensions: 4.6 x 1.3 and 4.0 x 1.0 mm.

Sonchus

Samples II-16-H, V-18-G and VII-22-K yielded each one fruit of *Sonchus asper*, a species from fields and waste places. The surface of the oblong, flat achenes is smooth. Dimensions: 2.8 x 1.0 and 2.7 x 1.1 mm.

One *Sonchus* fruit has been identified as *S. palustris* and one damaged specimen as probably *S. palustris*. The achenes of this species have 4 pronounced ribs; the fruit surface is transversely wrinkled. *S. palustris* is found in marsh vegetations.

3.8. Cornaceae

Cornus sanguinea

One non-carbonized fruit stone of *Cornus sanguinea* was recovered from a sieve residue. The almost globular fruit stone (greatest diameter c. 4 mm) shows longitudinal grooves which extend from the base to the apex of the stone. *C. sanguinea* is found in forest and shrub vegetations (Prunetalia spinosae, Fagetalia sylvaticae).

3.9. Cruciferae

Capsella bursa-pastoris

Cruciferae are represented at Swifterbant by altogether two seeds of *Capsella bursa-pastoris* (1.0 x 0.6 mm), a species from disturbed habitats, such as fields, waste places and roadsides.

3.10. Cyperaceae

Various cyperaceous species are represented at Swifterbant. The style base is not included in the measurements.

Carex

Carex fruits are conspicuously scarce in the samples examined, suggesting that sedges did not play a prominent part in the plant cover of the Swifterbant area. The following species or types have been distinguished:
Carex disticha in 1 sample (1.7 x 1.0 mm);
Carex nigra-type in 2 samples;
Carex paniculata in 2 samples (1.7 x 1.2 mm);
Carex pseudocyperus in 2 samples (1.8 x 1.0, 1.5 x 1.0 mm);
Carex riparia in 2 samples (3.5 x 2.3 mm);
Carex rostrata/vesicaria in 5 samples (2.6 x 1.2, 2.5 x 1.4, 2.2 x 1.3, 2.1. x 1. 5 mm);
Carex serotina-type in 1 sample (1.1. x 0.8 mm).
One *Carex* fruit remained unidentified.
For descriptions of the *Carex* fruits listed above the reader is referred to Körber-Grohne (1967, pp. 248-255) and Van Zeist (1974, pp. 276-280). *Carex nigra*-type corresponds to the *Carex acuta*-type in Van Zeist (1974, pp. 276-277). *Carex riparia* has not been reported for the coastal sites examined previously (Van Zeist, 1974). This fruit is noticeably larger than most other *Carex* fruits.

Cladium mariscus

This species is represented in a rather great number of samples, but generally only by one or a few fruits. The dimensions of *Cladium* fruits are shown in table 6. They do not differ noticeably from those obtained for *Cladium* fruits from Sneek (2.05(1.7-2.3) x 1.48(1.2-1.7) mm; Van Zeist, 1974, p. 280). *Cladium mariscus* is a predominantly fresh-water species, but it is also found in a slightly brackish environment.

Eleocharis palustris

This species, represented in only three samples, must have played a very minor part in the marsh vegetations of the Swifterbant area. This in contrast to the fresh-water coastal sites of Vlaardingen and Schiedam, dated to the pre-Roman Iron Age, which yielded rather great numbers of *Eleocharis* fruits (Van Zeist, 1974, p. 280-81).

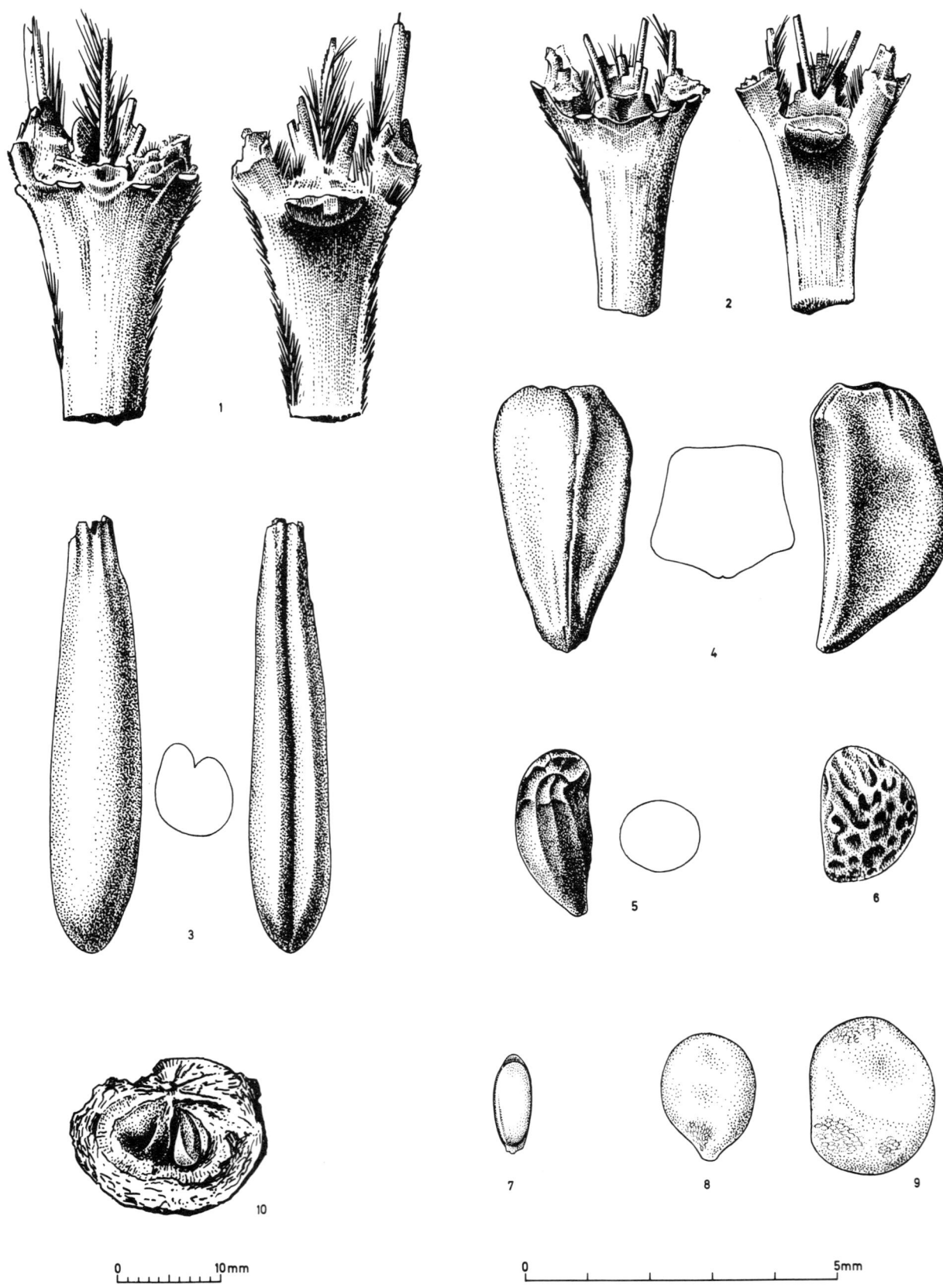

0 —————— 10mm

0 ————— 5mm

Scirpus

Of the *Scirpus* species established for Swifterbant, the fruits of *S. tabernaemontani (S. lacustris* ssp. *glaucus)* are most numerous. The bi-convex fruits of *S. tabernaemontani* are markedly smaller than those of *S. maritimus* (table 6). One may assume that this *Scirpus* species was a common constituent of the marsh vegetations in the Swifterbant area.

Scirpus maritimus, on the other hand, probably played a much more modest part in the local vegetation. This species is represented in a fairly great number of samples, but always by small numbers of fruits.

A third *Scirpus* species which could be established for Swifterbant is *S. lacustris* ssp. *lacustris*. Although in subfossil *Scirpus* fruits it may be difficult to determine whether they originate from *S. maritimus* or from *S. lacustris* ssp. *lacustris* (cf. Körber-Grohne, 1967, pp. 256-257), a few unmistakable fruits of the latter species were found. In *S. maritimus* the dorsal side is either domed or roof-shaped with a rounded median edge. In *S. lacustris* ssp. *lacustris* the median edge of the roof-shaped dorsal side is much sharper. Judging from the numbers of fruits, *S. lacustris* ssp. *lacustris* was not common in the vicinity of Swifterbant. For three fruits the dimensions have been obtained: 3.0 x 2.2, 3.1 x 2.0 and 2.9 x 1.8 mm.

Scirpus maritimus is common in brackish water. *S. tabernaemontani* occurs in fresh as well as in brackish water. *S. lacustris* ssp. *lacustris* is a predominantly fresh-water species, but it can tolerate slightly saline conditions.

3.11. Gentianaceae

Menyanthes trifoliata

The sediment samples yielded one seed of *Menyanthes trifoliata*, while a few more speci-

mens were recovered from the sieve residues. Dimensions of four seeds: 3.0 x 2.5, 2.5 x 1.9, 2.7 x 2.3, 2.9 x 2.4 mm. *Menyanthes trifoliata* is a constituent of marsh vegetations (Magnocaricion, Caricion curto-nigrae).

3.12. Gramineae

The crop-plant species *Hordeum vulgare* var. *nudum* and *Triticum dicoccum* will be discussed in section 5.

Among the caryopses of wild grasses, those of *Phragmites australis (P. communis)* are most numerous. Reed is represented in the majority of the samples, a few times by a fairly great number of fruits. Carbonized stem fragments of reed were recovered from the sieve residues and reed stems were frequently observed in the field. The dimensions of *Phragmites* caryopses (greatest length) in three samples from different layers are shown in table 7. One may assume that reed played a prominent part in the marsh vegetation of the Swifterbant area.

Poa pratensis/trivialis caryopses were found fairly regularly, but, except for one sample (XXIV-19-i), only in small numbers. The lengths of the Swifterbant caryopses are of the same order of magnitude as those obtained for *Poa pratensis/trivialis* in other sites (table 7).

In two samples a few *Bromus mollis/secalinus*-type caryopses were found (5.3 x 1.7, 4.6 x 1.7, 5.2 x 1.9, 4.5 x 1.5 mm). *Bromus secalinus*, a weed in grain fields, as well as *B. mollis*, which occurs in grassland, fields and ruderal habitats, come into consideration for Swifterbant.

Of *Agrostis* only two caryopses were recovered, suggesting that this grass was by no means common in the Swifterbant area. The same applies to *Alopecurus* represented by one damaged caryopsis (*A.* cf. *geniculatus*). One carbonized, slightly damaged fruit has been attributed to *Glyceria fluitans* (c. 3.4 x 1.3 mm). It is possible that *Glyceria* grains were collected intentionally by the Swifterbant people.

A few caryopses could not be identified (unident. Gramineae).

Fig. 5. 1-2: *Hordeum vulgare* var. *nudum,* rachis internodes (VIII-21-i); 3: *Anthriscus sylvestris* (XXIV-19-i); 4: *Rosa canina/rubiginosa* (XXIX-18-F); 5: *Caltha palustris* (V-18-K); 6: *Rubus fruticosus* (IX-19-F); 7: *Artemisia vulgaris* (XXIV-19-i); 8: *Solanum nigrum* (XVIII-19-i); 9: *Solanum dulcamara* (XXIV-19-i); 10: *Pyrus malus* (no. 23416).

Table. 6. Dimensions of Cyperaceae from Swifterbant.

		L	B	100 L:B
Scirpus tabernaemontani				
II-18-G	min.	1.9	1.3	114
N = 15	aver.	2.13	1.50	143
	max.	2.6	1.8	169
XIV-23-G	min.	1.7	1.1	133
N = 33	aver.	2.05	1.35	152
	max.	2.4	1.6	176
3 samples	min.	1.8	1.2	130
F and i layers	aver.	2.18	1.45	151
N = 21	max.	2.4	1.6	181
Scirpus maritimus				
various samples	min.	2.5	1.7	
N = 7	aver.	2.78	1.98	
	max.	3.2	2.3	
Cladium mariscus				
various samples	min.	1.4	1.0	
N = 16	aver.	1.88	1.33	
	max.	2.2	1.4	

Table 7. Greatest length of caryopses of *Phragmites* and *Poa* from Swifterbant and other sites.

	min.	*aver.*	*max.*
Phragmites australis			
Swifterbant IV-17-G (N = 50)	1.1	1.42	1.8
II-16-H (N = 50)	1.0	1.30	1.8
VIII-24-H (N = 50)	1.0	1.24	1.7
Ouddorp (N = 26)	1.1	1.39	1.7
Sneek (N = 15)	1.2	1.39	1.7
Elisenhof (N = 50)	0.65	1.18	1.6
Poa pratensis/trivialis			
Swifterbant XXIV-19-i (N = 50)	1.3	1.57	1.8
Paddepoel (N = 100)	1.1	1.76	2.3
Tritsum (N = 126)	1.2	1.47	2.3
Elisenhof (N = 50)	0.9	1.29	1.8

3.13. Juncaceae

Juncus

The only *Juncus* seed recovered from the Swifterbant samples is of *Juncus gerardii*, which is characteristic of salt-marsh vegetations. Whether *Juncus gerardii* actually occurred in the Swifterbant area at that time, be it very seldom, or whether the seed concerned had been carried in from the coastal salt marshes, *e.g.* by birds or by water during extremely high floods, must remain undecided.

The near-absence of *Juncus* at Swifterbant is very striking and not easy to explain. One may expect that habitats suitable for *Juncus bufonius, J. effusus* and *J. articulatus* were found in the vicinity of the site.

3.14. Labiatae

Galeopsis tetrahit/speciosa

This fruit type was met with in only two samples. One specimen has been measured: 3.6 x 3.0 mm. *Galeopsis tetrahit* and *G. speciosa* are both species from fields and ruderal sites.

Lycopus europaeus

A small number of mostly damaged fruits of *Lycopus europaeus* was recovered. Two fruits were suitable for measuring: 1.4 x 1.0 mm and 1.4 x 1.0 mm. *Lycopus* occurs in marsh vegetations.

Mentha cf. aquatica

Of the Labiatae, *Mentha* cf. *aquatica* is best represented at Swifterbant, viz. in 11 of the 46 samples, be it only occasionally by more than one fruit per sample. The subfossil fruits of *M. aquatica* and *M. arvensis* cannot be distinguished one from the other. It is most likely that *M. aquatica*, a species from marsh vegetations and wet grasslands, is concerned here. For 9 *Mentha* fruits the dimensions have been obtained: length 1.04 (0.9-1.1) mm, breadth 0.74 (0.7-0.8) mm.

3.15. Nymphaeaceae

Nymphaea alba

The seeds of *Nymphaea alba* are ovate in outline and more or less circular in cross-section. The base is rounded and the apex is pointed. A single, weakly developed longitudinal ridge is present. The upper layer of the seed wall consists of longitudinal rows of rectangular cells with thick, wavy cell walls. In subfossil seeds the surface pattern is often more distinct than in modern specimens.

Nymphaea alba is remarkably well represented. Most of the water-lily seeds were more or less seriously damaged. Only a few of them were complete and suitable for measuring: length 2.78 (2.3-3.2) mm; breadth 1.81 (1.6-2.1) mm for 8 specimens.

3.16. Papilionaceae

Trifolium repens

One *Trifolium* petal, recovered from sample III-14-F, has been attributed to *T. repens*. It is true that the subfossil petal is rather small, but the shape as well as the nervation match perfectly small specimens of *T. repens* petals from Paddepoel (Van Zeist, 1974, p. 295). *T. repens* is a plant from grasslands which is associated with grazing in prehistoric times.

3.17. Plantaginaceae

Plantago major

Compared to the Iron Age coastal settlement sites, *Plantago major* is rather poorly represented at Swifterbant. One or a few seeds of this plantain species were found in a comparatively small number of samples (8 of the 46 samples). In view of the good seed production of *Plantago major* one must assume that this species from trodden places and other disturbed habitats was not particularly abundant in the vicinity of the site. Five seeds measure: 1.32 (1.2-1.4) x 0.74 (0.7-0.8) mm.

Table 8. *Polygonum* species from Swifterbant.

		L	B	100L:B
Polygonum aviculare				
II-16-H	min.	2.4	1.2	128
N = 50	aver.	2.78	1.85	155
	max.	3.1	2.2	213
I-17-i	min.	2.4	1.5	130
N = 49	aver.	2.98	1.88	155
	max.	3.4	2.2	173
V-23-i	min.	2.4	1.5	134
N = 44	aver.	2.96	1.96	152
	max.	3.4	2.3	190
XI-23-K	min.	2.4	1.4	127
N = 46	aver.	2.80	1.86	151
	max.	3.1	2.2	177
Polygonum lapathifolium				
IV-20-H	min.	1.8	1.4	123
N = 50	aver.	2.33	1.64	142
	max.	2.7	2.1	160
I-18-i	min.	1.8	1.2	119
N = 50	aver.	2.28	1.58	145
	max.	2.8	1.8	171
VII-26-K	min.	1.8	1.2	119
N = 50	aver.	2.07	1.49	140
	max.	2.5	1.8	165
XI-23-K	min.	1.7	1.2	109
N = 50	aver.	2.21	1.60	139
	max.	2.6	2.1	160
Polygonum persicaria				
IV-20-H	min.	2.6	1.7	142
N = 23	aver.	2.89	1.88	155
	max.	3.2	2.0	174
I-17-i	min.	2.1	1.4	136
N = 38	aver.	2.80	1.84	153
	max.	3.4	2.2	178
I-18-i	min.	1.9	1.3	104
N = 49	aver.	2.65	1.76	148
	max.	3.2	2.2	181

3.18. Polygonaceae

Polygonum

Three *Polygonum* species, viz. *P. aviculare, lapathifolium* and *persicaria*, have been established for Swifterbant. Dimensions of the fruits of the *Polygonum* species are shown in table 8.

Polygonum aviculare is represented in more than half of the samples examined, sometimes by rather great numbers of fruits, suggesting that this species was common in the vicinity of the site. There are no significant differences in the dimensions and index values between samples from various levels. *P. aviculare* fruits from Tritsum are somewhat more slender (average L:B index of 182); those of Elisenhof are slightly smaller, on an average, than the Swifterbant fruits (table 9). *P. aviculare* is not only characteristic of the vegetation of frequently trodden places, but it occurs also in various other disturbed habitats.

Polygonum lapathifolium, a species from fields, waste places and other nitrate-rich habitats, must likewise have occurred frequently in the Swifterbant area. In sample III-20-F the distinction between *P. lapathifolium* and *P. persicaria* was not always satisfactory because the fruits were unripe. Some variation occurs in the average size of the *P. lapathifolium* fruits in the four samples for which measurements have been carried out. The fruits from Paddepoel and Vlaardingen (table 9) are of about the same length as those from Swifterbant, but they are broader (lower L:B index values). *P. lapathifolium* fruits from Elisenhof and Feddersen Wierde are larger as well as plumper than those from Swifterbant. In view of the differences in the local environments and of the presence of various subspecies of *P. lapathifolium* the variation in size and shape of the fruits is not surprising.

Polygonum persicaria is likewise well represented at Swifterbant, although somewhat less frequently and by smaller numbers of fruits than *P. lapathifolium*. The dimensions of the *Polygonum persicaria* fruits seem to be fairly large, on an average. They are larger than the *P. lapathifolium* fruits from Swifterbant

and also larger than the *P. persicaria* fruits from Vlaardingen and Paddepoel (table 9). The rather large size of the fruits makes one wonder whether *P. hydropiper* and not *P. persicaria* is concerned here. *P. hydropiper* fruits are usually larger than those of *P. persicaria*. However, the shape of the fruits as well as the glossy wall surface (in *P. hydropiper* dull and striate) point to *P. persicaria*. Moreover, the calyx remains which still adhere to some of the fruits do not show the glandular dots characteristic of *P. hydropiper*, whereas the venation corresponds wholly with that of the calyx of modern *P. persicaria* fruits. Consequently, there can be no doubt that the fruits are of *P. persicaria*. In this respect mention should be made of the *P. persicaria* fruits from Elisenhof which are of about the same average length as those from Swifterbant, but markedly broader. The *P. persicaria* fruits from Paddepoel and Vlaardingen are likewise comparatively broader (lower L:B index values) than those from Swifterbant. As in *P. lapathifolium*, also in *P. persicaria* the subfossil fruits display a rather great variation in size and shape.

The absence of *P. hydropiper* in the seed record of Swifterbant is striking. One must assume that wet, nitrate-rich habitats, in which *P. hydropiper* is found, were quite common near the Swifterbant S3 site. *Polygonum convolvulus*, a weed in fields the fruits of which often occur in charred grain samples, is equally not represented at Swifterbant.

Rumex

Rumex is scarcely represented at Swifterbant. One fruit from III-20-G enclosed by the valves (the inner three sepals) has been attributed to *Rumex conglomeratus*. Each of the three oblong, tongue-shaped valves bears a well-developed grain-like tubercle. The margin of the valves is entire. *R. conglomeratus* is a species from wet habitats, such as the sides of ditches.

Of the naked *Rumex* fruits, two specimens could be identified to the species level on the basis of the shape and the size: *R. maritimus* (1.6 x 0.8 mm) in IX-15-F and *R. hydrola-*

pathum in II-16-H. The species of the other naked *Rumex* fruits, measuring c. 1.9 x 1.3 mm, could not be determined.

3.19. Potamogetonaceae

Potamogeton

Only two *Potamogeton* fruits were found. One fruit (in VIII-24-H) has been identified as *Potamogeton crispus*. The greater part of the elongate acuminate beak, which is characteristic of *P. crispus*, is still preserved. Dimensions of the subfossil fruit (beak not included): 2.5 x 1.9 mm. *P. crispus* occurs in fresh as well as in brackish water.

The other fruit (XV-24-G) is too poorly preserved to allow a species identification.

Zannichellia palustris

One damaged fruit of *Zannichellia palustris* has been recovered (2.5 x 0.8 mm). Of the bristles on the dorsal side only the bases are still preserved. *Zannichellia palustris* is a species from brackish water (Ruppion maritimae), but occasionally it is also found in fresh water.

3.20. Ranunculaceae

Caltha palustris

One fruit of *Caltha palustris*, a species from fresh-water marsh vegetations, was found. The fruit is obliquely ovate in outline, pointed at the base and rounded at the apex. The upper part of the fruit is wrinkled (fig. 5:5). Dimensions: 2.8 x 1.3 mm.

Ranunculus

Two *Ranunculus* species have been established for Swifterbant, viz. *Ranunculus sceleratus* and *R. acris*. The fruits of *R. sceleratus* are characterized by the thickened margin of spongy tissue. They are rather small: length 1.06 (0.9-1.2) mm, breadth 0.86(0.8-1.0) mm for 7 specimens from various samples. The identification of the *R. acris* fruits is based upon the shape of the fruits (they have a more sharply pointed lower end than in *R. repens*) and by the absence of a reticulate surface pattern. Four fruits have been measured: 2.4 x 1.8, 2.3

x 1.6, 2.0 x 1.4, 1.7 x 1.2 mm. *Ranunculus acris* is a species from grasslands as well as from roadsides.

3.21. Rosaceae

Crataegus monogyna

Mainly carbonized fruit stones of *Crataegus monogyna* were found in the sediment samples as well as in the sieve residues. The rather great number of hawthorn fruit stones must partly be ascribed to the fact that because of their size they were easily recognized with the naked eye in the sieve residues. The thick-walled fruit stones are oval in outline and more or less circular in cross-section. Characteristic are the longitudinal grooves (fig. 4:10). In the carbonized fruits a hole is sometimes present in one of the grooves. The surface of the fruit wall is rough. Dimensions for 13 carbonized fruit stones from various samples: length 6.16 (5.1.-7.8) mm, breadth 4.86(4.3-5.3) mm.

Crataegus monogyna is a constituent of shrub and forest vegetations (Prunetalia spinosae, Fagetalia sylvaticae).

Pyrus malus

The sediment samples as well as the sieve residues yielded pips of *Pyrus malus*, mostly non-carbonized. Only few pips were suitable for measuring: 4.9 x 2.7, 5.0 x 2.6, 6.2 x 3.4 and 7.4 x 4.5 mm. In addition, three carbonized crab apples were recovered (sample 23416, fig. 5:10). Dimensions: 2.3 x 2.1, 1.85 x c. 1.6 and 2.55 x 2.4 cm. *Pyrus malus* must have formed part of the deciduous forest which constituted the natural vegetation of the levees bordering the major creeks or/and of the river-bank dunes. For a discussion of the possible economic importance of wild apple see 5.4.

Rosa

A small number of *Rosa* fruit stones were recovered. The angular fruit stones are irregularly shaped: ovate to obliquely ovate in outline, triangular to quadrangular in cross-section (fig. 5:4). For 5 fruit stones the dimensions have been obtained: 4.4 x 2.2 x 2.2, 4.6 x 3.1 x 2.5, 5.1 x 2.1 x 2.6, 4.6 x 2.4 x 3.0, 5.6 x 2.4 x 2.7 mm. The subfossil fruit stones correspond to those of *R. canina* and *R. rubiginosa*. Both rose species are found in forest and shrub vegetations.

One fruit stone fragment is reminiscent of *Rosa pimpinellifolia*. However, a satisfactory species identification is not possible, so this specimen is listed as *Rosa* spec.

Rubus fruticosus

Altogether 4 fruit stones of *Rubus fruticosus* were found. Blackberry fruit stones can be distinguished from those of *R. idaeus* by the shape. *R. fruticosus* fruit stones are rounded triangular to semi-circular in outline, the ventral side is straight to convex (fig. 5:6). *R. idaeus* fruit stones are more slender and the ventral side is straight to concave; they are more crescent-shaped in outline. For three *Rubus* fruit stones from Swifterbant the dimensions were obtained: 2.2 x 1.7, 2.5 x 1.8 and 2.2 x 1.7 mm.

3.22. Rubiaceae

Galium

The fairly large fruits of *Galium aparine* were recovered from sediment samples as well as from sieve residues. Only carbonized fruits of *Galium aparine* have been met with, which makes one wonder whether they became carbonized together with the cereal grains, implying that cleavers occurred as a weed in the grain fields of the Swifterbant farmers. Dimensions for 13 fruits from various samples: 3.19 (2.6-3.7) x 2.84(2.5-3.2) mm.

One small, nearly globular *Galium* fruit (1.4 x 1.2 mm) has been attributed to *G. palustre*, a species from wet habitats. The ventral side is largely taken up by a hole with a sharp edge: the indented hilum. The fruit surface is finely wrinkled.

3.23. Solanaceae

Solanum

Both *Solanum nigrum* and *S. dulcamara* have been established for Swifterbant. The seeds of *S. nigrum* differ from those of *S. dulcamara* by the shape. In *S. dulcamara* seeds the lower end with the hilum is rounded, whereas in *S. nigrum* seeds the lower end is obliquely pointed (fig. 5:8 and 9). Moreover, as is clear from table 10, *S. dulcamara* seeds are, on an average, larger than those of *S. nigrum*. The differences in shape find also expression in the L:B index values.

Solanum nigrum, a species from disturbed habitats, such as waste places and fields, is represented by much greater numbers of seeds than *S. dulcamara*, which is particularly found in alder-brook forest and shrub vegetations of wet places.

3.24. Typhaceae

Typha spec.

Small, cylindrical fruits, tapering towards the base. In the subfossil fruits the apex is truncated. Two *Typha* fruits were found: 1.0 x 0.3 and 1.0 x 0.3 mm. It is doubtful whether the subfossil fruits of *Typha latifolia* and *T. angustifolia* can be distinguished one from the other. Both *Typha* species occur in fresh-water marsh vegetations (Phragmitetalia).

3.25. Umbelliferae

Anthriscus sylvestris

Linear-cylindrical fruits. The lower end is rounded; the apex is truncated (the thickened style base (stylopodium) is not preserved in the subfossil fruits from Swifterbant). The ventral side shows a deep groove, the dorsal side is semi-circular in cross-section. There are no longitudinal ribs or oil-tubes (fig. 5:3). The fruit wall is smooth, with an epidermis pattern of isodiametric cells.

Anthriscus sylvestris fruit remains were found in three samples. Two specimens have been measured: 7.1 x 1.4 and c. 6.0 x 1.6 mm. *Anthriscus sylvestris* is a species from moist forests and grasslands.

Conium maculatum

This species is represented in a rather great

Table 9. Average dimensions of *Polygonum* from Swifterbant and other sites.

	L	B	100L:B
Polygonum persicaria			
Paddepoel (N = 50)	2.49	1.97	127
Vlaardingen (N = 60)	2.52	2.03	124
Elisenhof (N = 32)	2.75	2.20	
Swifterbant IV-20-H (N = 23)	2.89	1.88	155
I-17-i (N = 38)	2.80	1.84	155
I-18-i (N = 49)	2.65	1.76	148
Polygonum lapathifolium			
Paddepoel (N = 45)	2.27	1.96	116
Vlaardingen (N = 100)	2.40	1.99	121
Elisenhof (N = 50)	2.75	2.21	
Feddersen Wierde (N = 20)	2.6	2.2	
Swifterbant IV-20-H (N = 50)	2.33	1.64	142
I-18-i (N = 50)	2.28	1.58	145
VII-26-K (N = 50)	2.07	1.49	140
XI-23-K (N = 50)	2.21	1.60	139
Polygonum aviculare			
Tritsum (N = 100	3.00	1.66	182
Elisenhof (N = 50)	2.71	1.56	
Feddersen Wierde (N = 50)	2.6	1.0	
Swifterbant II-16-H (N = 50)	2.78	1.85	155
I-17-i (N = 49)	2.98	1.88	155
V-23-i (N = 44)	2.96	1.96	152
XI-23-K (N = 46)	2.80	1.86	151

Table 11. *Urtica dioica* from Swifterbant.

		L	B	100L:B
V-13-F	min.	1.0	0.7	130
N = 50	aver.	1.16	0.77	148
	max.	1.4	1.0	167
I-17-i	min.	1.0	0.6	127
N = 50	aver.	1.20	0.81	148
	max.	1.4	1.0	175
XII-15-G	min.	1.0	0.6	118
N = 50	aver.	1.12	0.76	145
	max.	1.4	1.0	175
XVIII-19-i	min.	1.0	0.6	127
N = 50	aver.	1.20	0.83	144
	max.	1.4	1.0	164
V-13-K	min.	1.0	0.6	125
N = 50	aver.	1.14	0.78	147
	max.	1.4	1.0	167

Table 10. *Solanum* from Swifterbant

		L	B	100L:B
Solanum nigrum				
various samples	min.	1.5	1.2	114
N = 30	aver.	1.85	1.42	131
	max.	2.0	1.8	150
Solanum dulcamara				
various samples	min.	1.7	1.5	107
N = 10	aver.	2.21	1.90	116
	max.	2.5	2.2	125

number of samples. In the subfossil fruits, the outer fruit wall with the longitudinal ribs is not preserved. The inner fruit wall shows long rows of transversally elongated cells. Behre (1976, pp. 107-108) has pointed out that the cell pattern of the inner fruit wall of *Aegopodium podagraria* resembles that of *Conium*. However, in *Conium* the short longitudinal cell walls are thicker than the transversal ones, whereas in *Aegopodium* no differences in thickness of the cell walls are observed. Moreover, in *Conium* the cells of the inner fruit wall are much narrower than in *Aegopodium*.

Three subfossil fruits were suitable for measuring: 3.3 x 1.8, 3.4 x 1.9 and 3.4 x 1.8 mm. *Conium maculatum* is found in waste places (Artemisietalia vulgaris).

Other Umbelliferae
Other umbelliferous species are scarcely represented at Swifterbant. Of *Oenanthe aquatica* two damaged split fruits were found, one carbonized and the other non-carbonized. *Sium latifolium* is represented by one somewhat deformed fruit. Three fruits of *Sium erectum* were recovered. One of the *Sium erectum* fruits, which are ovate in outline, with a pointed apex and with little pronounced longitudinal ribs, was suitable for measuring: 1.6 x 1.2 mm. The three species mentioned above are found in fresh-water marsh vegetations.

Finally mention should be made of two fruits of *Hydrocotyle vulgaris*. Both *Hydrocotyle* fruits are from samples which yielded predominantly poorly preserved seeds.

3.26. Urticaceae

Urtica
Urtica dioica, a species from waste places rich in nitrates, must have been a very common plant on and near the Swifterbant site. It is represented in all samples but one, often by great numbers of fruits. In various samples it is the dominant seed type (up to 72.9%). The dimensions and index values of *Urtica dioica* fruits from various levels show no significant differences (table 11).

In contrast to the frequent occurrence of *Urtica dioica*, *Urtica urens* is conspicuously absent at Swifterbant. In the Iron Age and early-medieval sites in the coastal area of the Netherlands and Northwest Germany *Urtica urens* fruits do occur, sometimes in fairly great numbers (Behre, 1976, p. 89; Körber-Grohne, 1967, p. 263; Van Zeist, 1974, p. 311).

4. THE RECONSTRUCTION OF THE VEGETATION

4.1. Introduction

The reconstruction of the vegetation in the vicinity of the Swifterbant S3 site is largely based upon the results of the examination of seeds and fruits presented in tables 1 to 3. In addition, the data provided by the examination of charcoal and uncarbonized wood carried out by W.A. Casparie and P.C. Struijk (Casparie *et al.*, 1977) play a prominant part in the reconstruction of the forest vegetations. The moss remains identified by H.J. During allow some very interesting conclusions concerning the trees in the Swifterbant area. The charcoal and wood samples are from squares V-13/25, VI-13/25 and VII-13/25; layers H, i and K. The seeds, fruits and moss remains were retrieved from samples more widely scattered over the excavated area (see figs. 25-33).

One may assume that in general the seeds, fruits and wood remains recovered from the site originate from herbs, shrubs or trees which were present in the vicinity of the settlement. For a few plant remains this may not be true, *e.g.* for those of the mosses *Sphagnum imbricatum* and *S. papillosum* for which it is suggested that they are derived from peat layers deposited before the time of the settlement (cf. 3.1.). It is superfluous to emphasize that it is not justified to convert relative frequencies of the plant macrofossils into a quantitative reconstruction of the vegetation. The seed production shows large differences among the various plant species. Moreover, the seeds of some wild species may have been collected intentionally, or at least the plants may have been brought into the settlement on purpose, as a result of which the seeds concerned occur in large numbers in samples from the cultural fill. The seeds of various other species, on the other hand, must have arrived in the settlement by accident. Further, the chances of remaining preserved in a non-carbonized state are not the same for all seeds and fruits. The same applies to the wood remains (cf. Casparie *et al.*, 1977, p. 37).

For the reconstruction of the vegetation the results of all samples have been combined. One wonders whether this is justified, because in the course of the habitation, the vegetation in the vicinity of the site may have changed to a greater or lesser extent as a result of the activity of man and his domestic animals. Man-induced ruderal vegetation types would not have been found there before the arrival of the inhabitants of the site. Moreover, changes in the physical environment, *e.g.* changes in mean water-level, could have effected changes in the natural vegetation. On the other hand, the seed contents of samples from lower and upper levels show at most some quantitative differences (see table 20, section 6.2.2.), suggesting that during the period covered by the samples examined the vegetation did not change markedly. This, combined with the fact that the habitation of the site must have lasted 75 years at most, led us to the opinion that under the given circumstances the lumping together of all palaeobotanical data for a reconstruction of the plant cover is justifiable.

The reconstruction of the vegetation types is based upon the phytosociological affinity of the species demonstrated for the Swifterbant S3 site. A discussion of this method of determining vegetation types in the vicinity of prehistoric habitation sites is presented in an earlier paper (Van Zeist, 1974, pp. 331-32). Suffice it to remark the following. The basic syntaxonomic unit is the plant association or plant community, the name of which is indicated with

the suffix *-etum* behind the root of the generic name of the plant chosen as an index species. Syntaxonomically related plant associations are grouped into alliances (suffix *-ion*), which in turn are united into orders (suffix *-etalia*) and classes (suffix *-etea*).

In table 12 the syntaxonomic units arrived at for Swifterbant are shown. A plus sign indicates that at present the species concerned is found in the syntaxonomic unit at the top of the table. Preferably, the presence of a vegetation type should only be concluded if a greater number of species characteristic of or common in the syntaxonomic unit concerned is represented in the site. However, in some instances it was not possible to meet this requirement: we are concerned here with syntaxonomic units which are poor in species. For the reconstruction of the vegetation types in the Swifterbant area the work of Westhoff & Den Held (1969) on plant communities in the Netherlands was extensively consulted. To that publication the reader is also referred for further details.

In attempting to reconstruct the vegetation of more than 5000 years ago one should not yield to the temptation to go too far into detail. Through natural causes, such as a change in climate or a slow rate of migration, plants may have appeared or disappeared from the area, in this case the Netherlands, while man is responsible for the intentional or unintentional introduction of a great many species. Consequently, the composition of the vegetation units deduced for Swifterbant would not have been exactly the same as that of the present-day.

Although the reconstruction of the vegetation types is primarily based upon the botanical data, the former topography of the Swifterbant area is also taken into consideration. Moreover, the topography gives indications where the prehistoric vegetation types may have been found. The area was dissected by larger and smaller creeks bordered by natural levees. Behind the levees low-lying back-swamps were found. At a distance of 1 km and more from the site river-bank dunes emerged as islets above the surroundings (Ente, 1976; Hacquebord, 1976; see also 1.1.).

As for the figures and other indications given in front of the plant names in table 12 the following should be remarked. The kind of palaeobotanical evidence is indicated with a letter: S for seeds and fruits, W for wood and Ma for other macrofossil remains. The sample frequency (second column) is the number of samples in which the species concerned was found, expressed as a percentage (total number of samples is 46). No sample frequency is given for species which are represented by wood or charcoal only. One fruit-stone of *Cornus sanguinea* was recovered from the sieve residues. The mean percentage was obtained by dividing the sum of the percentages of a species by the number of samples in which the species concerned is represented. For instance, the sum of the *Cladium mariscus* percentages is 30.9% and the number of samples in which seeds of this type were found in 13. This gives a mean percentage of 2.38% (30.9% divided by 13). The "weighed" percentage (fourth column) for *Cladium mariscus* is 0.67%, viz. 30.9% divided by 46 (the total number of samples). In the "weighed" percentage the relative importance of the species with respect to the whole of the plant macrofossil evidence should find expression. No mean and "weighed" percentages are given for mosses. The relative importance of wood remains is indicated with a symbol (after Casparie *et al.*, 1977, table 5). This symbol is placed in brackets if of the tree concerned also seeds or fruits were found. No mean percentage of *Corylus* is shown because only fragments of hazelnut shells were recovered.

4.2. Forest vegetations

The palaeobotanical evidence indicates that deciduous forest vegetations were found in the vicinity of the site. Oak (*Quercus*), elm (*Ulmus*), ash (*Fraxinus*), linden (*Tilia*), crab apple (*Pyrus malus*) and poplar (*Populus* spec.) must have formed part of the tree canopy, whereas of the understorey shrubs, in addition to hazel (*Corylus avellana*), dogwood (*Cornus sanguinea*), rose (*Rosa canina/ rubiginosa*) and hawthorn (*Crataegus monogyna*) are represented at Swifterbant. Alder (*Alnus glutinosa*) was very likely also found in this deciduous forest type, although this small tree must have had its main distribution in the alder carr vegetations (see below). Without being more specific the deciduous forest inferred for the Swifterbant area is attributed to the order of the Alno-Padion, which syntaxonomic unit includes forest vegetations along rivers and rivulets, on young soils rich in nutrients which are periodically flooded (Westhoff & Den Held, 1969, p. 263). The highest parts of the levees bordering the major creeks may have been covered by Alno-Padion forests. The river-bank dunes must likewise have supported deciduous forest, but this forest type must have been of different composition. At least, the soil conditions on the dunes (sand and no periodical flooding) differed markedly from those on the levees.

Some doubt has been cast on the assumption that well developed forest would have occurred in the Swifterbant area. The posts and pegs found in the site have only small diameters. If they were of the trunks, and not of branches, the felled trees would at most have been 50 years old, and in most cases they would have reached an age of only 10 to 20 years (Casparie *et al.*, 1977, p. 39). Moreover, in spite of the favourable conditions for preservation no remains (trunks, root systems) or other indications of the former presence of large trees were observed underneath the settlement deposits. Prior to the occupation no well developed forest would have grown at the locality of the S3 site. This leaves one with the question whether perhaps only young deciduous forest was found in the area, suggesting that it was not until some decades prior to the habitation that conditions had become favourable for the establishment of this forest type. However, in the case of exclusively pioneer forest vegetations one would not have expected the great variety in arboreal species established for Swifterbant.

Table 12. Reconstruction of vegetation units in the Swifterbant area.

type of remains	sample frequency	mean percentage	'weighed' percentage		Alno-Padion	Alnion glutinosae	Salicion albae	Molinietalia	Magnocaricion	Scirpo-Phragmitetum	Scirpetum maritimi	Potametea	Ranunculo-Rumicetum maritimi	Lolio-Plantaginetum	Artemisietalia	Polygono-Chenopodietalia	Sisymbrietalia
W			xx	Quercus spec.	+	·	·	·	·	·	·	·	·	·	·	·	·
W			x	Fraxinus excelsior	+	·	·	·	·	·	·	·	·	·	·	·	·
W			xx	Ulmus spec.	+	·	·	·	·	·	·	·	·	·	·	·	·
W			x	Tilia spec.	+	·	·	·	·	·	·	·	·	·	·	·	·
W			r	Populus spec.	+	·	·	·	·	·	·	·	·	·	·	·	·
WS	15.2	0.13	0.02(x)	Pyrus malus	+	·	·	·	·	·	·	·	·	·	·	·	·
WS	17.4	+	(xx)	Corylus avellana	+	·	·	·	·	·	·	·	·	·	·	·	·
S	( )			Cornus sanguinea	+	·	·	·	·	·	·	·	·	·	·	·	·
S	15.2	3.91	0.60	Crataegus monogyna	+	·	·	·	·	·	·	·	·	·	·	·	·
S	10.9	0.10	0.01	Rosa canina/rubiginosa (+ spec.)	+	·	·	·	·	·	·	·	·	·	·	·	·
S	6.5	0.3	0.02	Rubus fruticosus	+	·	·	·	·	·	·	·	·	·	·	·	·
S	2.2	3.6	0.08	Moehringia trinervia	+	·	+	·	·	·	·	·	·	·	·	·	·
S	8.7	0.93	0.08	Anthriscus sylvestris	+	·	+	·	·	·	·	·	·	·	·	·	·
WS	10.9	2.68	0.29(xxx)	Alnus glutinosa	+	+	·	·	·	·	·	·	·	·	·	·	·
Ma	10.9	+		Anomodon viticulosus	+	·	·	·	·	·	·	·	·	·	·	·	·
Ma	2.2	+		Antitrichia curtipendula	+	·	·	·	·	·	·	·	·	·	·	·	·
Ma	2.2	+		Brachythecium cf. populeum	+	·	·	·	·	·	·	·	·	·	·	·	·
Ma	4.4	+		Eurhynchium striatum	+	·	·	·	·	·	·	·	·	·	·	·	·
Ma	10.9	+		Homalothecium sericeum	+	·	·	·	·	·	·	·	·	·	·	·	·
Ma	8.7	+		Hypnum cupressiforme	+	·	·	·	·	·	·	·	·	·	·	·	·
Ma	6.5	+		Isothecium myosuroides	+	·	·	·	·	·	·	·	·	·	·	·	·
Ma	6.5	+		Leucodon sciuroides	+	·	·	·	·	·	·	·	·	·	·	·	·
Ma	15.2	+		Neckera complanata	+	·	·	·	·	·	·	·	·	·	·	·	·
Ma	2.2	+		Neckera crispa	+	·	·	·	·	·	·	·	·	·	·	·	·
Ma	6.5	+		Brachythecium cf. rutabulum	+	+	+	·	·	·	·	·	·	·	·	·	·
Ma	2.2	+		Brachythecium cf. velutinum	+	·	+	·	·	·	·	·	·	·	·	·	·
S	2.2	0.5	0.01	Rumex conglomeratus	+	·	·	·	·	·	·	·	·	·	·	·	·
S	4.4	1.3	0.06	Carex pseudocyperus	·	+	·	·	·	·	·	·	·	·	·	·	·
S	4.4	0.3	0.01	Carex paniculata	·	+	·	·	·	·	·	·	·	·	·	·	·
S	23.9	1.19	0.28	Solanum dulcamara	·	+	·	·	·	·	·	·	·	·	·	·	·
WS	2.2	0.1	0.002(x)	Betula spec.	·	+	+	·	·	·	·	·	·	·	·	·	·
S	10.9	0.78	0.08	Lycopus europaeus	·	+	·	·	+	+	·	·	·	·	·	·	·
W			xx	Salix spec.	·	·	+	·	·	·	·	·	·	·	·	·	·
S	6.5	0.37	0.02	Sium erectum	·	·	+	·	+	+	·	·	·	·	·	·	·
S	2.2	0.5	0.01	Sium latifolium	·	·	+	·	+	+	·	·	·	·	·	·	·
S	2.2	0.1	0.002	Rumex hydrolapathum	·	·	+	·	+	+	·	·	·	·	·	·	·
S	6.5	1.23	0.08	Alisma plantago-aquatica	·	·	+	·	+	+	·	·	·	·	·	·	·
Ma	4.4	+		Sphagnum palustre	·	·	?	·	·	·	·	·	·	·	·	·	·

Table 12 (continued).

type of remains	sample frequency	mean percentage	'weighed' percentage		Alno-Padion	Alnion glutinosae	Salicion albae	Molinietalia	Magnocaricion	Scirpo-Phragmitetum	Scirpetum maritimi	Potametea	Ranunculo-Rumicetum maritimi	Lolio-Plantaginetum	Artemisietalia	Polygono-Chenopodietalia	Sisymbrietalia
S	2.2	2.2	0.05	Carex serotina-type	·	·	·	+	·	·	·	·	·	·	·	·	·
S	23.9	4.05	0.97	Poa pratensis/trivialis	·	·	·	+	·	·	·	·	·	·	·	·	·
S	4.4	5.85	0.25	Ranunculus acris	·	·	·	+	·	·	·	·	·	·	·	·	·
Ma	2.2	+		Trifolium repens	·	·	·	+	·	·	·	·	·	·	·	·	·
S	4.4	0.5	0.02	Agrostis spec.	·	·	·	+	·	·	·	·	·	+	·	·	·
S	4.4	0.75	0.03	Hydrocotyle vulgaris	·	·	·	+	·	·	·	·	·	·	·	·	·
S	2.2	0.8	0.02	Galium palustre	·	·	·	+	+	·	·	·	·	·	·	·	·
S	4.4	0.35	0.02	Sonchus palustris	·	·	·	+	·	+	·	·	·	·	·	·	·
S	2.2	0.6	0.01	Lychnis flos-cuculi	·	·	·	+	·	+	·	·	·	·	·	·	·
S	2.2	0.3	0.006	Caltha palustris	·	·	·	+	·	+	·	·	·	·	·	·	·
S	4.4	0.6	0.03	Carex nigra-type	·	·	·	?	·	·	·	·	·	·	·	·	·
S	2.2	0.1	0.002	Alopecurus cf. geniculatus	·	·	·	?	·	·	·	·	·	·	·	·	·
S	4.4	0.4	0.02	Carex riparia	·	·	·	·	+	·	·	·	·	·	·	·	·
S	28.3	2.38	0.67	Cladium mariscus	·	·	·	·	+	·	·	·	·	·	·	·	·
S	10.9	0.8	0.09	Carex rostrata/vesicaria	·	·	·	·	+	·	·	·	·	·	·	·	·
S	4.4	0.3	0.01	Menyanthes trifoliata	·	·	·	·	+	·	·	·	·	·	·	·	·
S	2.2	0.1	0.002	Carex disticha	·	·	·	·	+	+	·	·	·	·	·	·	·
S	26.1	0.49	0.13	Mentha aquatica	·	·	·	·	·	+	+	·	·	·	·	·	·
S	6.5	0.53	0.03	Eleocharis palustris	·	·	·	·	·	+	+	·	·	·	·	·	·
MaS	67.4	9.85	6.64	Phragmites australis	·	·	·	·	·	+	+	·	·	·	·	·	·
Ma	6.5	+		Calliergonella cuspidata	·	·	·	·	·	+	+	·	·	·	·	·	·
Ma	2.2	+		Drepanocladus cf. aduncus	·	·	·	·	·	+	+	·	·	·	·	·	·
S	2.2	0.6	0.01	Glyceria fluitans	·	·	·	·	·	+	·	·	·	·	·	·	·
S	6.5	0.33	0.02	Scirpus lacustris ssp. lacustris	·	·	·	·	·	+	·	·	·	·	·	·	·
S	4.4	0.35	0.02	Typha angustifolia/latifolia	·	·	·	·	·	+	·	·	·	·	·	·	·
S	4.4	0.35	0.02	Oenanthe aquatica	·	·	·	·	·	+	·	·	·	·	·	·	·
S	37.0	0.61	0.22	Scirpus maritimus	·	·	·	·	·	·	+	·	·	·	·	·	·
S	60.9	2.70	1.64	Scirpus tabernaemontani	·	·	·	·	·	·	+	·	·	·	·	·	·
S	10.9	0.48	0.05	Aster tripolium	·	·	·	·	·	·	+	·	·	·	·	·	·
S	2.2	2.3	0.05	Juncus gerardii	·	·	·	·	·	·	?	·	·	·	·	·	·
S	2.2	0.1	0.002	Salicornia europaea	·	·	·	·	·	·	?	·	·	·	·	·	·
S	43.5	1.0	0.43	Nymphaea alba	·	·	·	·	·	·	·	+	·	·	·	·	·
S	6.5	0.47	0.03	Potamogeton crispus/spec.	·	·	·	·	·	·	·	+	·	·	·	·	·
S	2.2	0.1	0.002	Ceratophyllum submersum	·	·	·	·	·	·	·	+	·	·	·	·	·
S	2.2	0.6	0.01	Zannichellia palustris	·	·	·	·	·	·	·	+	·	·	·	·	·
S	4.4	0.6	0.03	Rumex maritimus	·	·	·	·	·	·	·	·	+	·	·	·	·
S	13.0	2.87	0.37	Ranunculus sceleratus	·	·	·	·	·	·	·	·	+	·	·	·	·
S	19.6	0.86	0.17	Plantago major	·	·	·	·	·	·	·	·	·	+	·	·	·

Table 12 (continued).

type of remains	sample frequency	mean percentage	'weighed' percentage		Alno-Padion	Alnion glutinosae	Salicion albae	Molinietalia	Magnocaricion	Scirpo-Phragmitetum	Scirpetum maritimi	Potametea	Ranunculo-Rumicetum maritimi	Lolio-Plantaginetum	Artemisietalia	Polygono-Chenopodietalia	Sisymbrietalia
S	58.7	15.67	9.20	Polygonum aviculare	·	·	·	·	·	·	·	·	·	+	·	·	·
S	6.5	0.17	0.01	Capsella bursa-pastoris	·	·	·	·	·	·	·	·	·	+	·	+	+
Ma	2.2	+		Barbula convoluta	·	·	·	·	·	·	·	·	·	+	·	·	·
S	26.1	1.12	0.29	Conium maculatum	·	·	·	·	·	·	·	·	·	·	+	·	·
S	4.4	1.15	0.05	Artemisia vulgaris	·	·	·	·	·	·	·	·	·	·	+	·	+
S	17.4	0.38	0.07	Arctium cf. lappa	·	·	·	·	·	·	·	·	·	·	+	·	·
S	97.8	28.62	28.00	Urtica dioica	+	·	·	·	·	·	·	·	·	·	+	·	·
S	13.0	0.63	0.08	Galium aparine	+	·	·	·	·	·	·	·	·	·	+	·	·
S	8.7	0.45	0.04	Carduus crispus	·	·	·	·	·	·	·	·	·	·	+	·	·
S	4.4	4.2	0.18	Lapsana communis	·	·	·	·	·	·	·	·	·	·	+	·	·
S	58.7	20.00	11.74	Polygonum lapathifolium	·	·	·	·	·	·	·	·	·	·	·	+	·
S	43.5	3.12	1.36	Polygonum persicaria	·	·	·	·	·	·	·	·	·	·	·	+	·
S	4.4	0.05	0.002	Galeopsis tetrahit/speciosa	·	·	·	·	·	·	·	·	·	·	·	+	·
S	6.5	0.37	0.02	Cirsium arvense	·	·	·	·	·	·	·	·	·	·	·	+	·
S	82.6	14.76	12.20	Chenopodium album	·	·	·	·	·	·	·	·	·	·	·	+	+
S	80.4	12.22	9.83	Stellaria media	·	·	·	·	·	·	·	·	·	·	·	+	+
S	6.5	0.2	0.01	Sonchus asper	·	·	·	·	·	·	·	·	·	·	·	+	+
S	82.6	2.15	1.78	Solanum nigrum	·	·	·	·	·	·	·	·	·	·	·	+	+
S	8.7	0.8	0.07	Chenopodium rubrum/glaucum	·	·	·	·	·	·	·	·	·	·	·	·	+
S	84.8	11.41	9.68	Atriplex hastata/patula	·	·	·	·	·	·	·	·	·	·	·	·	+
S	4.4	9.95	0.43	Bromus secalinus/mollis	·	·	·	·	·	·	·	·	·	·	·	·	+

Convincing evidence for the presence of full-grown deciduous trees at not too great a distance from the site is provided by the moss remains. A relatively great number of the moss species demonstrated for Swifterbant grow on tree trunks (see table 2), and for some of these mosses it is specifically mentioned in the literature (Landwehr, 1966) that they are found on old trees, e.g. *Anomodon viticulosus, Antitrichia curtipendula, Neckera complanata* and *Leucodon sciuroides*. One may assume that the mosses were brought into the settlement adhering to the bark which must have been collected on purpose. Small pieces of bark were frequently observed in the culture layer.

There is hardly any doubt that well developed forest with full-grown trees was found in the Swifterbant area, although the area covered by this vegetation type was of very limited extent. The locality chosen for the foundation of the S3 site, on a levee along a secondary creek, may not have supported bigger trees because it was too wet. This levee had not silted up high enough. The absence of big trees may have been one of the factors which determined the choice of the site for habitation. Subsequently, the wet conditions forced the inhabitants to raise the level of occupation and/or to achieve some insulation from the damp subsoil by carrying in substantial amounts of plant material, such as reed and stems.

Alderwood vegetations (Alnion glutinosae) must particularly have been found in the transitional zone between levee and back-swamp. Alder is by far the most common wood type in the site (Casparie *et al.*, 1977, fig. 6) suggesting that *Alnus glutinosa*, and hence alder carr, were quite common in the area. Of the herbs represented at Swifterbant, *Carex pseudocyperus, C. paniculata* (two sedge species), *Solanum dulcamara* (bittersweet) and, to a lesser extent, *Lycopus europaeus* are characteristic of Alnion glutinosae vegetations.

The presence of willow carr (Salicion albae) is concluded on the basis of the rather frequent occurrence of wood and charcoal of *Salix* in the site. Of the other species with a plus sign in the column of Salicion albae, none grows exclusively or predominantly in willow carr. Willow shrub must have covered a part of the back-swamp areas, the other part of which was the domain of open vegetations (4.3.).

4.3. Marsh and aquatic vegetations

In addition to willow carr, Magnocarion and Scirpo-Phragmitetum vegetations were found in the back-swamps. These syntaxonomic units are attested by fairly great numbers of species, but as is clear from table 12 most of the species are poorly represented in the site (low sample frequencies and low weighed percentages). *Cladium mariscus* and *Mentha aquatica* show somewhat higher sample-frequency percentages, but it is reed (*Phragmites australis*) that is really well represented in the seed record, particularly if one takes into consideration that this species is very likely seriously under-represented by its fruits. The many reed-stem fragments observed in the occupational fill do not find expression in the weighed percentage value. The poor representation of most back-swamp plants can be explained in two ways. It is possible that dense willow and reed vegetations dominated the back-swamps, preventing a more luxurious growth of other marsh plants. On the other hand, willow and reed were exploited by the inhabitants of the site. Willow wood and reed stems were carried in on purpose, whereas most of the other species from the back-swamps must have arrived in the settlement by accident. In the natural succession, the Scirpo-Phragmitetum, which is found in places with shallow water, is succeeded by Magnocaricion vegetations.

Scirpo-Phragmitetum vegetations may also have bordered on to the creeks. However, the frequent occurrence of fruits of *Scirpus tabernaemontani (S. lacustris* ssp. *glaucus)* and the fairly good representation of *Scirpus maritimus* suggest that the Scirpetum maritimi, another vegetation type found in and near creeks, was not rare in the Swifterbant area. This vegetation points to brackish water in the creeks, at least temporarily. At high storm floods, salt water must have penetrated into the area through the creek system which was in open connection with the sea. Marine influence in the Swifterbant area is also suggested by the presence of *Aster tripolium, Juncus gerardii* and *Salicornia europaea.* The plant evidence does not suggest that the area was more or less regularly flooded with brackish water. On the contrary, brackish-water conditions must generally have been confined to the creeks.

It was more or less a surprise to find that a fairly great number of species demonstrated for Swifterbant are found in Molinietalia vegetations. Except *Poa trivialis/pratensis,* all possible Molinietalia species are poorly represented. The order of the Molinietalia includes wet meadows which owe their origin and their continued existence to the activity of man, viz. to the cutting of tree stands and a subsequent non-intensive exploitation of the vegetation. It is feasible that at Swifterbant Molinietalia vegetations developed in places where alder carr had been cut.

A few species from open water are attributed to the Potametea, which syntaxonomic unit includes vegetations in fresh and slightly saline water. Of these aquatics, only *Nymphaea alba* (water-lily) is well represented at Swifterbant, although the distribution of its seeds poses some questions (6.2.1.). *Zannichellia palustris, Potamogeton crispus* and *Ceratophyllum submersum* occur in fresh as well as in brackish water. *Nymphaea alba,* on the other hand, is a typical fresh-water species which may have been found in cut-off secondary creeks and in the lowermost parts of the back-swamp areas with permanent open water.

4.4. Synanthropic vegetations

A great number of species from Swifterbant is most common in vegetations which are due to the interference of man. Of synanthropic vegetations not only many species could be established for Swifterbant, but quite a few species from this group of syntaxonomic units show high sample frequencies and weighed percentage values, such as *Urtica dioica, Polygonum aviculare, Polygonum lapathifolium, Chenopo-*

dium album, Stellaria media and *Atriplex hastata/patula.* The high frequencies of seeds and fruits of these species must have been the result of the activity of the inhabitants who dumped these weeds in the site (see 6.1.1.). Nevertheless, the great numbers of seeds indicate clearly that the species concerned must have been quite common in the immediate vicinity.

Plantago major, Polygonum aviculare and *Capsella bursa-pastoris* point together to the presence of the Lolio-Plantaginetum, which vegetation type is found in frequently trodden places. The moss species *Barbula convoluta,* which in table 12 has been assigned to the Lolio-Plantaginetum, grows in various disturbed habitats. It is, of course, no great surprise to find indications of a vegatation from more or less heavily trodden places. Such places must have been common in the vicinity of settlement sites.

The evidence for the presence of the Ranunculo-Rumicetum maritimi is rather meagre. The only species of this plant association from wet places rich in nitrates found in Swifterbant are *Ranunculus sceleratus* and *Rumex maritimus,* and both are scarcely represented. One must assume that this vegetation type was not particularly common in the Swifterbant area, which is somewhat surprising in view of the wet environment. The absence of *Bidens tripartitus,* another species from wet, nitrate-rich habitats, should be mentioned. This species is well represented in the Iron Age sites of Vlaardingen and Schiedam, both situated in a moist, fresh-water environment. The differences in soil conditions, clay at Swifterbant and predominantly peaty soils at Vlaardingen and Schiedam, could perhaps provide an explanation. *Bidens tripartitus* seems to prefer more peaty soils. These differences in soil conditions cannot be adduced to explain the absence of *Juncus bufonius* at Swifterbant, whereas at Vlaardingen and Schiedam this species is very well represented. *Juncus bufonius,* which is a characteristic species of vegetations on moist to wet, bare soils is rather undemanding as to the kind of soil it grows on. One wonders whether in the fourth millennium B.C. this species did not yet occur in the Netherlands.

Mention should also be made of the absence of the Chenopodio-Urticetum urentis. At least *Urtica urens* and *Chenopodium ficifolium*, which are the most characteristic species of this plant community from muckheaps and other nitrate-rich habitats, are conspicuously absent at Swifterbant (see also 3.6. and 3.26.).

The presence of Artemisietalia vulgaris vegetations is suggested particularly by *Artemisia vulgaris, Conium maculatum, Arctium* cf. *lappa* and *Carduus crispus*. This syntaxonomic unit includes vegetations from a large variety of natural and man-made nitrate-rich habitats, *e.g.* along streams, roads and fields, in farmyards and in neglected gardens. Artemisietalia vulgaris vegetations are quite common in the vicinity of habitation sites. The relatively good representation of *Conium maculatum* (Poison Hemlock) in the seed record is striking. At present this highly poisonous plant is far from common. It cannot wholly be excluded that this species was collected on purpose. From *Conium* not only a powerful poison can be prepared, but in lower concentrations the alkaloids extracted from the plant act as a sedative.

Of the Polygono-Chenopodietalia and Sisymbrietalia, relatively small numbers of species have been established for Swifterbant, but quite a few of them show high to very high frequencies, suggesting that the plants concerned were found in great numbers. Both syntaxonomic units mentioned above belong to the Chenopodietea which class comprises exclusively synanthropic plant communities: weed associations in fields and other ruderal vegetations of predominantly annual and bi-annual species. One wonders whether it is justified to distinguish between Polygono-Chenopodietalia and Sisymbrietalia for vegetations from more than 5000 years ago. It may also be questioned whether the modern weed associations compare with those in Neolithic times. Consequently, no attempt has been made to arrive at a somewhat more detailed reconstruction of Chenopodietea weed vegetations in the vicinity of Swifterbant. It may only be concluded that

these vegetations were present and that very probably they were quite common.

From the above it will be clear that no suggestions concerning the composition of the flora in the grain fields of the Swifterbant farmers can be presented. All Chenopodietea species demonstrated for this site could have been found in the fields, but typical field weeds are almost completely lacking. Only if *Bromus mollis/secalinus* fruits are of *B. secalinus*, a typical grain-field weed would be represented. *Polygonum convolvulus*, the seeds of which are frequently found in charred grain samples, has not been established for Swifterbant.

5. THE PLANT HUSBANDRY

5.1. Cultivated plants

Charred cereal grains were recovered from the sediment samples and in particular from the sieve residues. At least two and probably three cereal crop plant species are represented at Swifterbant. The total numbers of grains found of each species are shown in table 13.

So far, the crop plant remains from Swifterbant constitute a rather isolated case. The Rössen sites in the Rhineland area of West Germany come most into consideration for a comparison. The habitation of the Swifterbant S3 site, dated to 3400-3300 B.C., falls within the time limits of the Rössen culture: 3800-3350 B.C. (cf. Lanting & Mook, 1977, fig. 4). The Rhineland area is not too far away from Swifterbant, and, moreover, charred crop plant remains are reported for those sites (Knörzer, 1971; Schiemann, 1954).

5.1.1. *Hordeum vulgare var. nudum*

Fairly large numbers of barley grains, altogether nearly 2000 specimens (table 13), were recovered. The kernels show the characteristics of naked barley: they are rounded in cross-section, and in various grains a fine transverse wrinkling on the surface can be observed (fig.

4:1-3). Many grains are more or less distinctly lop-sided indicating that six-rowed barley is concerned here. Thus, the morphology of the grains points clearly to *Hordeum vulgare* var. *nudum*. This conclusion is supported by the rachis internode remains.

A few fragmentary rachis internodes were recovered from the sediment samples and the sieve residues, but of particular interest are the internode remains found in a small hand-picked sample of 'vegetable material' from VIII-21-i. This sample turned out to consist of charred threshing remains of barley. In addition to unidentifiable ashes and fragments of glumes, bracts and rachillas, the sample yielded a small number of grains and rachis internodes of barley. One should be grateful to the keen-eyed, unknown volunteer who noticed the plant remains at "his/her" square metre parcel. On the other hand, it is to be regretted that it was not until many months afterwards that the nature of the vegetable material was discovered. Otherwise a much larger sample could perhaps have been taken.

The internode remains (fig. 5:1-2) are characteristic of free-threshing, six-rowed barley. At the distal end of the internodes the basal parts of three spikelets (a median spikelet and two lateral ones) can be observed. Remains of the glumes, the lemma (the lower bract) and the hairy rachilla are still present on some of the internodes. It should be mentioned that the lateral spikelets were not sessile as in modern six-rowed barley, but that they were pedicellate. The Swifterbant rachis internodes show short, stout stalks which formed the base of the lateral spikelets. Similar pedicellate lateral spikelets are described and depicted by Knörzer (1971, fig. 3:1b) for naked barley from the Rössen site near Langweiler, in the Rhineland area. Villaret-von Rochow (1967) pays special attention to the phenomenon of the short-stalked lateral spikelets in hulled and naked barley from Neolithic Burgäschisee-Süd in Switzerland. This author raises the question to what extent the pedicellate lateral spikelets should be considered as a primitive feature.

The Swifterbant internode remains confirm the conclusion arrived at on the basis of the grains, viz. that naked, six-rowed barley is concerned here. A few almost complete internodes point to the lax-eared variety (the nodding spike type). This is of course no proof that the dense-eared variety would not have been cultivated. Moreover, plump kernels (short and broad specimens) suggest that the dense-eared variety is equally represented at Swifterbant. It is likely that a mixture of both varieties was grown, the lax-eared type being the predominant one. The latter conclusion is based upon the fact that slender grains are significantly more numerous than plump ones.

Of four comparatively large barley samples, grains have been measured. As a large proportion of the grains had been more or less seriously affected by the carbonization, only relatively small numbers of kernels were suitable for measuring. For that reason, in table 14 the dimensions and index values of grains from two adjacent squares, viz. from VIII-21-i and VIII-22-i and from XVI-24-i and XVI-25-i, respectively, have been taken together. It is likely that in both cases the grains originated from the same supply. The grains from XVI-24/25-i are, on an average, slightly smaller than those from VIII-21/22-i, but the index values do not show significant differences. Below (5.2.) the dimensions of the Swifterbant barley will be compared with those from other sites.

5.1.2. *Triticum dicoccum*

Of *Triticum dicoccum* only small numbers of grains were found. There is a rather great variety in shape; fairly slender as well as plump specimens occur (fig. 4:4-6). The L:B index values of 20 grains from various samples (table 14) demonstrate that, on the whole, the grains are rather plump. Of emmer wheat grains from other prehistoric sites in the Netherlands, only those from Neolithic Vlaardingen show a mean L:B index value (182) which is as low as that obtained for the Swifterbant grains. The grains from the other sites listed in table 64 of Van

Zeist ((1968) 1970) are more slender, on an average. In addition to the grains, one spikelet base and one glume base were found in the sediment samples. There may have been many more spikelet forks and glume bases in the settlement layers, but they were not retained on the rather large-meshed sieve through which the soil was washed.

5.1.3. *Triticum cf. aestivum*

Among the *Triticum* grains, one specimen (4.3 x 2.6 x 2.4 mm) from sieve sample XII-15-i is strongly reminiscent of *T. aestivum*. As is clear from fig. 4:8 this grain shows the greatest width in its lower part, which feature is characteristic of *T. aestivum*. Moreover, the surface of the grain suggests a free-threshing wheat. It is clear that only one grain is no particularly firm evidence of bread wheat at Swifterbant. Thus, it cannot be excluded that this grain is a deformed emmer wheat kernel. On the other hand, Knörzer (1971) reports a small number of bread wheat (club wheat) grains for the Rössen site near Langweiler. In that case a few rachis remains confirm the presence of free-threshing wheat.

It is not likely that *T. aestivum* was grown intentionally, but at most it may have constituted an admixture to emmer wheat.

5.1.4. *Claviceps purpurea*

One carbonized sclerotium of *Claviceps purpurea* (ergot) was met with (fig. 4:7). The sclerotium is lanceolate, tapering at both ends, with longitudinal grooves. Dimensions: 7.2 x 1.9 mm. *Claviceps purpurea* is a parasitic fungus on cereals as well as on wild grasses. The Swifterbant specimen is of about the same size as the *Claviceps* sclerotia from the Rössen site near Langweiler (8.1 x 1.1 x 1.3 mm; Knörzer, 1971). Knörzer argues that these sclerotia are too small for *Claviceps* on barley or wheat, but that wild grasses with medium-sized grains come rather into consideration. In this connection it should be mentioned here that 10 non-carbonized sclerotia of *Claviceps* on *Triticum*

monococcum grown on an experimental plot at Orvelte measure 10.1 (9.0-12.0) x 2.5 (2.0-2.8) mm. They are considerably smaller than the *Claviceps* sclerotia on *Secale* (18.9 x 3.3 x 4.4 mm) to which Knörzer refers. As wild grasses with medium-sized grains are hardly represented at Swifterbant, it is tempting to assume that in spite of the rather small size of the sclerotium concerned barley or wheat was nevertheless the host plant.

5.2. Plant cultivation at Swifterbant

The barley and wheat grains from Swifterbant do not in themselves necessarily imply that these cereals were grown locally. In view of the local situation, with only very little potential arable land, one may wonder whether crop plants were actually grown there. One could imagine that the crops had been grown elsewhere, on the higher soils of the Land van Vollenhove, the Veluwe or the coastal dune area, and that the people who spent a part of the year at Swifterbant had brought the grains with them. However, as has already been mentioned before (Casparie *et al.*, 1977), the threshing remains of barley from VIII-21-i indicate that the crop plants were grown in the vicinity of the site. If the plants had been cultivated at a great distance from the site, threshed grains would have been transported because they are much less bulky than unthreshed ears.

Thus, one must assume that plant cultivation was practised locally. Fields can only have been laid out on the highest parts of the levees bordering the major creeks and possibly on the river-bank dunes, at a distance of at least 1 km from the site. In spite of the rather extreme edaphic conditions the quality of the crop was not exactly poor. At least, the average size of the Swifterbant barley grains is larger than that obtained for naked barley from other sites in the Netherlands and from the Rössen site near Langweiler (table 15). The size of the emmer wheat grains equally does not suggest a poor crop (cf. Van Zeist, (1968) 1970, table 64). It is striking that the *Triticum dicoccum*

Table 13. Total numbers of cereal grains from Swifterbant.

	Hordeum vulgare nudum	Triticum dicoccum	Triticum cf. aestivum
sediment samples	179 (99.4%)	1 (0.6%)	—
sieve residues	1788 (96.1%)	71 (3.8%)	1 (0.05%)
totals	1967 (96.4%)	72 (3.5%)	1 (0.05%)

Table 14. Dimensions in mm and index values of cereal grains from Swifterbant.

		L	B	T	100 L:B	100 T:B
Hordeum vulgare nudum						
VIII-21/22-i	min.	4.9	2.9	2.2	144	71
N = 50	aver.	5.86	3.32	2.77	170	82
	max.	6.8	4.2	3.3	208	91
XVI-24/25-i	min.	4.8	2.7	1.9	143	68
N = 46	aver.	5.55	3.25	2.55	172	79
	max.	6.8	4.2	3.0	197	90
Triticum dicoccum						
various samples	min.	4.8	2.4	2.0	155	70
N = 20	aver.	5.62	3.08	2.59	184	84
	max.	6.8	3.9	3.2	227	100

Table 16. Samples from inside or near flint concentration areas (group I) and from outside those areas (group II).

Group I in/near activity areas		Group II outside activity areas	
no. 15	V-21-G	no. 1	I-17-i
no. 19	III-17-F	no. 2	I-18-i
no. 20	VIII-17-F	no. 3	II-16-H
no. 21	VIII-23-F	no. 4	III-14-G
no. 26	IX-14-G	no. 5	III-20-F
no. 27	IX-15-F	no. 6	III-20-G
no. 28	IX-19-F	no. 7	IV-13-F
no. 30	IX-29-F	no. 8	IV-17-G
no. 31	X-15-G	no. 9	IV-20-H
no. 32	X-22-F	no. 10	IV-24-F
no. 33	XI-16-F	no. 11	IV-24-G
no. 35	XII-15-G	no. 12	V-13-K
no. 36	XIV-17-i	no. 13	II-18-G
no. 37	XIV-23-G	no. 14	V-18-K
no. 39	XV-24-G	no. 16	V-23-i
no. 40	XV-27-G	no. 17	VI-25-K
no. 41	XVI-25-G	no. 18	VII-22-K
no. 45	XXIX-18-F	no. 22	VIII-23-H
no. 46	XXXIII-18-F	no. 23	VIII-24-H
		no. 24	VIII-26-H
		no. 25	VIII-26-K
		no. 29	IX-25-H
		no. 34	XI-23-K
		no. 38	XV-19-K
		no. 42	XVIII-19-i
		no. 43	XX-19-i
		no. 44	XXIV-19-i

Table 15. Average dimensions in mm and index values for *Hordeum vulgare nudum* from various sites.

	L	B	T	L:B	T:B
Ur-Fulerum (N = 50)	5.8	3.1	2.5	187	81
Zandwerven (N = 8)	4.3	2.5	1.9	178	77
Eeserveld (N = 100)	4.58	2.58	1.99	179	77
Emmerhout (N = 25)	4.46	2.41	1.80	188	76
Angelsloo (N = 100)	4.36	2.36	1.66	186	70
Elp 98 (N = 100)	5.34	2.64	1.93	204	73
Elp 69 (N = 69)	4.78	2.74	2.15	176	79
Bovenkarspel (N = 49)	4.69	2.82	2.13	166	76
Langweiler (N = 17)	4.66	2.56	2.01		

grains from Neolithic Vlaardingen, where the edaphic conditions must have been comparable to those of Swifterbant, are even somewhat larger than those of the latter site. It is true that only some information on the quality, that is on the size, of the grains is available and not on the yields in kilogrammes per unit area (and on the frequency of crop failures), but nevertheless it seems that the conditions for grain growing were not particularly unfavourable, although the cultivated acreage must have been of very limited extent. It is unlikely that cereals could have constituted the main food of the inhabitants of the Swifterbant site; they may at most have been a welcome addition to a diet of predominantly fish and meat (Clason & Brinkhuizen, 1978).

The Rössen site near Langweiler (Knörzer, 1971, table 1) shows about the same ratio between naked barley and emmer wheat as the Swifterbant S3 site, viz. c. 25 to 1. In both sites *Triticum aestivum* is represented, but at Langweiler free-threshing wheat seems to have been of nearly equal importance as emmer wheat. At Langweiler, *Triticum monococcum* is the predominant wheat species: the ratio between einkorn and emmer wheat is nearly 4 to 1. At Swifterbant, on the other hand, *Triticum monococcum* is conspicuously absent. It should be emphasized that an identical crop-plant assortment was not to be expected for two settlement sites of different archaeological cultures and with different environmental conditions. Moreover, both Rössen sites in the Rhineland area, for which archaeobotanical data have been published, show already striking differences. Thus, for Ur-Fulerum (Schiemann, 1954) no *Triticum monococcum* is reported, whereas there the barley was predominantly of the hulled type.

5.3. The distribution of cereal grains in the site

Only once were charred grains observed *in situ* during the excavations (VIII-21-i). The density of the cereal grains was generally low. The sediment samples consisting of 3 litres of soil yielded mostly only a few charred grains. On the other hand, from a few samples, viz. from IV-20-H, IX-19-F and VIII-22-K, somewhat greater numbers of cereal grains were recovered, suggesting that the grains were not very evenly distributed through the archaeological deposits. This suggestion is supported by the numbers of charred grains recovered from the sieve residues. From the latter category of samples it appears that small concentrations of charred grains must have occurred.

In figs. 6-10, for levels F to K, the numbers of charred grains found in the sieve residues of the square metre quadrants are shown. A cross indicates that other plant remains but no cereal grains were recovered from that particular quadrant.

In evaluating the data presented in fig. 6-10 the following should be taken into consideration. The quality of the data varies considerably because the recovery of botanical remains depended in no small measure upon the interest and the attentiveness of the people who on that particular day were carrying out the soil-washing operation. If greater numbers of vegetable remains were present on the sieve they should certainly have been observed by the person who was in charge of recovering the archaeologically important items from the sieve (see 2.1.). However, if only a few charred grains or other plant remains, such as hazelnut shell fragments and *Galium* seeds, were present they may easily have been overlooked if the person concerned did not pay particular attention to possible archaeobotanical material.

The greatest uncertainty concerns the quadrants for which no vegetable remains are reported. In some instances there may actually have been no plant material on the sieve, but in many others it must have remained unnoticed. If the majority of a consecutive series of quadrants yielded plant remains, but for a few quadrants no mention is made of seeds one may assume that, indeed, no seeds were present. In that case it is very likely that the same person or group of persons handled the whole series of quadrants. In the other instances one is left with the uncertainty to what extent a blank in figs. 6-10 is due to the absence of

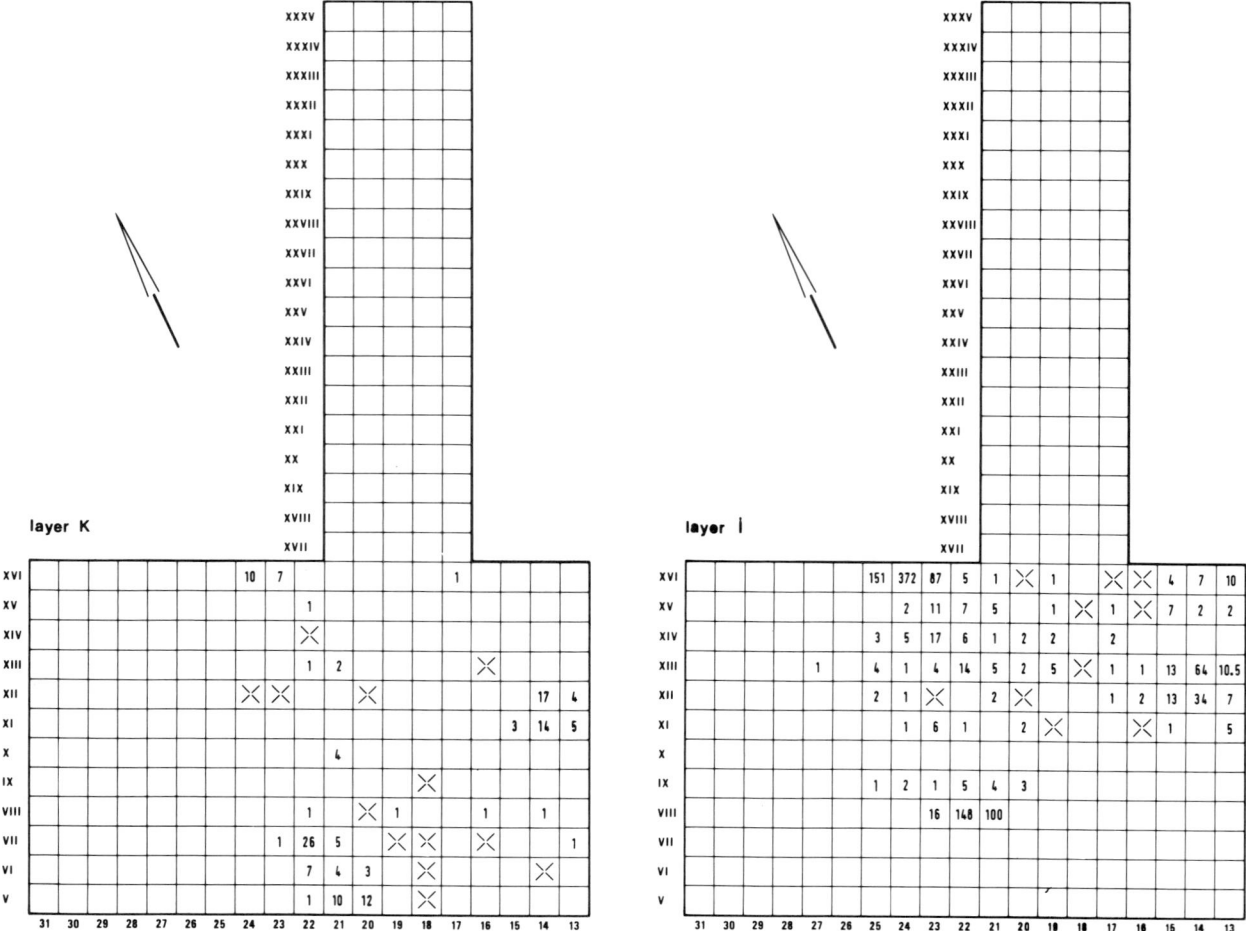

Fig. 6. Numbers of cereal grains recovered from sieve residues, layer K. For discussion of figs. 6-10, see 5.3.

Fig. 7. Numbers of cereal grains recovered from sieve residues, layer i.

plant remains or to the attitude of the excavation volunteer concerned.

Nevertheless, in spite of the many uncertainties and particularly of the fragmentary nature of the data presented in figs. 6-10, the numbers demonstrate at least that concentrations of charred grains occurred, that is to say, that in some areas charred grains were considerably more numerous than in the rest of the archaeological deposit. Obvious concentrations have been established for XI-14-F, XI-21/22-G, VIII-21/22-i, XII/XIII-14-i and XVI-23/24/25-i. One may assume that these concentrations are due to specific activities of the inhabitants of the site. In one instance it is clear to

what kind of activity a charred grain concentration must be ascribed. We are concerned here with the concentration in quadrants VIII-21-i and VIII-22-i. From one of the quadrants (VIII-21-i) the sample with the threshing remains of naked barley originates (5.1.1.). Other concentrations may likewise have come from places where charred threshing refuse had been dumped, but the sieve residues are not informative on this matter. The small rachis and glume fragments must practically all have been washed through the sieve, and the few remains which may already have stayed on the screen would generally have remained unnoticed in the field.

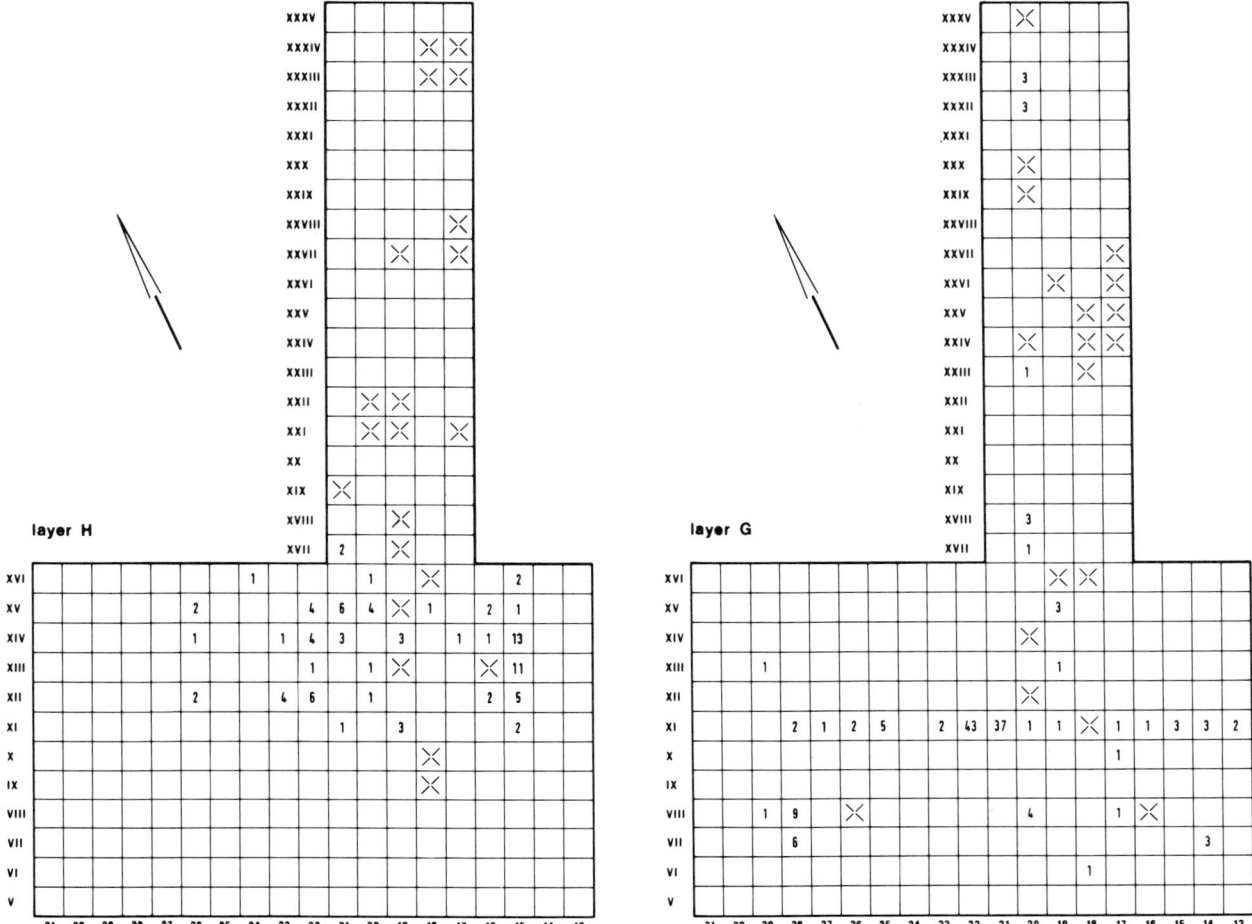

Fig. 8. Numbers of cereal grains recovered from sieve residues, layer H.

Fig. 9. Numbers of cereal grains recovered from sieve residues, layer G.

It is likely that threshing remains were used as fuel. It is true that these remains burn up rapidly, but they give high temperatures. Threshing refuse could also have been used to light or to revive the hearth fire. However, cereal grain concentrations certainly do not always represent threshing remains. Thus it is possible that charred grains accumulated near the fire place where the meals were prepared. One could also imagine that if the people were tidy, the rubbish including occasional charred grains was swept out of the room and deposited at a place outside the house. As the archaeological remains do not provide indications of the location and the outline of houses

or huts, it cannot be tested whether there is, indeed, any relation between charred grain concentrations and house sites.

It has been examined whether the charred grain accumulations suggested by the sieve residues possibly correlate with distinct features observed in the field, such as fire places and ash patches. In this connection it should be mentioned that a feature map and an accompanying list with information on the type, the shape and the size of the feature, the location within the grid system and the level has been composed (by Mr. D. Kielman). Corresponding information on the location within the site is available for the sieve residues, so that it was

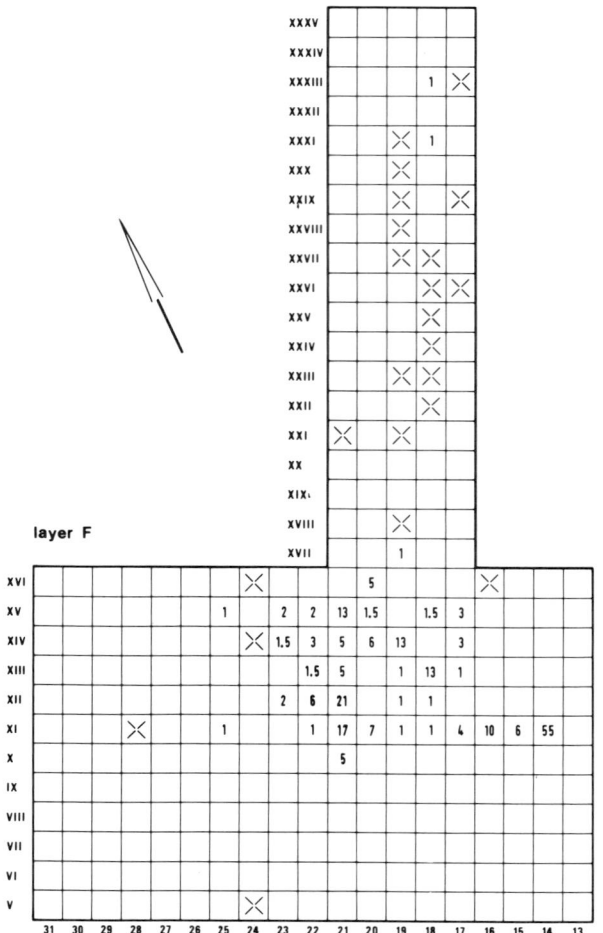

Fig. 10. Numbers of cereal grains recovered from sieve residues, layer F.

great number of cereal grains. None of the charred grain accumulations is from an area with a fire place.

As will come up for discussion in chapter 6 the flint material shows distinct distribution patterns in the successive 10 cm levels. Areas of concentration of flint material suggest activity areas. No correlation between the charred grain accumulations and the distribution of the flint material could be established. Fairly great numbers of grains were recovered from metre squares in or near areas of flint concentrations as well as from those outside these activity areas, viz. 16 and 11 samples, respectively.

In summary, it may be remarked that at least one cereal grain concentration represents threshing residues. The origin of the other accumulations remains unclear. Verly likely none of the concentrations constitutes *in situ* remains of a grain supply. In that case one would have expected a cache of charred seeds which had certainly been observed in the field. One must assume that most if not all of the cereal grain concentrations are from refuse that had been dumped in the settlement area.

5.4. Food collecting

The conclusion that plant cultivation was probably of secondary economic importance at Swifterbant (5.2.) leads more or less automatically to the question of the possible role of wild plants in the diet of the inhabitants of the site. The factual evidence indicates that various fruits and nuts were collected: *Corylus avellana* (hazelnut), *Pyrus malus* (crab apple), *Crataegus monogyna* (hawthorn), *Rosa* (rose-hips) and *Rubus fruticosus* (blackberry). In addition to the uncarbonized pips, three carbonized crab apples were found (dimensions: 2.3 x 2.1, 1.85 x c. 1.6, 2.55 x 2.4 cm). There are no indications of the drying of apples, a preservation method that has been reported for various prehistoric sites in Europe (cf. Helbaek, 1952). It is true that the specimen depicted in fig. 5:10 is half an apple, but this it not due to the intentional cutting into two halves. Schiemann (1954) reports five carbonized, whole crab

possible to determine whether charred grain concentrations are from places where a feature was observed. This has been done for 12 samples with more than 20 charred grains (21 to 372 kernels) and for 15 samples with 11-20 specimens. The result is distinctly negative. Only the concentration in VIII-21-i is from a square for which a patch of carbonized seeds and ashes is reported. In the same metre quadrant a concentration of fish remains was observed. It is from this feature that the hand-picked sample of threshing remains was recovered. It is likely that the charred threshing remains were not confined to VIII-21-i, but that they extended over the adjoining square VIII-22-i which likewise yielded a relatively apples for the Rössen site of Ur-Fulerum in the Rhineland area. One wonders whether the presence of carbonized apples could point to the roasting of these fruits, which treatment may have improved the taste. Hawthorn fruits, rosehips, blackberries and (non-roasted) apples were a source of vitamin C but of only little caloric value. On the other hand, hazelnuts, with a high content of fat (c. 60%), must have constituted a valuable additional food. Remains of acorns were not recovered at Swifterbant. Acorns, which in addition to starch contain a fairly high content of fat, must have been collected for human nutrition in prehistoric times as is attested by the finds of the carbonized fruits in various sites (cf. Jørgensen, 1978). The absence of acorn remains at Swifterbant cannot be ascribed to a non-availability of this potential source of food. For the rather frequent occurrence of oak charcoal at Swifterbant (Casparie *et al.*, 1977) indicates that the trees must have been found at not too great a distance from the site.

Much more difficult than in the case of wild fruits and nuts is the evaluation of the possible exploitation of other potential wild food plants, even if one considers only those species the seeds of which could have been used for human consumption. Thus, the great numbers of seeds of *Chenopodium album, Polygonum lapathifolium* and *Polygonum aviculare* in various samples could be interpreted as human food waste. There is convincing evidence that prehistoric man harvested the seeds of *Chenopodium album* and *Polygonum* species, in particular for Iron Age Denmark (Helbaek, 1951, 1960). However, *Stellaria media* and *Urtica dioica* are likewise represented by considerable numbers of seeds although it is unlikely that these seeds were eaten by the inhabitants of the site. At most the leaves of these species were used as vegetables. The seeds of *Nymphaea alba* (water-lily) occur rather regularly, albeit never in great numbers. However, it is unlikely that these fairly large seeds were collected on purpose. Of water-lily, it is the rhizomes, rich in starch, that have been eaten in northern Europe.

It will be clear that the seed evidence from Swifterbant does not provide clues as to the possible role of wild food plants in the diet of the inhabitants. It may only be concluded that various plants which could have contributed to the demand for carbohydrates were quite common in the Swifterbant area. Whether or not these species were exploited as such by the inhabitants of the site must remain unanswered. Another possible explanation for the great numbers of seeds of various plants will be discussed below (6.1.1.).

6. THE DISTRIBUTION OF SEEDS AND FRUITS IN THE SITE

6.1. The distribution of the seed frequencies

6.1.1. *Seed frequency distribution per layer*

In section 2 it has been stated that during the campaigns of 1974-1977, of each square metre parcel one litre of soil was examined for seeds in the field. (In the following the term seeds includes also fruits although from a morphological point of view this is not correct.) Depending on the numbers of seeds recovered from the samples four categories were distinguished. This procedure which had been set up with the aim of selecting samples for possible examination in the laboratory had an interesting additional result. It turned out not only that the numbers of seeds in the occupational fill vary considerably, as was already known, but also that areas of concentrations of seeds can be distinguished.

In figs. 11-15 the seed frequencies are plotted per layer. The seed frequencies are classified here into three groups, viz. no or few seeds (less than 10), fairly numerous (10-19 seeds) and numerous (20 and more seeds). A blank means that from that particular square no sample was examined for seeds. Figs. 11-15 point to the presence of distinct concentrations of seeds. Squares with a rather high seed content are not randomly distributed,

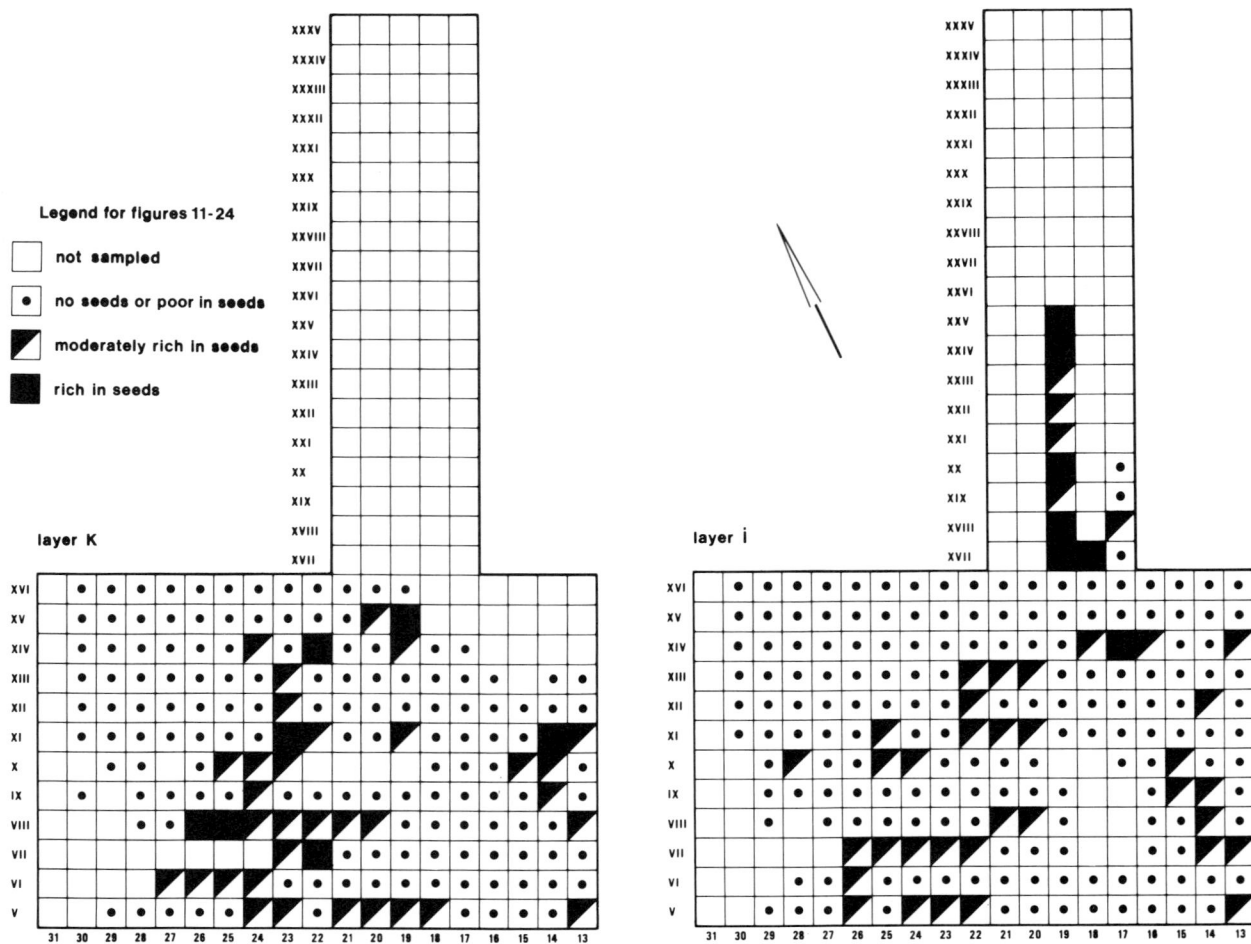

Fig. 11. Seed frequency distribution chart for layer K. For discussion of figs. 11-15, see 6.1.1.

Fig. 12. Seed frequency distribution chart for layer i.

but they show a kind of clustering. In each level areas of predominantly fairly numerous to numerous seeds constrast with those in which seeds are absent or scarce.

Although a fairly great number of seed types has been established for Swifterbant, of only rather few species were considerable quantities of seeds found: *Atriplex hastata/patula, Chenopodium album, Polygonum aviculare, Polygonum lapathifolium, Stellaria media, Urtica dioica*. It is the seeds of these species which largely determined the seed frequencies in the one litre samples. As for the explanation of the presence of the great numbers of seeds, in an earlier paper (Casparie *et*

al., 1977, p. 53) it has been advocated that the plants concerned had grown on the spot, that is to say, in the settlement itself. This would only have been possible if the site was not inhabited each year. In the case of seasonal habitation, the above would imply a non-annual return of the inhabitants to the summer occupation site. Leaving aside questions concerning seasonal or year-round habitation and possible temporary abandonment of the settlement, we now take another line for the explanation of the great numbers of seeds. If in periods of non-habitation the whole of the site was covered by weed vegetations mainly consisting of one or more the species mentioned above, one

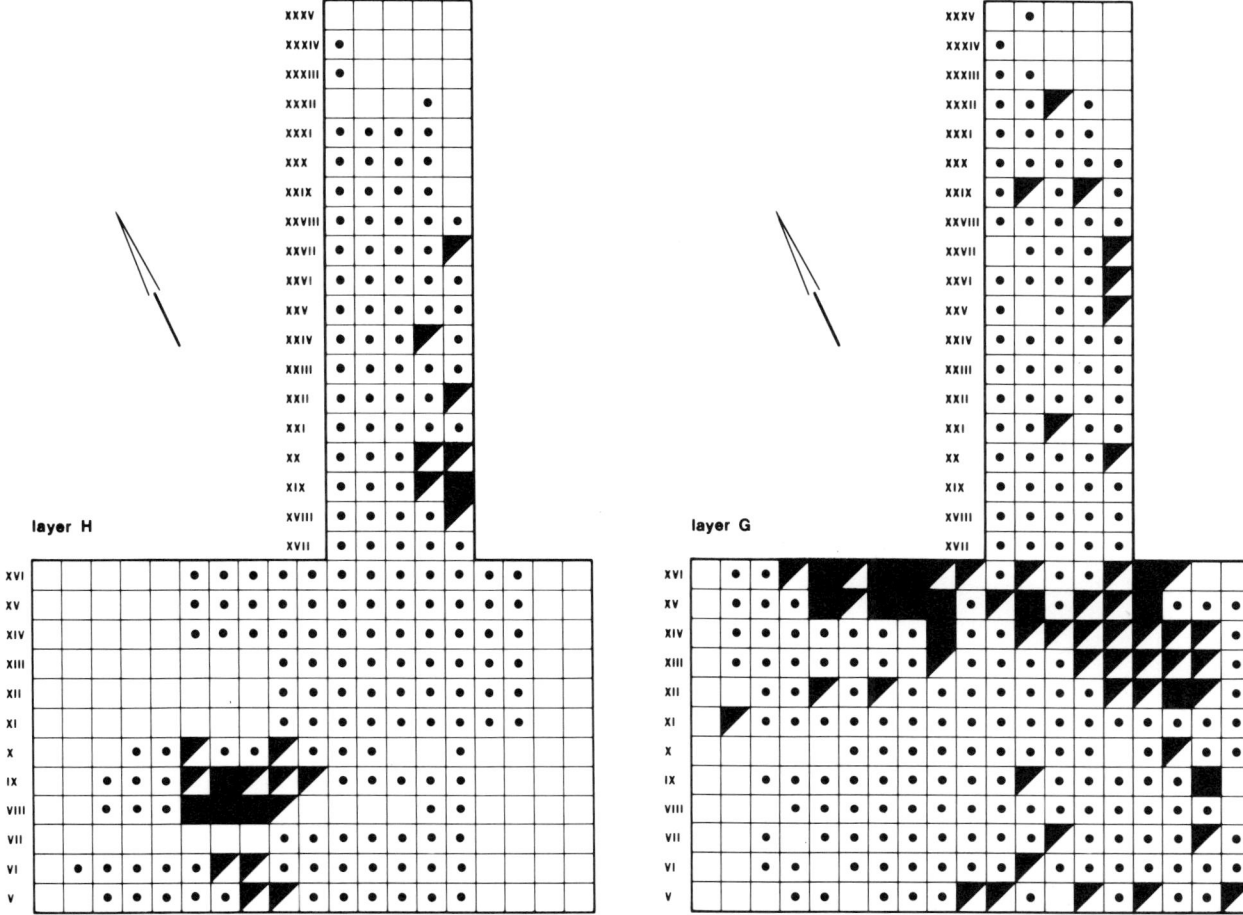

Fig. 13. Seed frequency distribution chart for layer H.

Fig. 14. Seed frequency distribution chart for layer G.

would have expected a less patchy distribution of the squares with higher seed concentrations. The seed frequency distribution patterns shown in figs. 11-15 indicate rather that man must have been responsible for the great quantities of these seed types. These seeds do not so much constitute the remains of human food, in other words, the seeds may not have been collected on purpose, but the plants concerned were dumped on the spot. In this connection it should be remembered that the culture layer consisted of clay with much organic material. Huge quantities of vegetable material must in the course of time have been carried in by the inhabitants of the site. In the field, fragments

of reed stems were frequently observed. It is self-evident that not only the plants of which great numbers of seeds were found were brought to the site, but also many other plant species. One may safely assume that most of the seeds and fruits arrived in the site adhering to the plants. At places where plants with a prolific seed production had been dumped the cultural fill is comparatively rich in seeds.

6.1.2. *Seed frequency distribution per 10 cm level*

If the inhabitants of the site were primarily responsible for the differences in the seed fre-

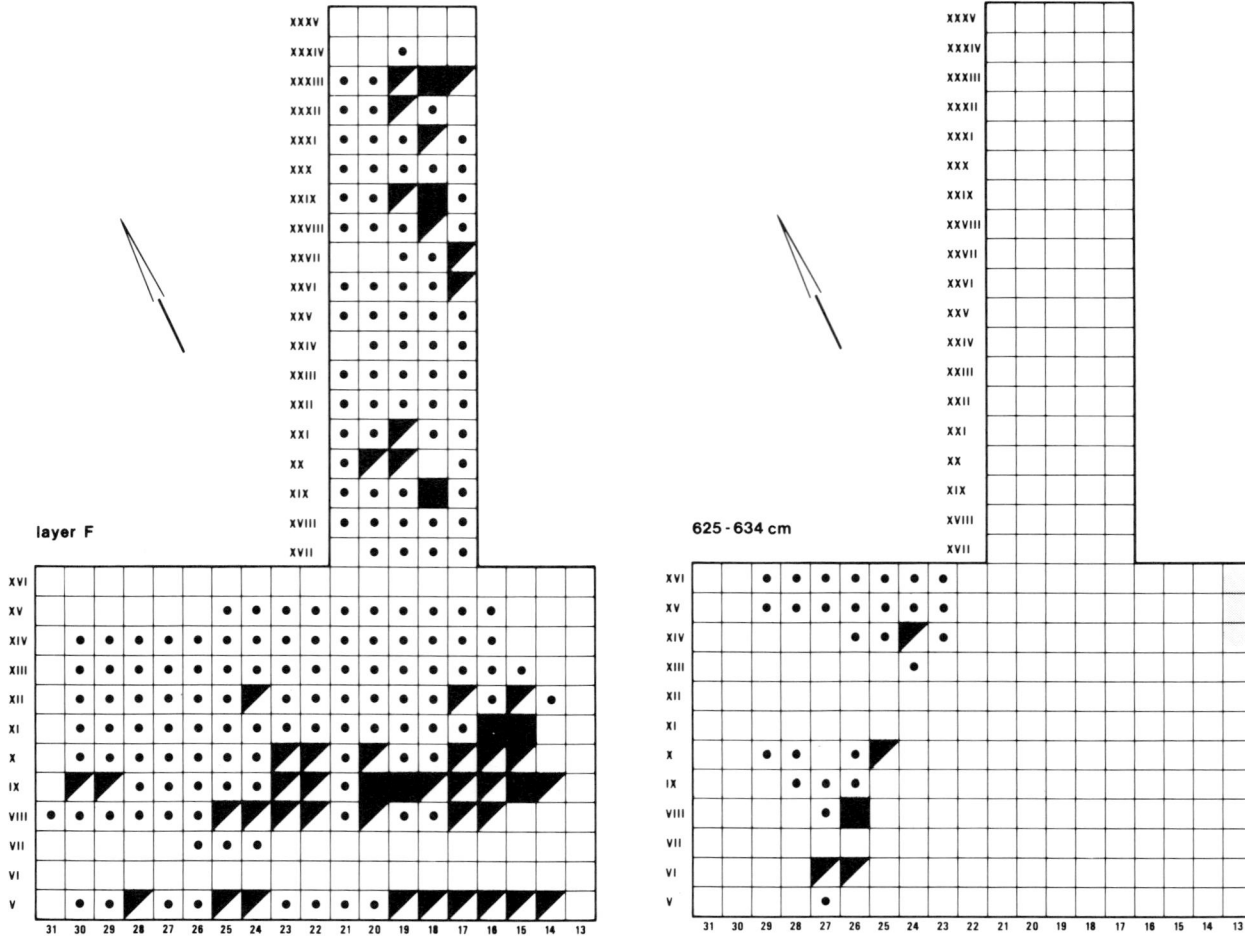

Fig. 15. Seed frequency distribution chart for layer F.

Fig. 16. Seed frequency distribution chart for 625-634 cm level. The contour lines delimit areas of concentrations of unburnt flint material (so-called activity areas). For discussion see 6.1.2.

quencies, albeit unintentionally, the question arises to what extent the seed frequency distribution patterns correlate with other indications of human activity. Although a great number of posts or remains thereof have been excavated it has not been possible to reconstruct ground-plans of houses. Consequently, no possible relations between house sites and areas of seed concentration can be determined. Better prospects in this connection are offered by the artifacts. For the flint material as well as for the pottery sherds, studied by Mr. P.H. Deckers and Mrs. J.P. de Roever, respectively, distribution maps have been prepared. The distribu-

tion of the unburnt flint material, which largely corresponds to the distribution of the burnt flint material and of the pottery, has been chosen for comparison with the botanical data. Mr. Deckers kindly placed the flint distribution data, which have not yet been published, at our disposal.

The distribution of the artifacts is given per 10 cm level, e.g. for the levels of 565-574, 575-584 cm etc. below N.A.P. (mean sea-level). This was possible because all finds had been registered three-dimensionally (Van der Waals & Waterbolk, 1976). The seed frequency distributions are per layer. Under ideal circum-

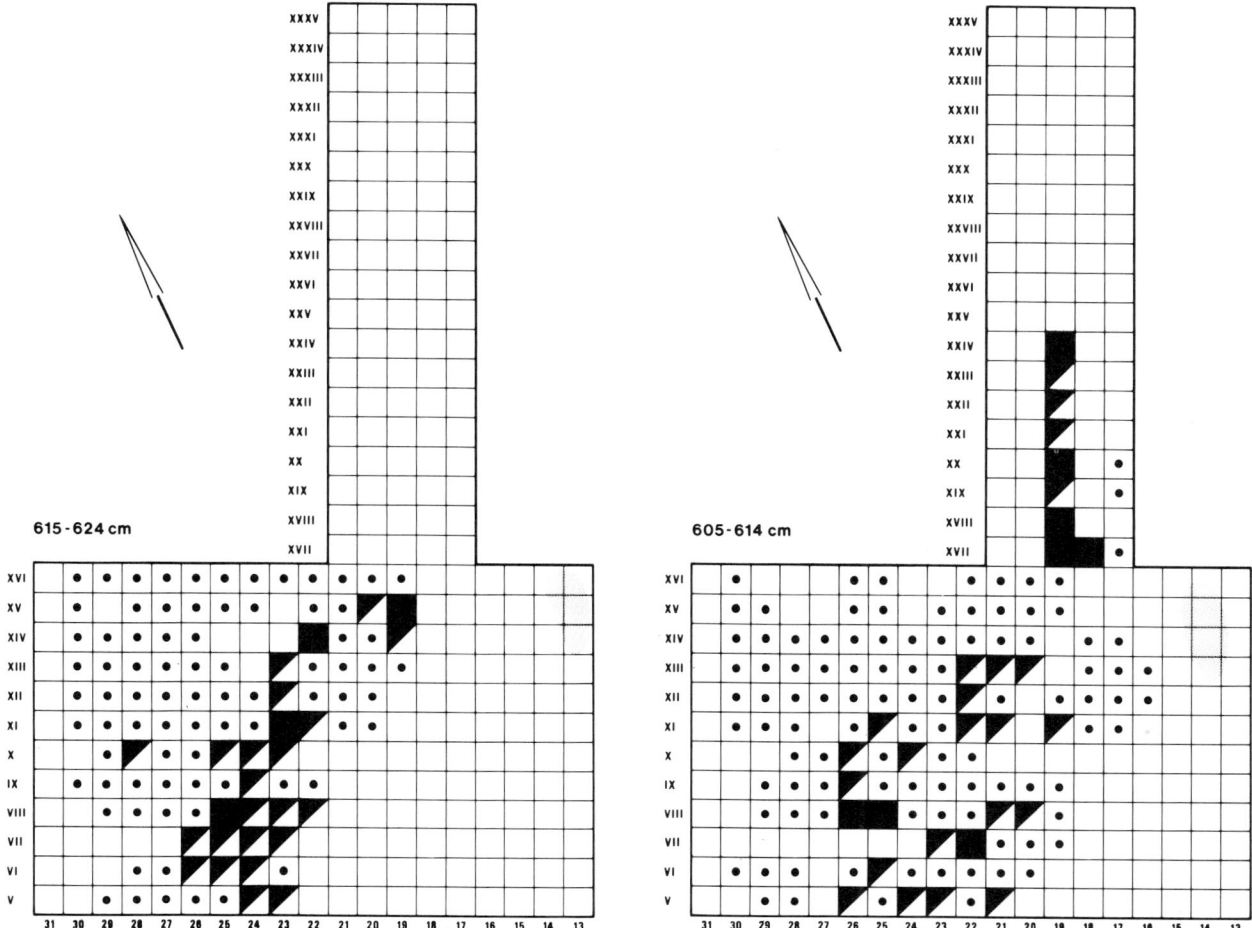

Fig. 17. Seed frequency distribution chart for 615-624 cm level. See caption of figure 16.

Fig. 18. Seed frequency distribution chart for 605-614 cm level. See caption of figure 16.

stances, the 10 cm levels could have been made corresponding with the layers employed in the field. However, these layers are often more than 10 cm thick and, moreover, the absolute height of one and the same layer may vary quite considerably over the excavated area. In this respect it should be mentioned that per metre square the upper and lower limits of each layer were levelled. For a comparison of the distribution of seed frequencies with that of the flint material it was necessary to convert the botanical data obtained for layers F to K to the 10 cm levels mentioned above. The re-

arrangement was performed on the basis of the average height (the mean of upper and lower limit) of each layer per metre square. Thus, this average height determines to which 10 cm level the sample concerned has been attributed. If of a layer only one height measurement is available (of the lowermost layer in general only the upper limit has been levelled), an estimation of the average height has been made. The seed frequency distributions per 10 cm levels are presented in figs. 16-24.

With respect to this re-arrangement a few remarks should be made. This procedure inevi-

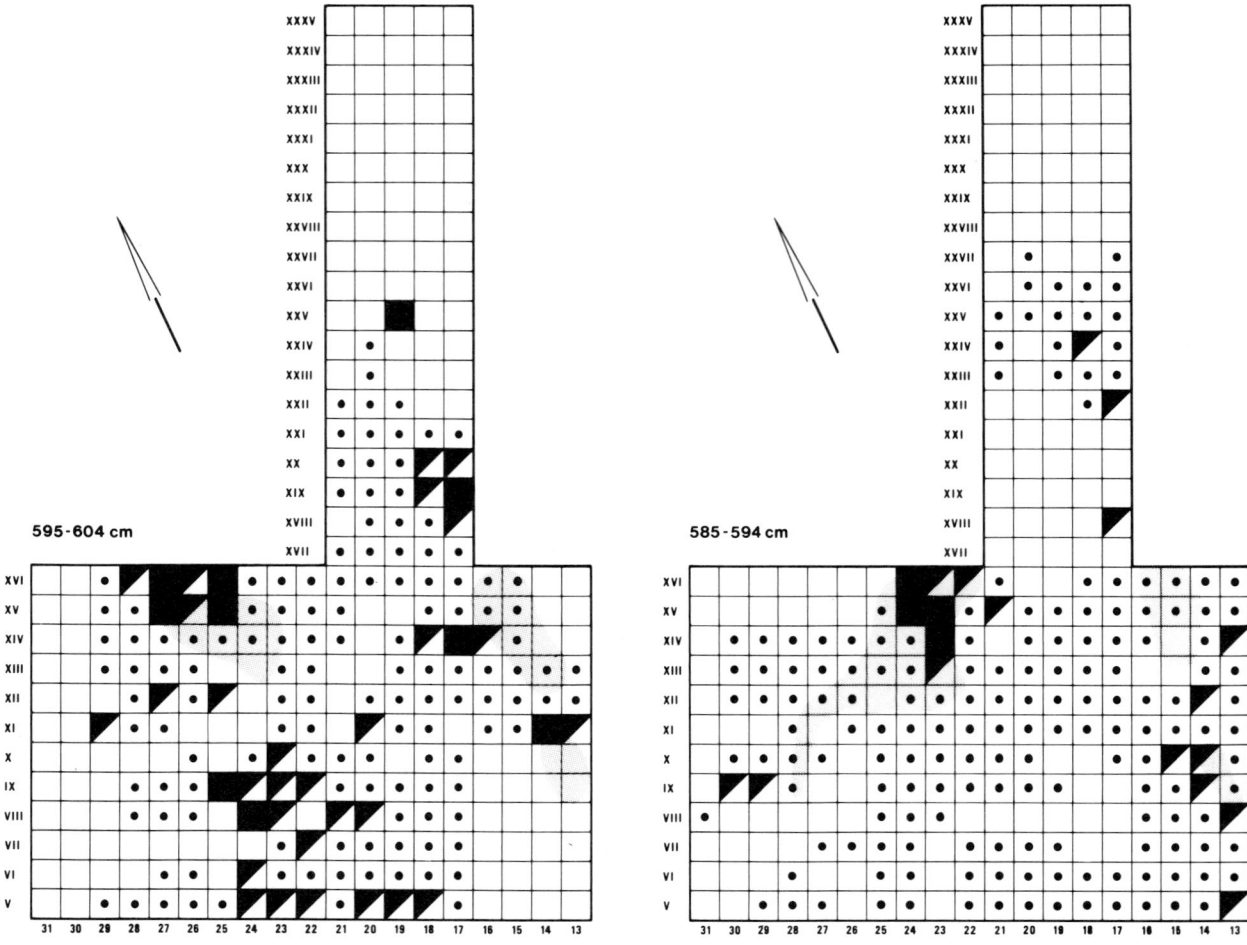

Fig. 19. Seed frequency distribution chart for 595-604 cm level. See caption of figure 16.

Fig. 20. Seed frequency distribution chart for 585-594 cm level. See caption of figure 16.

tably leads to inaccuracies and gaps in the seed record. Thus, it could happen that of two samples from adjacent squares in the same layer one ended up in another 10 cm level than the other, although the difference in average height was only 1 or 2 cm. The absolute height of the top of the subsoil and the thickness of the culture layer vary quite markedly. As a result, ten 10 cm levels are distinguished, whereas only five layers were employed in the field. It is self-evident that the splitting up must result in a considerable increase in the numbers of blank squares on the distribution charts. In some cases this has led to distribution patterns of examined and of non-examined metre squares which are real and quite informative (see 6.1.3.). In other instances blank squares constitute distinct gaps in the seed record, although happily they do not drastically reduce the significance of the distribution charts. Examples of the latter category are level 555-564 cm, squares 16-19, VI-XI (fig. 23) and level 565-574 cm, squares VI-VII, 20-25, VIII-XI, 19-21 (fig. 22). The data for levels 535-544 and 545-554 cm have been combined (fig. 24).

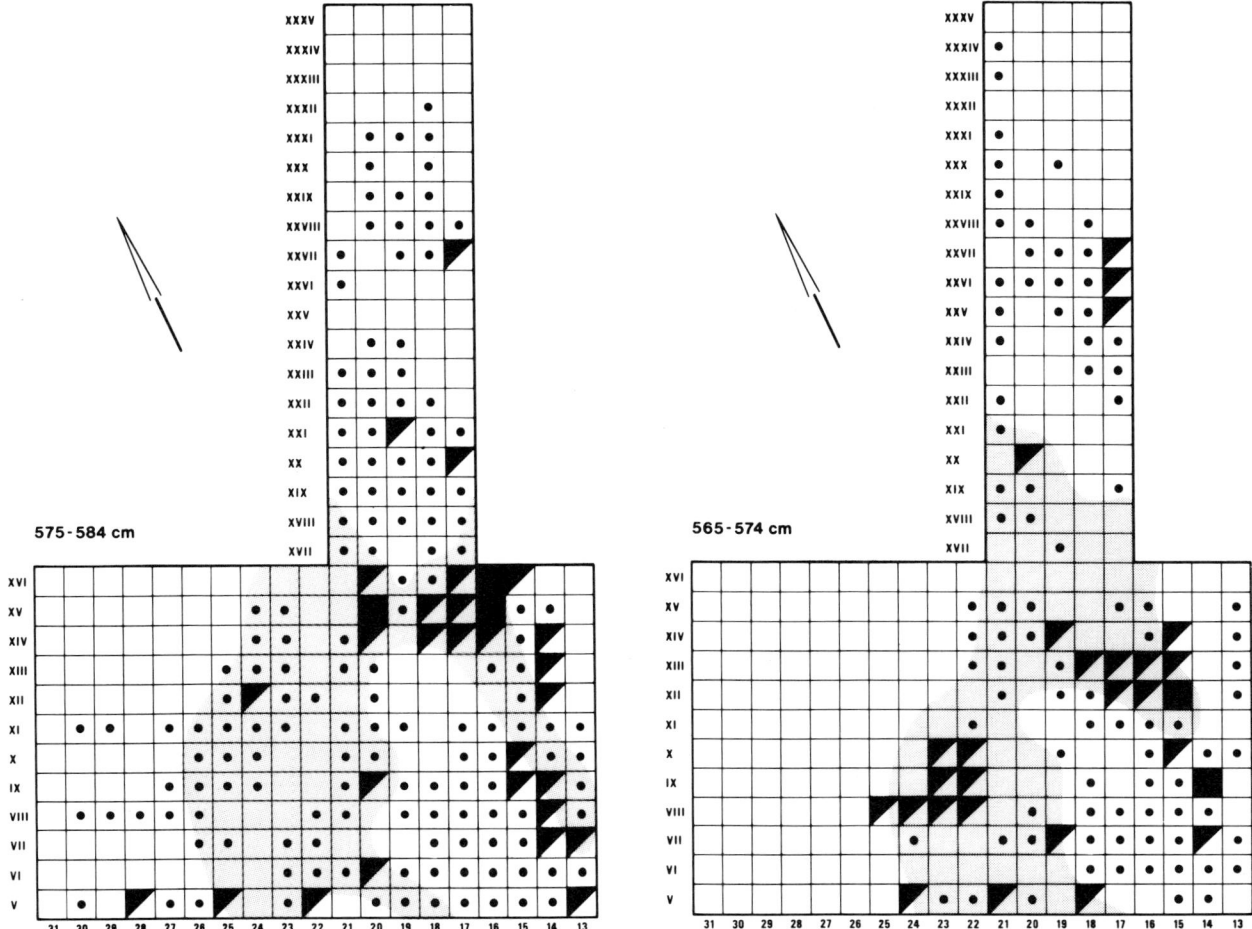

Fig. 21. Seed frequency distribution chart for 575-584 cm level. See caption of figure 16.

Fig. 22. Seed frequency distribution chart for 565-574 cm level. See caption of figure 16.

6.1.3. *Seed frequency distribution and activity areas*

The contour lines of the areas in which unburnt flint material was found, the so-called activity areas, are indicated in figs. 16-24 which show the seed frequency distributions. As for these activity areas, not all flint material is confined to these areas, but scattered specimens were also recovered outside them. However, inside the areas delimited by the contour lines flint material was distributed more or less continuously, and one can speak of flint concentration areas. From figs. 16-24 it appears that there is no correlation between activity areas and areas of high seed frequencies. In figs. 21-24 squares with high seed frequencies fall predominantly within the limits of the activity areas, whereas in figs. 17-19 distinct concentrations of seeds occur outside the areas of regular distribution of flint material.

With regard to the distribution of seed frequencies and flint material per 10 cm level, no correlation, either negative or positive, seems to exist. However, this by no means implies

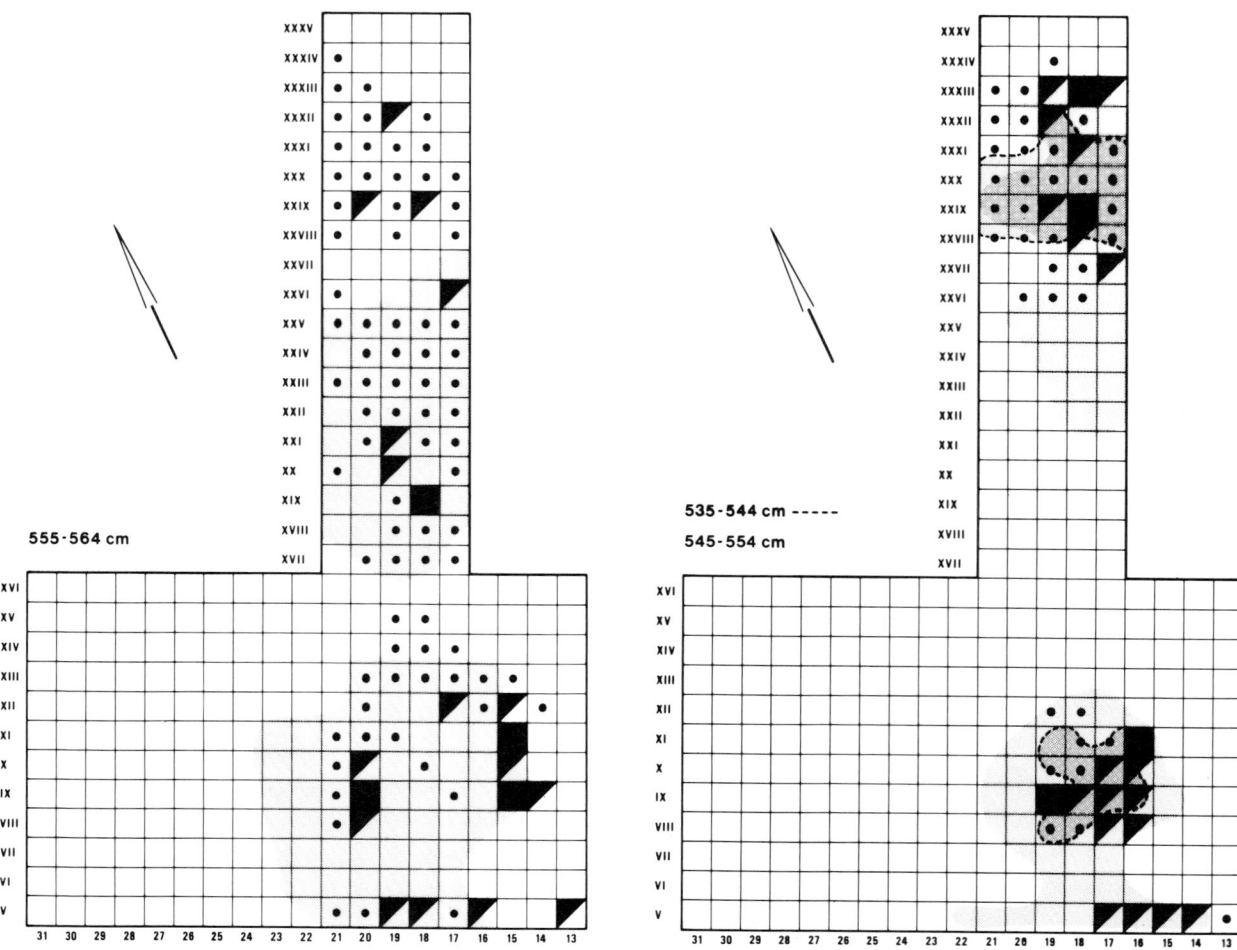

Fig. 23. Seed frequency distribution chart for 555-564 cm level. See caption of figure 16.

Fig. 24. Seed frequency distribution chart for 535-544 and 545-555 cm levels. See caption of figure 16.

that there would be no relation at all between the occurrence of seed concentration areas and the activity of man. Above (6.1.1.) it has been pointed out that the high seed frequencies must have been the result of the carrying in of plant material by the inhabitants of the site. However, all vegetable remains in the culture layer, irrespective as to whether seeds are frequent or not, have been brought in by man. For that reason, all squares examined in the field for seeds, and not only those with high to fairly high seed frequencies, will be considered in the following discussion. In this connection

it should be mentioned that all samples examined for plant remains are from dark occupational soil layers.

At the level of 625-634 cm (fig. 16) the distribution of the metre squares examined for seed frequencies shows two patches of culture soil. At the succeeding level (615-624 cm, fig. 17) the occupational fill covers a much greater area. The fact that no metre squares were examined from the southeastern part of the excavation area indicates that there the surface of the subsoil was higher; it was there that the highest part of the levee bordering the north-south oriented gully was found. At the lower-most levels factual evidence of the activity of man has, indeed, only been established for the presumably highest part of the creek ridge, although the flint concentration areas at the levels of 625-634, 615-624 and 605-614 cm (see figs. 16-18) are quite small. Moreover, according to Mr. P.H. Deckers (oral communication) the flints would not have lain there *in situ*. However, it is conceivable that the people who dumped plant material and perhaps other refuse in the areas indicated in figs. 16-18 nevertheless had their work-floor on the levee along the creek. For it was in this area that in an early stage of the habitation a part of the site had been washed away (Van der Waals, 1977; Deckers *et al.*, 1980; cf. section 1.3.). Be this as it may, the deposition of organic material must have resulted in a gradual filling up of the lower-lying area to the west of the levee. At least this is suggested by the expansion of the area covered by squares examined for seeds in an easterly direction, towards the crest of the levee.

At the level of 595-604 cm (fig. 19) accumulation of organic material had started also on the highest part of the levee. A flint concentration in squares XIV-XV/24-26 indicates that the inhabitants had extended their activities to the formerly low-lying area which by now had been filled up. In the succeeding level, 585-594 cm (fig. 20), the latter activity area had already expanded quite markedly. It is likely that the activity areas have always protruded to some extent above the rest of the site (see below), implying that within a 10 cm level the soil inside and near an acitvity area must have been deposited somewhat earlier than that at some distance from the flint concentration areas. At the level of 575-584 cm (fig. 21) the activity area in the southern section of the excavation area had reached its greatest extent. One may assume that from here the deposition of organic material in the northern section (XVII-XXXII/17-21) was carried out. In fig. 21 the blank squares in the middle of the northern section are mainly due to the fact

that here a layer of culture soil more than 10 cm thick was excavated per operation. The distribution of the metre squares examined for seeds in fig. 20, representing the underlying level of 585-594 cm, is nearly complementary to the area of blank squares in fig. 21.

The raising of the surface in the northern section of the excavation area was followed by a partial shift of the activity area to that section, which finally resulted in two separate flint concentration areas (figs. 23 and 24).

In the uppermost levels, metre squares examined for seeds are confined to both activity areas (fig. 24). This demonstrates clearly that these activity areas emerged above the rest of the site. At the levels of 545-554 and 535-544 cm (fig. 24), outside the areas of flint concentration the sediment consisted of clay deposited by the floods which brought to an end the habitation of the S3 site. In the southwest section of the area shown in figs. 16-24 this clay was found at the level of 555-564 cm (fig. 23) and perhaps already at the level of 565-574 cm (fig. 22). The differences in height of the base of the clay sealing off the culture layers suggest that in the final phases of the habitation the level of the work floors was 20 to 30 cm above that of the rest of the site. In this figure of 20-30 cm no allowance has been made for the possible effects of differential soil compaction and of local erosion of top layers of occupational fill.

6.2. The distribution of seed types

Table 1 shows that the seed composition of the samples examined in the laboratory varies considerably. There are great differences in relative frequencies as well as in absolute numbers of seeds. This variation in seed content gives occasion to two questions. In the first place one wonders whether there is any relation between seed composition and origin of the sample. Secondly, the question arises whether differences in seed content correlate with the stratigraphical position of the samples.

6.2.1. *Seed composition and origin of samples*

As for the question to what extent the seed
content of the culture soil may have depended
on the location within the site, it has been
examined whether a possible relation exists
with the activity areas discussed above. In figs.
25-33 the location of the samples is indicated
on the 10 cm level charts with the contour
lines of the activity areas (concentrations of
unburnt flint material). From these charts it
appears that some samples are clearly from
outside activity areas, particularly but not
exclusively in the lower levels. Other samples
originate from inside or close to areas of flint
concentration. It may be needless to say that
for the choice of the samples to be analysed in
the laboratory, the position with regard to
activity areas played no part. It was not until
the botanical examination had been finished
that this point of view was taken into consideration. For the discussion below the samples
are divided into two groups. The samples taken
from and near activity areas (group I) and
those from well outside these areas (group II)
are listed in table 16.

In re-arranging the results of the botanical
examination into both groups mentioned
above, some striking differences become
obvious. These differences are best visualized
in table 17, in which for both groups mean and
weighed percentages and sample frequencies
(see 4.1.) of a selected number of species are
shown. Only species represented in more than
10 (of the 46) samples are included in table 17,
except *Arctium* cf. *lappa* which was found in
only 8 samples. From this table it is clear that
the seeds of some species are most frequent in
samples from outside the activity areas, whereas other species are much better represented in
the samples from inside or near flint concentrations. The species which show a clear preference for one of both groups of samples are
framed by a solid line. In determining whether
a species may be considered to be more or less
"characteristic" of one of both groups of
samples, the weighed percentage (the relative
importance) played a decisive part. In this

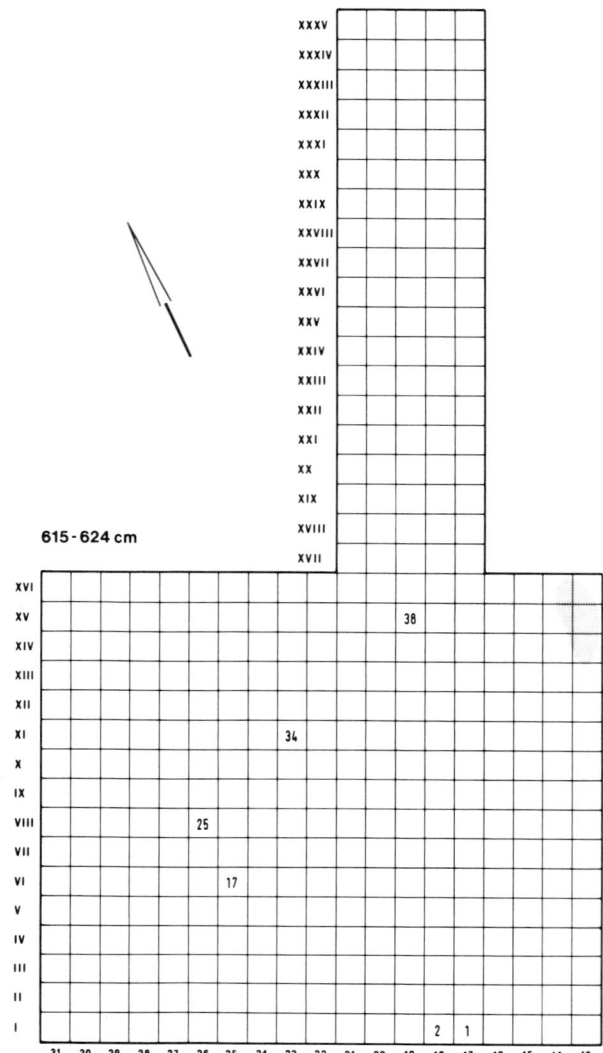

Fig. 25. 615-624 cm level. Location of samples examined
(table 1) and flint concentration areas. For discussion of
figures 25-33, see 6.2.1. and 6.2.2.

connection it should be mentioned that in
general no weighed percentages between
species should be compared, because this figure
depends, among other things, on the production and dispersal of seeds which vary greatly
among the various taxa. On the other hand,
weighed percentages of the same species may
provide useful information on possible differences between groups of samples.

The differences between both groups of

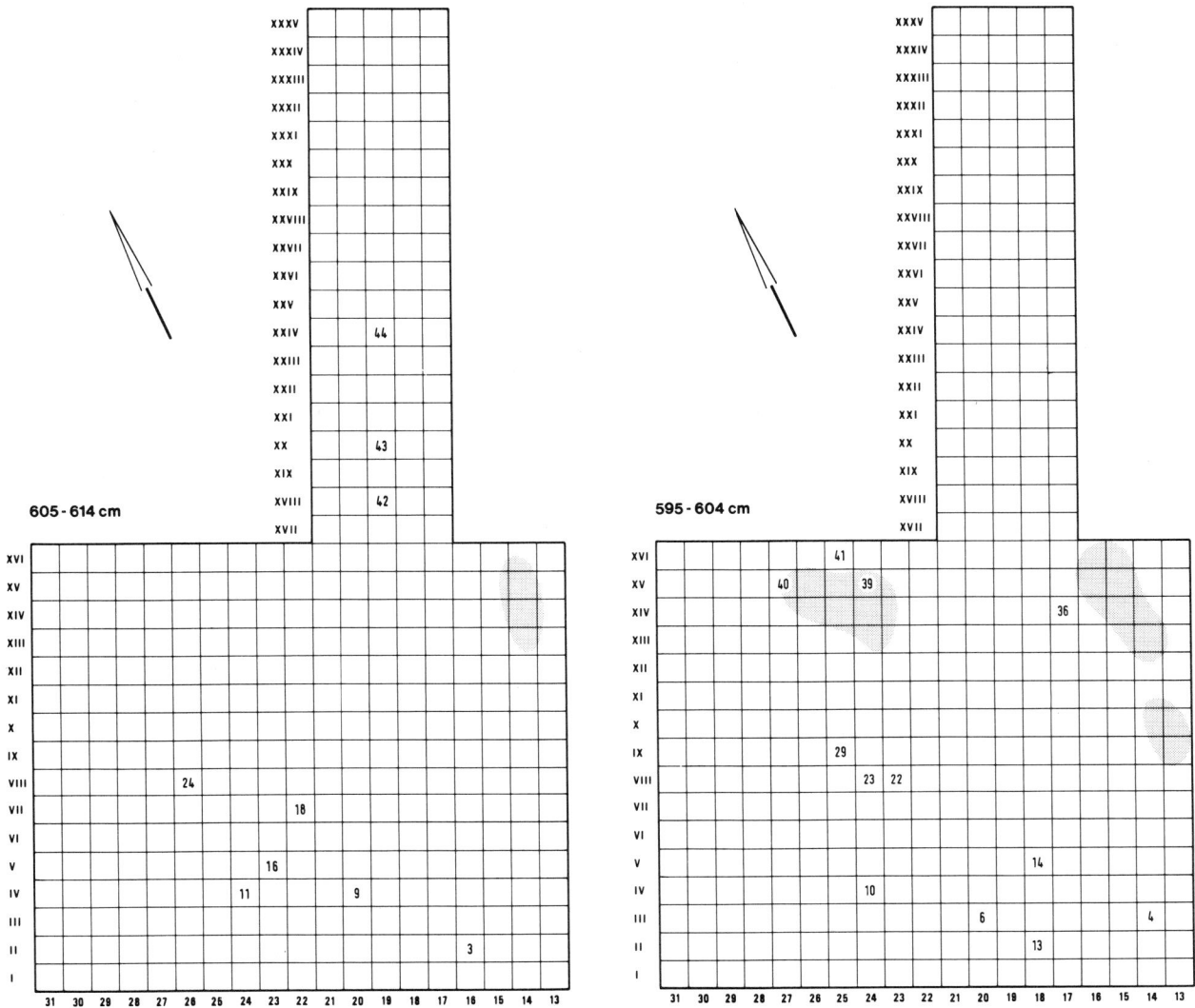

Fig. 26. 605-614 cm level. Location of samples examined (table 1) and flint concentration areas.

Fig. 27. 595-604 cm level. Location of samples examined (table 1) and flint concentration areas.

samples in table 17 are not always of identical nature. Thus, *Chenopodium album*, *Solanum nigrum* and *Urtica dioica*, the seeds of which are by far most numerous in the samples from activity areas, show high sample frequency values also in group II, the differences being determined by the much lower mean and weighed percentages. On the other hand, the representation of *Polygonum* species in both groups of samples differs not only in terms of

mean and weighed percentages, but these species score also low sample frequencies in group I. Very striking is the behaviour of *Nymphaea alba* which is fairly well represented in group II, but not a single seed of which was found in the samples from activity areas.

The numbers of seeds per 3 litres of soil are, on an average, significantly greater in the samples from outside activity areas than in those from group I. The same applies to the numbers

Table 17. Representation of some selected species in samples from inside or near flint concentration areas (group I) and from outside these areas (group II). For discussion see 6.2.1.

	Group I			Group II		
	sample frequency N = 19	mean percentage	weighed percentage	sample frequency N = 27	mean percentage	weighed percentage
Chenopodium album	100	27.70	27.70	70	1.83	1.29
Hordeum vulgare nudum	89	2.49	2.23	48	0.89	0.43
Solanum nigrum	84	3.61	3.04	81	1.09	0.88
Urtica dioica	95	46.01	43.63	100	17.00	17.00
Cladium mariscus	32	4.30	1.36	26	0.73	0.19
Scirpus maritimus	21	1.48	0.31	48	0.34	0.16
Scirpus tabernaemontani	53	4.71	2.48	67	1.58	1.05
Conium maculatum	47	1.43	0.68	11	0.17	0.02
Mentha aquatica	21	0.55	0.12	30	0.46	0.14
Poa pratensis/trivialis	11	4.50	0.47	33	1.75	0.59
Solanum dulcamara	32	1.78	0.56	19	0.48	0.09
Atriplex hastata/patula	74	5.06	3.73	93	14.97	13.86
Polygonum aviculare	11	0.50	0.05	93	16.88	15.63
Polygonum lapathifolium	21	3.35	0.71	85	22.90	19.51
Polygonum persicaria	5	0.6	0.03	70	3.25	2.29
Stellaria media	84	6.74	5.68	78	16.39	12.75
Arctium cf. lappa	—	—	—	30	0.38	0.11
Nymphaea alba	—	—	—	74	1.00	0.74
Phragmites australis	37	3.25	1.20	89	11.78	10.47
	min.	mean	max.	min.	mean	max.
number of seeds	23	1269	10317	109	3431	10402
number of species	8	11.4	17	7	15.5	28

Table 18. Dominance of one or two species (for explanation see 6.2.1.).

Group	I	II
Phragmites	•	2
Atriplex	•	1
Atriplex/Polygonum aviculare	•	1
Polygonum aviculare	•	1
Polygonum lapathifolium	•	6
Stellaria	•	1
Stellaria/Urtica	•	3
Urtica	8	1
Urtica/Chenopodium	6	•
no dominance	5	11

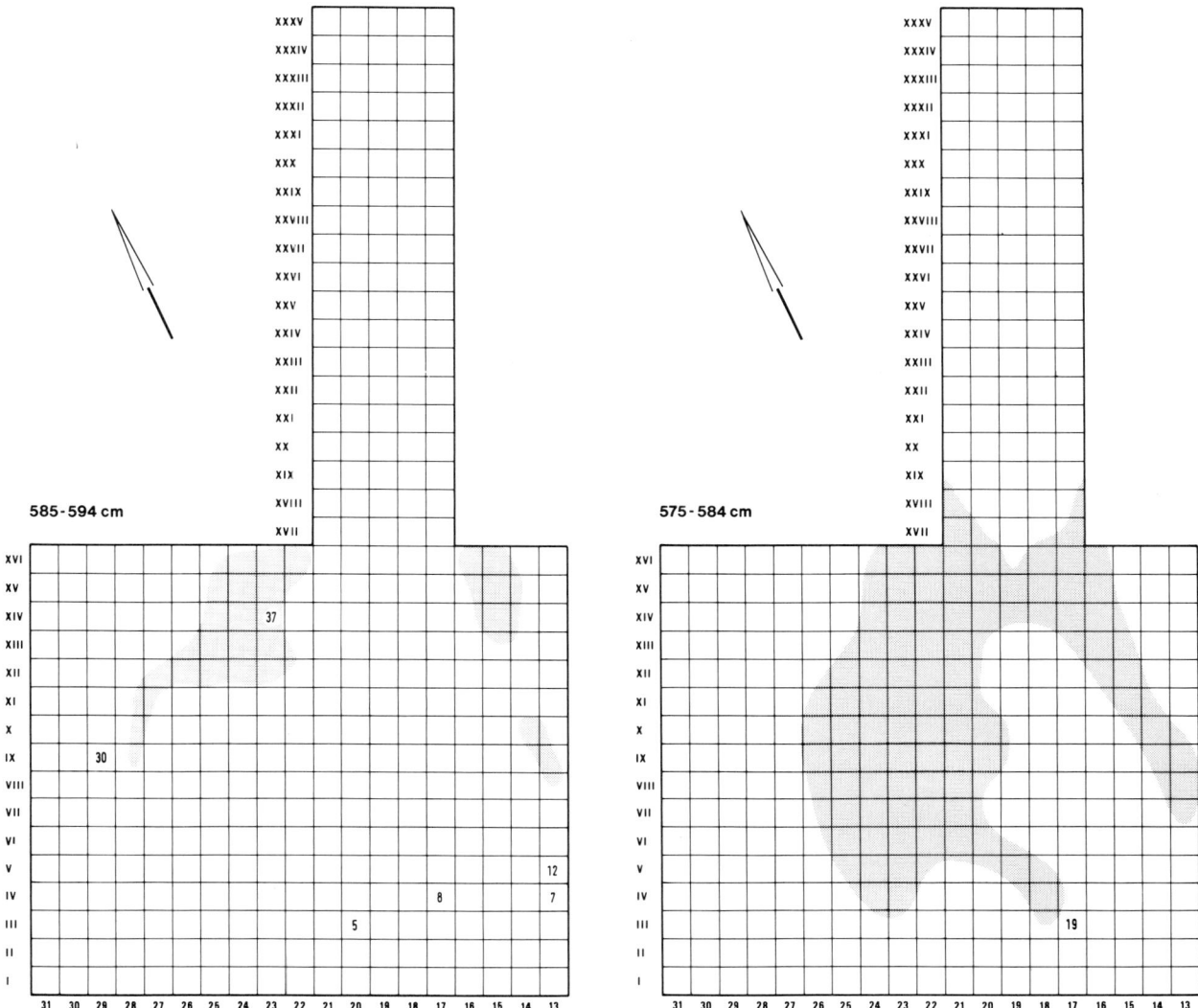

Fig. 28. 585-594 cm level. Location of samples examined (table 1) and flint concentration areas.

Fig. 29. 575-584 cm level. Location of samples examined (table 1) and flint concentration areas.

of species represented per sample. Thus, the soil from flint concentration areas is poorer in absolute numbers of seeds as well as in species. Differences between both groups of samples also find expression in the following approach. It has been determined for which samples there is a question of dominance of one or at most two species. If one species accounts for more than 50% of the seeds in a particular sample, this sample is considered to be dominated by

that species. If no species is represented by more than 50%, but two species score together more than 70%, the sample is dominated by these two species. As is shown in table 18 various species are dominant in one or more samples, either alone or in combination with another species. Table 18 shows also very marked differences in dominant species or species combinations between both groups. As a matter of fact, only the *Urtica* dominance

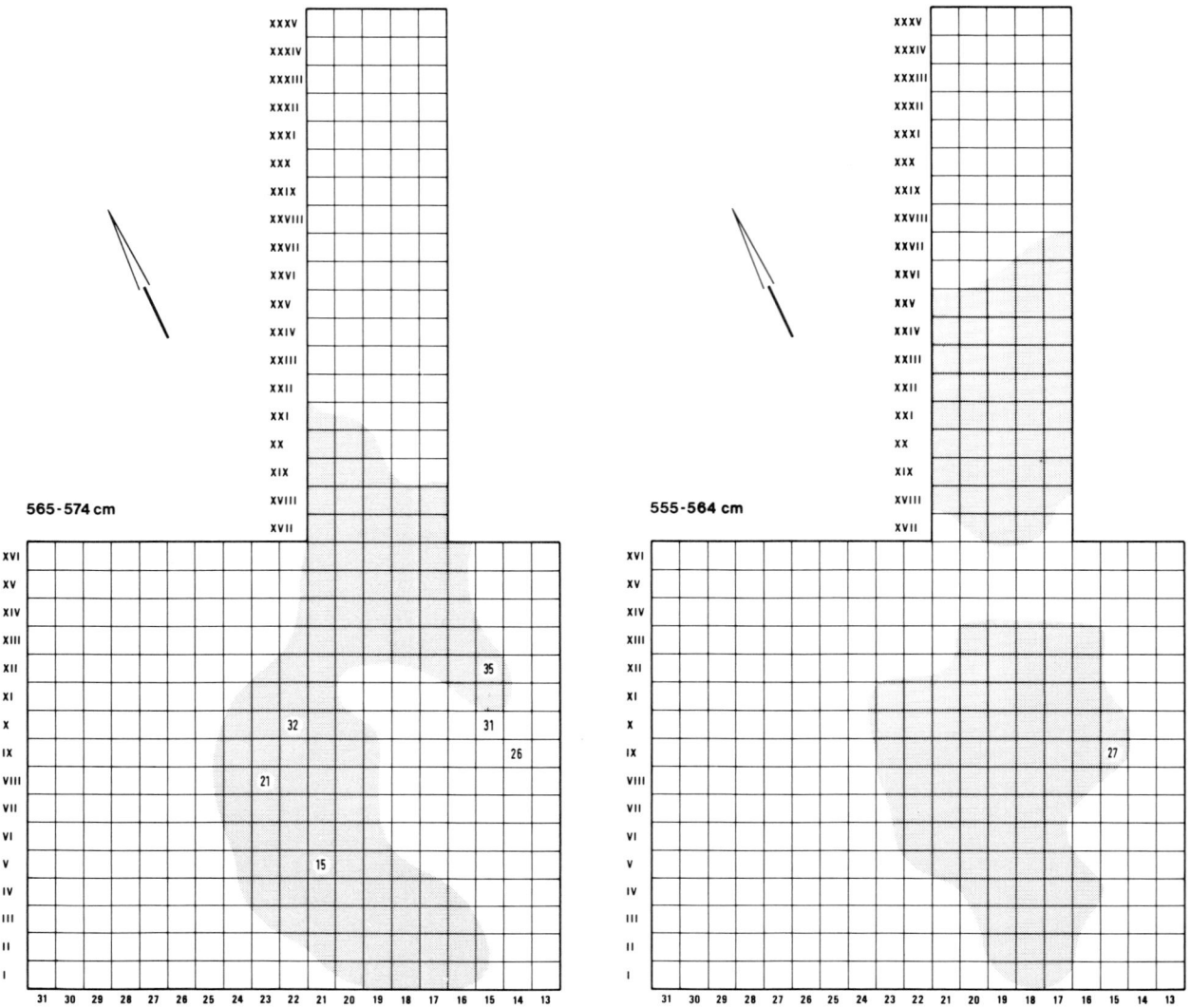

Fig. 30. 565-574 cm level. Location of samples examined (table 1) and flint concentration areas.

Fig. 31. 555-564 cm level. Location of samples examined (table 1) and flint concentration areas.

is found in both groups, and here there is a conspicuous difference in frequency: common in group I and present only once in group II. Moreover, in more samples of group II (41%) no dominance of one or two species occurs than in those of group I (26%). From table 18 the predominant role of *Urtica* in the samples from activity areas is again clearly apparent.

Thus, tables 17 and 18 demonstrate, along different lines, the differences in seed content

between both groups of samples. Illustrative of these differences may also be the fact that of the samples of group II, only one (no. 12: V-13-K) should on the basis of the seed content rather belong to group I. Now, the question forces itself, as it were, as to how far the observed differences can be explained. The differences in the numbers of seeds per soil volume could be due to differences in the conditions for preservation. The activity areas

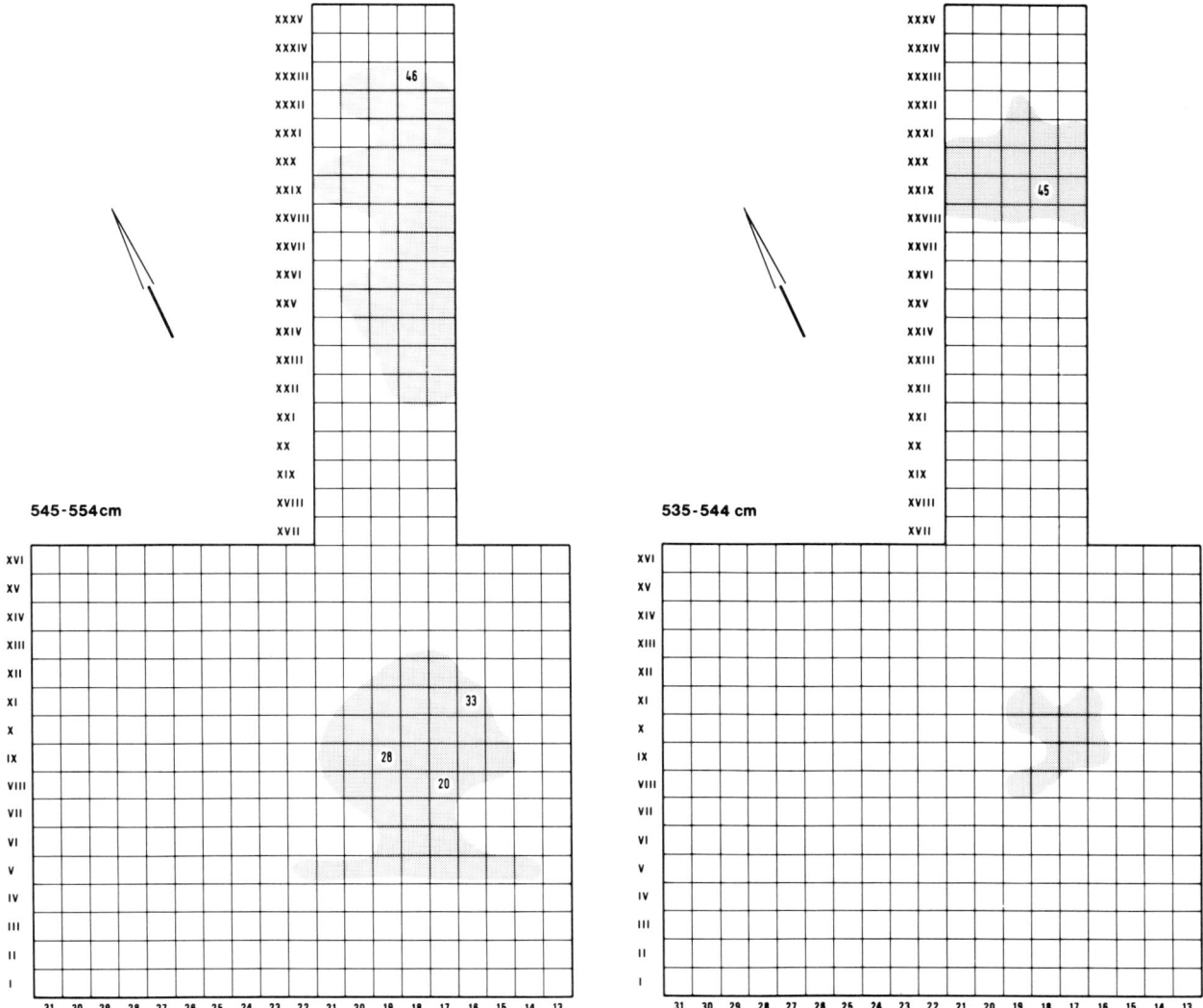

Fig. 32. 545-554 cm level. Location of samples examined (table 1) and flint concentration areas.

Fig. 33. 535-544 cm level. Location of samples examined (table 1) and flint concentration areas.

protruded above the rest of the site and were consequently better drained, implying slightly drier conditions. Moreover, in the activity areas treading must have been more severe than in other places. These factors could have resulted in much greater losses of seeds (breakage, corrosion, biological destruction) in activity areas than elsewhere within the site. However, in general the differences in the relative frequencies of the seeds cannot be explained in

this way. One could imagine that the very good representation of *Chenopodium album* in group I was due to the fact that even under rather poor conditions the seed wall of this species remains preserved. However, in that case it is difficult to understand why *Atriplex* does not equally show its highest frequencies in group I. The seeds of *Solanum nigrum* are not particularly sturdy, but nevertheless this species is best represented in group I. The seed wall of *Nymphaea alba* is rather thick, but curiously water-lily is not represented in group I. Only in the case of *Phragmites australis*, one could argue that the differences in mean frequencies between both groups were due to preservation. In the activity areas, the small, thin-walled seeds of reed could particularly have suffered from the assumed less favourable preservation conditions. Although it is very well possible that differential preservation of seeds had some effect, one must assume that the observed differences in the seed composition are largely of primary nature, due to the action of the inhabitants of the site.

Unfortunately, the above conclusion is not very helpful in understanding the why and wherefore. The apparent preference of barley grains (*Hordeum vulgare nudum*) for activity areas is not particularly surprising as this is an unmistakable food plant. One could perhaps argue that the high frequencies of *Chenopodium album* in group I samples are connected[1] with the intentional harvesting of these seeds for human consumption. However, in that case one wonders why knotweed (*Polygonum*) species, the seeds of which are likewise known as potential food of prehistoric man (cf. 5.4.), are by far better represented in the samples from outside activity areas. The berries of black nightshade (*Solanum nigrum*) are poisonous, and of stinging nettle (*Urtica dioica*) the leaves and not the seeds are consumed. It is striking that not a single seed of water-lily (*Nymphaea alba*) was recovered from the group I samples, whereas this plant is rather well represented in the samples of group II. It is true that of this species not the seeds but the root-stocks are a well-known wild food, but if water-lily had been prepared by the inhabitants of the site in the activity areas, one would have expected at least some seeds in group I samples.

As has been discussed above (6.1.1.) it is very likely that by far most of the seeds had arrived in the settlement adhering to the plants, which were deposited there to raise the level of habitation and to fill up low-lying areas. But this assumption leaves one with the question why, for instance, *Chenopodium album* and

Urtica dioica were so much preferred for raising the level of the activity areas, whereas *Atriplex* and various *Polygonum* species particularly were dumped in other parts of the settlement. In conclusion, no hypothesis can be brought forward to explain the striking differences in the seed content of the culture soil inside and outside the activity areas.

6.2.2. *Seed composition and stratigraphical position of samples*

One wonders to what extent the differences in seed composition discussed above (6.2.1.) relate to the stratigraphical position of the samples. Moreover, in the course of the habitation the vegetation in the vicinity of the site may have changed to some degree, which could have found expression in the composition of the plant material brought to the settlement. These considerations occasioned us to examine whether changes in the seed composition between lower and upper layers occur, irrespective of the origin of the samples (from an activity area or not). To that end, it was first necessary to establish the chronostratigraphical sequence of the samples examined. The (assumed) stratigraphical position of the samples is shown in table 19. As for the sequence suggested in this table, the following should be remarked. The basis of determining the most likely chronostratigraphical sequence is formed by the 10 cm level charts of figs. 25-33 on which the locations of the samples are indicated. The stratigraphical order of the samples from outside flint concentration areas (to the left in table 19) is based upon the successive 10 cm levels from which the samples originate. This may lead to inaccuracies due to the fact that the layers employed in the field do not coincide with the 10 cm levels (6.1.2.). Also if samples are scattered over a large area they may not all be of the same period although they are from the same 10 cm level. Samples at some distance from most of the others in a 10 cm level are shown in brackets.

The same procedure was applied to the samples from or near activity areas (table 19,

Table 19. Chronostratigraphical sequence of samples.

		Group II outside activity areas		Group I in/near activity areas	
younger				20, 28, 33, (46)	(545-554 cm)
				27	(555-564 cm)
				15, 21, 32, 26, 31, 35	(565-574 cm)
				19	(575-584 cm)
	(585-594 cm)	5, 8, 7, 12			
	(595-604 cm)	4, 6, 10, 13, 14, 22, 23, 29			
older	(605-614 cm)	3, 9, 11, 16, 18, 24, (42), (43), (44)		30, 37	(585-594 cm)
	(615-624 cm)	17, 25, 34, 38, (1), (2)		39, 40, 41, (36)	(595-604 cm)

to the right). One may assume that the stratigraphical sequences established for both groups of samples are, in broad outline, correct. However, how do both sequences correlate? A complicating factor in this respect is the fact that the activity areas are supposed to have emerged above the rest of the site, implying that there is no synchrony of the 10 cm levels in and outside the areas of flint concentrations. Fortunately, there is at least one reliable stratigraphic datum for correlating both sequences. At the levels of 575-584 and 565-574 cm (figs. 29 and 30) the activity area expanded considerably in the southern section of the site. Here the fill of the flint concentration area must definitely be younger than the underlying culture soil. This implies that samples 19 (575-584 cm) and 15, 21, 32, 26, 31, 35 (565-574 cm) are younger than samples 5, 8, 7, 12 (585-594 cm), from directly underneath the activity area fill. There remains the chronostratigraphical relation of activity area samples 36, 39, 40, 41 (595-604 cm) and 30, 37 (585-594 cm) to the samples at the left side of table 19. The assumption that the level of the activity areas was higher than that in the rest of the site implies that samples 30 and 37 (from a flint concentration area) must chronologically be placed below samples 5, 8, 7, 12 from the same level (585-594 cm). In their turn samples 30 and 37 lie stratigraphically above samples 39, 40 and 41, from the underlying level in the same activity area (fig. 27). The latter samples

must chronostratigraphically be placed below samples 4 to 29 from the same 10 cm level (595-604 cm). It is not possible to determine exactly where the 6 stratigraphically lowermost samples from activity areas should fit into the sequence established for the samples of group II. In table 19, this uncertainty finds expression by a broken line indicating the possible chronostratigraphical range of these 6 samples.

As appears from table 19, most of the samples from activity areas examined for plant remains are chronostratigraphically younger than those of group II. Only 6 samples of group I fall within the time range of the samples from outside flint concentration areas. It is clear that these 6 samples should be particularly informative with regard to the question of a possible relation between seed content and stratigraphical position of the samples. In this connection it must be remembered that it was to be examined whether the differences in seed composition between both groups of samples should rather be ascribed to the relative age of the samples and not so much to the origin (from an activity area or not).

To determine possible time-stratigraphic changes in the seed content of the occupational deposits, for both groups the samples have been divided into those from lower and upper levels. For group I it was obvious to combine the 6 lowermost samples to one sub-group (lower levels), although this implies that the other sub-group has a much greater number of

Table 20. See caption of Table 17. Within both groups of samples a subdivision is made between samples from lower and upper levels. For discussion see 6.2.2.

	Group I						Group II					
	upper levels			lower levels			upper levels			lower levels		
	N = 13 sample frequency	mean %	weighted %	N = 6 sample frequency	mean %	weighted %	N = 12 sample frequency	mean %	weighted %	N = 15 sample frequency	mean %	weighted %
Chenopodium album	100	31.01	31.01	100	20.53	20.53	75	1.83	1.38	67	1.82	1.21
Hordeum vulgare nudum	100	2.98	2.98	67	0.90	0.60	67	0.79	0.53	33	1.06	0.35
Solanum nigrum	85	3.76	3.18	83	3.28	2.67	75	1.59	1.12	87	0.75	0.65
Urtica dioica	92	44.16	40.76	100	49.85	49.85	100	20.21	20.21	100	14.43	14.43
Cladium mariscus	38	4.72	1.82	17	2.2	0.37	33	1.18	0.39	20	0.10	0.02
Scirpus maritimus	15	2.30	0.35	33	0.65	0.22	58	0.39	0.23	40	0.28	0.11
Scirpus tabernaemontani	46	1.40	0.66	67	9.63	6.42	100	1.97	1.97	40	0.80	0.32
Conium maculatum	62	1.56	0.96	17	0.4	0.07	8	0.3	0.03	13	0.10	0.01
Mentha aquatica	23	0.40	0.09	17	0.1	0.02	25	1.07	0.27	33	0.10	0.03
Poa pratensis/trivialis	15	4.50	0.70	–	–	–	33	0.40	0.13	33	2.84	0.95
Solanum dulcamara	38	1.94	0.75	17	1.0	0.16	17	0.40	0.07	20	0.53	0.11
Atriplex hastata/patula	77	3.92	3.02	67	7.93	5.28	83	19.59	16.33	100	11.89	11.89
Polygonum aviculare	8	0.2	0.02	17	0.8	0.13	83	12.35	10.29	100	19.91	19.91
Polygonum lapathifolium	31	4.95	1.52	–	–	–	92	6.69	6.13	80	37.76	30.21
Polygonum persicaria	8	0.6	0.05	–	–	–	75	2.71	2.03	67	3.74	2.49
Stellaria media	85	5.78	4.89	83	8.86	7.38	67	18.38	12.25	87	12.75	11.05
Arctium cf. lappa	–	–	–	–	–	–	25	0.60	0.15	33	0.24	0.08
Nymphaea alba	–	–	–	–	–	–	75	1.56	1.16	73	0.55	0.40
Phragmites australis	23	3.20	0.74	67	3.30	2.20	83	23.92	19.93	93	3.10	2.89
	min.	mean	max.	min.	mean	max.	min.	mean	max.	min.	mean	max.
numbers of seeds	23	1394	10317	150	999	2526	109	1848	8703	226	4698	10402
numbers of species	8	11.5	16	8	11.2	17	11	16.5	22	7	14.7	28

samples and covers a thicker culture layer. A subdivision of the samples of group II is laid between levels 605-614 and 595-604 cm (see table 19). In table 20, the sample frequencies, mean and weighed percentages are presented of the same species as in table 17, but now for the sub-groups. For group II, table 20 shows some differences in the seed composition of lower and upper levels. Most striking are the much higher frequencies of *Phragmites* in the samples from the upper levels. Does this mean that in later stages of the habitation more reed was brought in or could it have something to do with the time of the year when the reed was cut? Similarly no obvious explanation can be presented for the lower frequencies of *Polygonum aviculare* and *P. lapathifolium* in the upper levels. Another striking difference between the samples from lower and upper levels in group II is constituted by the mean numbers of seeds per unit of soil volume. The samples from the lower levels contain, on an average, 2½ times more seeds than those from the upper levels. Above (6.2.1.) it has been argued that the lower numbers of seeds in samples from activity areas, as compared to those from outside flint concentration areas, could be due to poorer preservation conditions. If this hypothesis is correct it would imply that in the upper levels of the culture soil outside the activity areas, conditions for the preservation of seeds were less favourable than in the lower levels. The fact that precisely in those upper levels *Phragmites* seeds show the highest frequencies invalidates in no small measure the suggestion that the poorer representation of reed in the samples from activity areas was due to poorer preservation conditions (6.2.1.). Again, more problems are raised than are solved in attempting to analyse the results of the botanical examination.

As for the samples from flint concentration areas (group I), no conspicuous differences are found between those from lower and upper levels, taking into account that the sub-group "lower levels" consists of only 6 samples. Particularly interesting is the fact that these 6 samples not only compare well with those of the other samples from activity areas, but also that they differ markedly from the largely synchronous samples from outside flint concentration areas. This demonstrates that the differences in seed content established between both groups of samples are related primarily to the location of the samples with respect to the activity areas. The chronostratigraphical position of the samples plays at most a minor part in the composition of the seeds. Consequently, one must assume that certain plants were preferred for raising the level of the activity areas, whereas the plant material deposited elsewhere in the site was of a different composition. It remains puzzling what may have determined the preference.

7. ACKNOWLEDGEMENTS

The authors kindly acknowledge the co-operation of various people in the study presented in this paper. First of all, we wish to express our gratitude to Mr. P.H. Deckers for his tremendous help in our attempts to integrate the palaeobotanical data into an archaeological context. He made the distribution maps of flint artifacts he had prepared available to us for comparison with the palaeobotanical results and he allowed us to present his so far unpublished data in this publication (figs. 16-33). We profited also a great deal from his comments on the first draft of section 6 of this paper.

The identifications of the moss remains by Dr. H.J. During (Utrecht) constitute a highly valuable contribution to the palaeobotany of Swifterbant. Mr. D. Kielman provided us with the list of archaeological features observed in the field. Dr. S. Bottema and Professor J.D. van der Waals critically read the manuscript, pointed at inaccuracies and inconsistencies, and made suggestions for improving the text. The drawings were executed by Mr. H.R. Roelink and Mr. J. Klein. Mrs. G. Entjes-Nieborg assisted in the preparation of the manuscript. The English text was improved by Mrs. S.M. van Gelder-Ottway.

8. REFERENCES

BAKELS, C., 1979. Linearbandkeramische Früchte und Samen aus den Niederlanden. *Archaeo-Physika* 7, pp. 1-10.

BEHRE, K.-E., 1976. *Die Pflanzenreste der frühgeschichtlichen Wurt Elisenhof.* Studien zur Küstenarchäologie Schleswig-Holsteins, Serie A: Elisenhof, Band 2. Bern & Frankfurt/M.

CASPARIE, W.A., BETTY MOOK-KAMPS, RITA M. PALFENIER-VEGTER, P.C. STRUIJK & W. VAN ZEIST, 1977. The palaeobotany of Swifterbant (Swifterbant contribution 7). *Helinium* 17, pp. 28-55.

CLASON, A.T. & D.C. BRINKHUIZEN, 1978. Swifterbant. Mammals, birds, fishes. A preliminary report (Swifterbant contribution 8). *Helinium* 18, pp. 69-82.

DECKERS, P.H., J.P. DE ROEVER & J.D. VAN DER WAALS, 1980. Jagers, vissers en boeren in een prehistorisch getijdengebied bij Swifterbant. *Z.W.O.-Jaarboek* 1980, pp. 111-145.

ENTE, P.J., 1976. The geology of the northern part of Flevoland in relation to the human occupation in Atlantic times (Swifterbant contribution 2). *Helinium* 16, pp. 15-35.

HACQUEBORD, L., 1976. Holocene geology and palaeogeography of the environment of the levee sites near Swifterbant (Swifterbant contribution 3). *Helinium* 16, pp. 36-42.

HELBAEK, H., 1951. Ukrudtsfrø som Naeringsmiddel i førromersk Jernalder. *Kuml* 1951, pp. 65-74.

HELBAEK, H., 1952. Preserved apples and *Panicum* in the prehistoric site at Nørre Sandegaard in Bornholm. *Acta Archaeologica* 23, pp. 107-115.

HELBAEK, H., 1980. Comment on *Chenopodium album* as a food plant in prehistory. *Ber. Geobot. Forschungsinst. Rübel, Zürich* 31, pp. 16-19.

JØRGENSEN, G., 1978. Acorns as a food-source in the Late Stone Age. *Acta Archaeologica* 48, pp. 233-238.

KNÖRZER, K.-H., 1972. Subfossile Pflanzenreste aus der bandkeramischen Siedlung Langweiler 3 und 6, Kreis Jülich, und ein urnenfeldzeitlicher Getreidefund innerhalb dieser Siedlung. *Bonner Jahrbücher* 172, pp. 395-403.

KNÖRZER, K.-H., 1974. Bandkeramische Pflanzenfunde von Bedburg-Garsdorf, Kreis Bergheim/Erft. In: *Beiträge zur Urgeschichte des Rheinlands*, Bd. 1 (= Rheinische Ausgrabungen Bd. 15), Bonn, pp. 173-192.

KNÖRZER, K.-H., 1977. Pflanzliche Grossreste des bandkeramischen Siedlungsplatzes Langweiler 9. In: R. Kuper, H. Löhr, J. Lüning, P. Stehli & A. Zimmermann, *Der bandkeramische Siedlungsplatz Langweiler 9, Gemeinde Aldenhoven, Kreis Düren* (= Rheinische Ausgrabungen, Bd. 18). Bonn, pp. 279-304.

KNÖRZER, K.-H., 1980. Pflanzliche Grossreste des bandkeramischen Siedlungsplatzes Wanlo (Stadt Mönchengladbach). *Archaeo-Physika* 7, pp. 7-20.

KÖRBER-GROHNE, U., 1967. *Geobotanische Untersuchungen auf der Feddersen Wierde.* Wiesbaden (Textband & Tafelband).

LANTING, J.N. & W.G. MOOK, 1977. *The pre- and protohistory of the Netherlands in terms of radiocarbon dates.* Groningen.

LANDWEHR, J., 1966. *Atlas van de Nederlandse bladmossen.* Amsterdam.

PRICE, T.D., 1981. Swifterbant, Oost-Flevoland, Netherlands: excavations at the river dune sites S-21-S-24, 1976. Final reports on Swifterbant III. *Palaeohistoria* 23, pp. 75-104.

SCHIEMANN, E., 1954. Die Pflanzenreste der Rössener Siedlung Ur-Fulerum bei Essen. *Jahrbuch des Römisch-Germanischen Zentralmuseums Mainz* 1, pp. 1-14.

VILLARET-VON ROCHOW, M., 1967. Frucht- und Samenreste aus der neolithischen Station Seeberg, Burgäschisee-Süd. *Acta Bernensia* 2, pp. 21-64.

WAALS, J.D. VAN DER & H.T. WATERBOLK, 1976. Excavations at Swifterbant – discovery, progress, aims and methods (Swifterbant contribution 1). *Helinium* 16, pp. 4-14.

WESTHOFF, V. & A.J. DEN HELD, 1969. *Plantengemeenschappen in Nederland.* Zutphen.

ZEIST, W. VAN, (1968) 1970. Prehistoric and early historic food plants in the Netherlands. *Palaeohistoria* 14, pp. 41-173.

ZEIST, W. VAN, 1974. Palaeobotanical studies of settlement sites in the coastal area of the Netherlands. *Palaeohistoria* 16, pp. 223-371.

PLANT REMAINS FROM IRON AGE NOORDBARGE, PROVINCE OF DRENTHE, THE NETHERLANDS

Willem van Zeist

CONTENTS

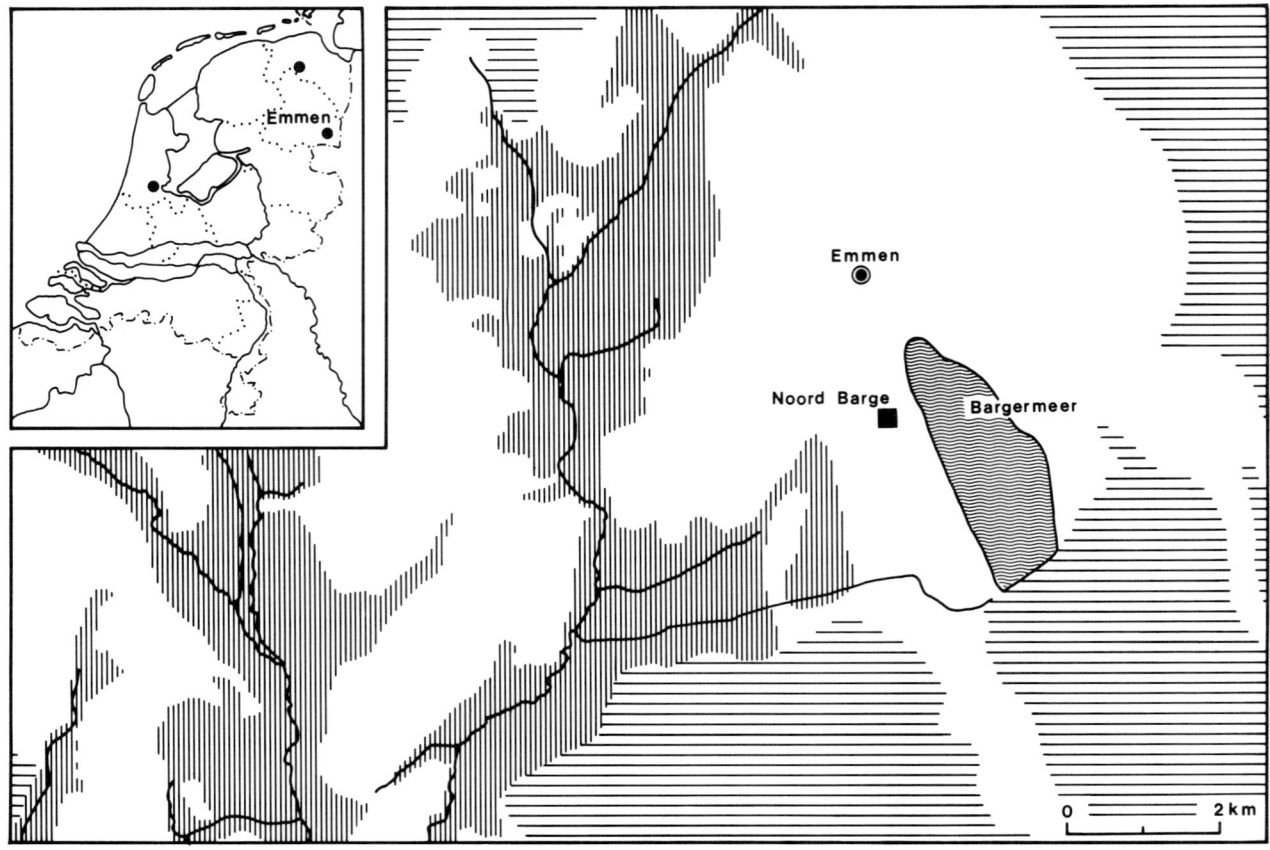

Fig. 1. Map of the Noordbarge area showing the maximum extent of the Bargermeer (after Kooi, 1979, fig. 160). Vertical hatching: stream-valley fen land; horizontal hatching: raised bog; white: higher sandy soils. Situation around A.D. 1850.

1. INTRODUCTION

In this paper the results will be discussed of the examination of charred plant remains recovered from the settlement site of Noordbarge, south of Emmen (fig. 1). On a terrain, called the "Hooge Loo" (52° 47' N, 6° 53' E) excavations have been carried out by the Biologisch-Archaeologisch Instituut in 1920, 1935, 1949 and 1972-1974 (cf. Kooi, 1979). These excavations revealed an urnfield, dated to the 9th to 5th centuries B.C., and settlement remains from various periods, consisting of houses of different type, sunken huts, granaries and fences. The urnfield and the associated finds have been described by Kooi (1979). Harsema

(1976) published a preliminary report on the results of the 1972-1974 settlement excavations. The following details have been taken from Harsema's report.

On the basis of the horizontal stratigraphy (intersecting house-plans) and of the orientation, distribution and typology of the houses, five phases of habitation are distinguished by Harsema. The schematic settlement plan of fig. 2 shows the sites of the houses.

Houses 28 and 21, in the southeastern part of the excavation area, are attributed to phase I. In contrast to all other houses, these two have approximately a N-S orientation. The phase I houses are dated to the Middle Bronze Age (1100-1000 B.C.) or the beginning of the

Table 1. Samples taken for the examination of seeds and fruits.

No.	Phase	Origin of sample	Remarks
159	IV?	pit inside house 7	
168	?	pit near house 8	
247	IV	sunken hut between houses 5 and 6	
275	IV	sunken hut beside house 5	
307	IV	post inside house 7	
308	IV	post inside house 7	^{14}C: 1930 ± 35 B.P. (GrN-7251)
312	IV	post inside house 5	
318	IV	central post, house 5	
336	IV	central post, house 5	broomcorn millet sample
338	IV	central post, house 5	
341	IV	post inside house 5	^{14}C: 2125 ± 50 B.P. (GrN-6865)
352	IV	sunken hut between houses 7 and 8	
612	?	pit near sunken hut and granary	
621	III	post, houses 15 or 16	
654	?	pit through house 12	
659	IV	sunken hut near house 5	no seeds
702	?	pit near house 17	
710	III	upright, house 17	
711	III	upright, house 17	
712	III	upright, house 17	
713	III	entrance trench, house 17	
714	III	upright, house 17	
715	III	central (?) post, house 17	
716	III	pit inside house 17	
717	?	foundation trench of fence	
720	III	upright, house 15	
721	III	post, houses 15 or 16	
819	III	post, granary near house 12	
833	IV	sunken hut near house 19	
848	IV?	post inside house 23	
849	IV?	post inside house 23	
856	IV?	entrance pit, house 18	
857	IV?	upright, house 18	
870	IV	fire place, house 19	
871	IV	post, house 19	
872	IV	foundation trench, house 19	
876	II	wall post, house 20	no seeds
877	II	wall post, house 20	
878	II	fire place, house 20	no seeds
879	II	wall post, house 20	
881	II	wall post, house 20	no seeds
885	II	corner post, house 20	
887	I	upright, house 21	no seeds
888	I	wall post, house 21	
889	I	wall post, house 21	
890	I	wall post, house 21	
892	I?	pit besides house 21	no seeds
895	I	upright, house 21	
896	I	wall post, house 21	
898	IV	fire place, house 22	
899	IV?	pit in house 22	no seeds
910	II?	wall post, house 20 (= 26)	
911	IV	wall post, house 25	
912	?	post in houses 26 and 28	
915	II	wall post, house 20 (= 26)	no seeds
920	IV	upright, house 23	
921	IV	upright, house 23	
922	IV	upright, house 23	
923	IV	wall post, house 23	

Table 1 (continued).

No.	Phase	Origin of sample	Remarks
722	III	upright, house 15	no seeds
723	III	upright house 16	
724	III	upright, house 16	
727	III	post, houses 15 or 16	
738	III	pit inside house 13	
748	III	post, house 13	
756	III	upright, house 14	14C house 14: 2175 ± 55 B.P. (GrN-7216)
761	III	wall post, house 13	no seeds
762	III	wall post, house 13	
766	III	upright, house 13	
767	III	upright, house 13	
768	III	wall post, house 14	
774	III	post, granary	
794	III	upright, house 11	14C: 2180 ± 50 B.P. (GrN-7217)
795	III	pit inside house 11	
804	III	wall post, house 11	
815	III	pit near granaries (near house 13)	
818	III	wall post, house 12	
931	IV	wall post, house 25	no seeds
932	II?	post in row of posts	no seeds
936	II?	post in row of posts	
941	IV	sunken hut over house 18	
952	IV	sunken hut over house 25	
954	IV	sunken hut near house 23	no seeds
956	II	wall post, house 29	
957	IV	foundation trench, house 25	
959	IV	foundation trench, house 25	
979	IV?	upright, house 18	no seeds
980	IV?	upright, house 18	
994	II	wall post, house 29	
995	?	pit besides foundation trench, house 19	
997	II	wall post, house 20 (= 26)	
999	IV	upright, house 25	
1004	?	upright?, house 20 (= 26)	
1029	III	post	
1041	III	post, granary	barley sample

Late Bronze Age. Phase II is also represented by only two houses, viz nos. 29 and 20 (=26), again situated in the southeastern part of the excavation area. Phase II was either contemporaneous with phase III, dated to the second century B.C. (see below), or it preceded phase III directly. Be this as it may, there must have been a very considerable difference in time between phases I and II (800-1000 years).

Nine houses, nos. 11-17, 31 and 32, in the western part of the excavation area, are attributed to phase III. The houses show a NE-SW orientation. Two radiocarbon determinations and the comparison with the pre-Roman Iron Age settlement of Hijken lead Harsema to a dating of around 200 B.C. for phase III.

The following house plans are attributed to phase IV: 1, 5, 6, 7, 8, 9, 19, 22, 23, 25 and 32. Most of the houses of this phase are located in the central part of the excavation area. Radiocarbon dates and comparisons with other sites suggest an age of 100 B.C. to A.D. 100 for phase IV.

Houses 27 and 30, a small part of which was unearthed at the eastern edge of the excavation area, are probably younger than phase IV.

Houses 2 and 4 (phase V) and 3 (probably phase IV) have been excavated by Professor A.E. van Giffen in 1949.

This study would not have been possible without the co-operation of various people. Dr. K.-H. Knörzer (Neuss, West-Germany) gave advice on the identification of some seed types. In the examination of the samples the present author was greatly assisted by Mrs. R.M. Palfenier-Vegter. Drs. O.H. Harsema procured the data on the origin of the samples, made the settlement plan available for publication and read the manuscript. Information on the site and its surroundings was provided by Dr. P.B. Kooi. The drawings of seeds and fruits and of the graphs were made by Mr. H.R. Roelink, while Mr. G. Delger prepared the map (fig. 1) and the settlement plan. The typing of the manuscript was carried out by Mrs. G. Entjes-Nieborg. The English text was improved by Mrs. S.M. van Gelder-Ottway.

2. THE SAMPLES

2.1. Sample processing

During the 1972-1974 excavations soil samples of a volume of 3000 cc were taken for botanical examination. The seeds occurred dispersed in the soil. Only in one case (sample 1041) were charred grains observed in the field. The location of the palaeobotanical samples is indicated on the plan of fig. 2. Further information on the origin of the samples (fill of post-hole, sunken hut, etc.) is given in table 1.

The plant remains were recovered in the laboratory by means of manual water flotation. The flotation residues were completely examined for seeds and fruits, the numbers of which are presented in table 2. Unidentified seeds or possible seeds and unidentifiable cereal grain fragments are not listed in table 2. From a small number of samples no seeds or fruits were retrieved. This is indicated in table 1. The local soil conditions allow the preservation of only carbonized plant remains. Occasional non-carbonized seeds and fruits have been discarded because they must have been due to modern intrusion (*e.g.* by animals). In all samples charcoal was present, sometimes in rather great quantities. Oak seems to be the dominant type of charcoal. Charred wood will not be considered in this paper.

Some of the results of the botanical examination have already been published in a paper in which particular attention is paid to the presence of rye at Noordbarge and in another site from the first centuries A.D. (Van Zeist, 1976). The final examination of the Noordbarge samples did not lead to any major changes in the data presented in table 2 of the 1976 paper. In the latter table, the phase indications for samples 856 (must be phase IV) and 1004 (must be of uncertain phase) have to be corrected.

2.2. The origin of the plant remains

The Noordbarge charred seed finds are from

Table 2. Numbers of seeds, fruits and other plant remains from Noordbarge. The numbers are not seldom estimates (broken seeds and fruits); this is not indicated in the table. cf. = identification uncertain; + = present. Species included in the palaeo-ecological histograms of fig. 7 are indicated with an asterisk.

Sample number	888	889	890	896	892	877	879	885	956	994	997	910	936	621	711	712	713	714	715	716	720
Phase	I	I	I	I	I?	II	II	II	II	II	II	II?	II?	III	III	III	III	III	III	III	III
Panicum miliaceum	.	.	.	.	5	.	.	.	.	.	.	1	.	6	1	.	½	1	5	7	.
Avena fatua/sativa	.	.	.	.	.	.	.	.	.	.	.	.	.	.	.	.	.	.	3	1½	.
Secale cereale	.	.	.	.	.	.	.	.	.	.	.	.	.	.	.	.	.	.	.	.	.
Hordeum vulgare	.	.	.	.	1	.	.	.	.	.	.	+	1	.	.	.	.	.	.	1	.
Triticum dicoccum	.	.	.	.	.	.	.	.	.	.	.	.	.	.	.	.	.	.	.	.	.
Triticum spikelet forks	.	.	.	.	.	.	.	.	.	.	.	.	.	.	.	.	.	.	.	.	.
Triticum glume bases	.	.	.	.	.	.	.	.	.	.	.	.	.	.	.	.	.	.	.	.	.
Camelina sativa	.	.	.	.	.	.	.	.	.	.	.	.	.	.	.	.	.	.	.	.	.
Linum usitatissimum	.	.	.	.	.	.	.	.	.	.	.	.	.	.	.	.	.	.	.	.	.
Vicia faba var. minor	.	.	.	.	.	.	.	.	.	.	.	.	.	.	.	.	.	.	.	.	.
Corylus avellana	.	+	.	.	.	+	.	.	.	.	.	.	.	.	.	.	.	.	.	.	.
Pyrus malus	.	.	.	.	.	.	.	.	.	.	.	.	.	.	.	.	.	.	.	.	.
Rubus idaeus	.	.	.	.	.	.	.	.	.	.	.	.	.	.	.	.	.	.	.	.	.
Humulus lupulus	.	.	.	.	.	.	.	.	.	.	.	1	.	.	.	.	.	.	.	.	.
*Bromus mollis/secalinus	.	.	.	.	.	.	.	.	.	.	.	.	.	2	.	.	.	.	1	1	.
*Echinochloa crus-galli	.	.	.	.	.	.	.	.	.	.	.	.	.	.	.	.	.	.	.	.	.
*Setaria viridis/italica	.	.	.	.	1	.	.	.	.	.	.	.	.	.	.	.	.	.	.	.	.
*Digitaria ischaemum	.	.	.	.	.	.	.	.	.	.	.	.	.	.	.	.	.	.	.	.	.
*Spergula arvensis	.	.	.	.	.	.	.	1	.	.	.	.	.	1	1	2	.	.	21	10	.
*Chenopodium album	3	.	3	.	.	.	.	.	.	.	.	.	.	1	.	.	.	.	5	.	1
*Chen. rubrum/glaucum	.	.	.	.	.	.	.	.	.	.	.	.	.	.	.	.	.	.	.	.	.
*Chenopodium ficifolium	.	.	.	.	.	.	.	.	.	.	.	.	.	.	.	.	.	.	.	.	.
Atriplex hastata/patula	.	.	.	.	.	.	.	.	.	.	.	.	.	.	.	.	.	.	.	.	.
*Rumex acetosella	.	1	2	1	.	.	.	.	1	1	3	.	.	6	1	.	.	.	33	7	.
*Rumex crispus	.	.	.	.	.	.	.	.	1	.	.	.	.	.	.	.	.	.	.	.	.
*Polygonum lapathifolium	.	½	1	.	.	.	.	.	.	.	.	.	.	cf.2	1	.	.	.	1½	2	.
*Polygonum persicaria	.	.	.	.	.	.	.	.	.	.	.	.	.	.	.	.	.	.	.	½	.
*Polygonum aviculare	.	.	2½	.	.	.	.	.	.	.	.	.	.	.	.	.	.	.	.	.	.
*Polygonum convolvulus	1	.	.	.	.	.	.	.	.	.	1	.	.	1	.	.	.	.	.	.	.
*Polygonum hydropiper	.	.	.	.	.	.	.	.	.	.	.	.	.	.	.	.	.	.	.	.	.
*Scleranthus annuus[1]	.	.	.	.	.	.	.	.	.	.	.	.	.	.	.	.	.	.	.	.	.
*Stellaria media	.	.	.	.	.	.	.	.	.	.	.	.	.	.	.	.	.	.	.	2	.
*Stellaria graminea	.	.	.	.	.	.	.	.	.	1	.	.	.	.	.	.	.	.	.	.	.
*Raphanus raphanistrum	.	.	.	.	.	.	.	.	.	.	.	.	.	.	.	.	.	.	.	.	.
*Stachys arvensis	.	.	.	.	.	.	.	.	.	.	.	.	.	.	.	.	.	.	.	.	.
*Prunella vulgaris	.	.	.	.	.	.	.	.	.	.	.	.	.	.	.	.	.	.	.	.	.
*Galium aparine	.	.	.	.	.	.	1	.	.	.	.	.	.	.	.	.	.	.	.	.	.
*Urtica dioica	.	.	.	.	.	.	.	.	.	.	.	.	.	.	.	.	.	.	.	.	.
*Solanum nigrum	.	.	.	.	.	.	.	.	.	.	.	.	.	.	.	.	.	.	.	.	.
*Capsella bursa-pastoris	.	.	.	.	.	.	.	.	.	.	.	.	.	.	.	.	.	.	.	.	.
Vicia	.	.	.	.	.	.	.	.	.	.	.	.	.	.	.	.	.	.	.	.	.
*Viola	.	.	.	.	.	.	.	.	.	.	.	.	.	.	.	.	.	.	.	.	.
*Plantago lanceolata	.	.	.	.	.	.	.	.	.	.	.	.	.	.	.	.	.	.	.	.	.
*Ranunculus repens	.	.	.	.	.	.	.	.	.	.	.	.	.	.	.	.	.	.	.	.	.
Lotus	.	.	.	.	.	.	.	.	3	1	.	.	.	.	.	.	.	.	.	.	.
*Trifolium repens	.	.	.	.	.	.	.	.	.	.	.	.	.	.	.	.	.	.	1	.	.
*Medicago lupulina	.	.	.	.	.	.	.	.	.	.	.	.	.	.	.	.	.	.	.	.	.
*Festuca rubra	.	.	cf.1	.	.	.	.	.	.	.	.	.	.	.	.	.	.	.	.	.	.
Phleum	.	.	1	.	.	.	.	.	.	.	.	.	.	.	.	.	.	.	.	.	.
Poa	.	.	.	.	.	.	.	.	.	.	.	.	.	.	.	.	.	.	.	.	.
*Potentilla cf. erecta	.	.	.	.	.	.	.	.	.	.	.	.	.	.	.	.	.	.	.	.	.
Calluna vulgaris[2]	.	.	.	.	.	.	.	.	.	.	.	.	.	.	.	.	.	.	.	.	.
Valeriana officinalis	.	.	.	.	.	.	.	.	.	.	.	.	.	.	.	.	.	.	.	.	.
Carex nigra-type	.	.	.	.	.	.	.	.	.	.	.	.	.	.	.	.	.	.	.	.	.
Carex serotina-type	.	.	.	.	.	.	.	.	.	.	.	.	.	.	.	.	.	.	.	.	.
Carex rostrata/vesicaria	.	.	.	.	.	.	.	.	2	.	.	.	.	.	.	.	.	.	.	.	.
Eleocharis palustris	.	.	.	.	.	.	.	.	.	.	.	.	.	.	.	.	.	.	.	.	.
Oenanthe aquatica	.	.	.	.	.	.	.	.	.	.	.	.	.	.	.	.	.	.	.	.	.
Cenococcum geophilum[3]	15	6	25	1	.	.	.	.	14	.	1	.	2	295	12	4	2	18	100	18	700

721	722	723	724	727	738	748	756	762	766	767	768	774	794	795	804	815	818	819	1029	1041	Sample number
III	III	III	III	III	III	III	III	III	III	III	III	III	III	III	III	III	III	III	III	III	Phase
1	8	2	4		8				2		1½		2	5		14				7	Panicum miliaceum
												11	13			4				29	Avena fatua/sativa
				1																	Secale cereale
				2½								10	12	1½		14				1980	Hordeum vulgare
				½							1	18	8	2		7				150	Triticum dicoccum
																1			½	18	Triticum spikelet forks
												3		2		4				16	Triticum glume bases
					25											165					Camelina sativa
																15					Linum usitatissimum
																					Vicia faba var. minor
																					Corylus avellana
																					Pyrus malus
			1																		Rubus idaeus
																					Humulus lupulus
		1										1									*Bromus mollis/secalinus
			1		27											11					*Echinochloa crus-galli
					9				1			1				17					*Setaria viridis/italica
									1							13					*Digitaria ischaemum
3	1	2		2	2½		8		2	1			1	5	1	32	2	4½	2		*Spergula arvensis
½	5	4	18	4	20	cf.1	cf.1	1	15	2½	1		1	1	6	820	½	6	11		*Chenopodium album
																					*Chen. rubrum/glaucum
1																					*Chenopodium ficifolium
																					Atriplex hastata/patula
14	8	12	35	16	3		1				2					70		1	1		*Rumex acetosella
					1											1				1	*Rumex crispus
	2	1	12	4	}45	1			14	1		4	2	2	3	18	1	8½	4		*Polygonum lapathifolium
cf.½						1			3				cf.1½	1½	5			3			*Polygonum persicaria
																					*Polygonum aviculare
	1				4		1	⅓	1	1			21			7		½		2	*Polygonum convolvulus
							1		2½		1										*Polygonum hydropiper
		1														1		1			*Scleranthus annuus[1]
			1		1																*Stellaria media
																					*Stellaria graminea
																					*Raphanus raphanistrum
														1½							*Stachys arvensis
			1																		*Prunella vulgaris
																					*Galium aparine
																					*Urtica dioica
			1		3			1							5						*Solanum nigrum
							1														*Capsella bursa-pastoris
			1	½	½				1							2					Vicia
	1																				*Viola
			1	1												2				1	*Plantago lanceolata
																					*Ranunculus repens
									1												Lotus
1																					*Trifolium repens
																					*Medicago lupulina
					2																*Festuca rubra
											1					3					Phleum
																					Poa
																					*Potentilla cf. erecta
									3												Calluna vulgaris[2]
					1																Valeriana officinalis
	1								1												Carex nigra-type
								1													Carex serotina-type
																					Carex rostrata/vesicaria
																		½			Eleocharis palustris
									1												Oenanthe aquatica
1730	3325	295	1565	695		6	15	5	7	3	3								2		Cenococcum geophilum[3]

Table 2 (continued).

Sample number	247	275	307	308	312	318	336	338	341	352	659	833	870	871	872	898	911	920	921	923	941
Phase	IV	IV	IV	IV	IV	IV	IV	IV	IV	IV	IV	IV	IV	IV	IV	IV	IV	IV	IV	IV	IV
Panicum miliaceum	4	2	25	105	2½	1	750	4	25	7	.	14	2	3	1	1	.	.	.	.	.
Avena fatua/sativa	.	.	.	.	½	.	1½	.	10	.	.	2	.	3	½	.	.	.	.	.	.
Secale cereale	28	½	290	50	15	.	.	6	.	.	.	2	.	.	.	.	.	.	.	.	.
Hordeum vulgare	5	1	.	.	5	1	2	.	1	.	.	2	.	.	.	26	.	.	.	.	1
Triticum dicoccum	.	½	.	.	.	.	.	.	.	.	.	.	.	.	.	58	.	.	.	.	.
Triticum spikelet forks	.	.	.	.	.	.	.	.	.	.	.	.	.	.	.	2	.	.	.	.	.
Triticum glume bases	.	.	.	.	.	.	.	.	.	.	.	.	.	.	.	.	.	.	.	.	.
Camelina sativa	.	.	.	.	.	.	.	.	.	.	.	.	.	.	.	.	.	.	.	.	.
Linum usitatissimum	.	.	.	.	.	.	.	.	.	.	.	.	.	.	.	.	.	.	.	.	.
Vicia faba var. minor	.	.	.	.	.	.	.	.	.	.	.	.	.	.	.	.	.	.	.	.	.
Corylus avellana	.	.	.	.	.	.	.	.	.	.	+	.	.	.	.	.	+	.	.	.	.
Pyrus malus	.	.	.	.	.	.	.	.	.	.	.	½	.	.	.	.	.	.	.	.	.
Rubus idaeus	.	.	.	.	.	.	.	.	.	.	.	.	.	.	.	.	.	.	.	.	.
Humulus lupulus	.	.	.	.	.	.	.	.	.	.	.	.	.	.	.	.	.	.	.	.	.
*Bromus mollis/secalinus	2	⅔	2½	.	.	.	1	.	35	.	.	.	.	½	.	.	.	.	.	.	2
*Echinochloa crus-galli	.	.	1	.	.	.	.	.	.	.	.	.	.	.	.	.	.	.	.	.	.
*Setaria viridis/italica	.	1	1	.	.	.	2	.	1	.	.	.	.	.	.	.	.	.	.	.	.
*Digitaria ischaemum	.	.	1	.	.	.	.	.	.	.	.	.	.	.	.	.	.	.	.	.	.
*Spergula arvensis	12	8	.	1	1	4	9	6	22	43	.	6	.	6	4	.	1	1	.	2	25
*Chenopodium album	26	13	1	2	.	1	7	12	11	.	.	1	3	.	1	6	.	.	1	5	12
*Chen. rubrum/glaucum	.	.	.	.	.	.	.	.	.	.	.	.	.	.	.	.	.	.	.	.	.
*Chenopodium ficifolium	.	.	.	.	.	.	.	.	.	.	.	.	.	.	.	.	.	.	.	.	.
Atriplex hastata/patula	.	.	.	.	.	.	.	.	.	.	.	.	.	.	.	.	.	.	.	.	.
*Rumex acetosella	18	16	4	1	.	.	3	20	10	18	.	6	8	4	3	.	2	1	.	8	11
*Rumex crispus	3	cf.½	.	.	1	.	.	1	.	.	.	.	.	.	1	.	1	.	.	.	.
*Polygonum lapathifolium	1	10	7	1½	.	.	3	2	7	1	.	.	.	2	1½	1	.	1	.	1	2
*Polygonum persicaria	.	2	1	.	½	.	.	.	.	.	.	.	.	.	.	.	.	.	.	.	.
*Polygonum aviculare	1	1½	.	.	.	.	.	.	.	.	.	.	.	.	.	.	.	.	.	.	.
*Polygonum convolvulus	1	.	.	.	.	.	1	.	.	.	.	.	.	.	½	.	.	.	.	.	3
*Polygonum hydropiper	.	.	.	.	.	.	.	.	.	.	.	.	.	.	.	.	.	.	.	.	.
*Scleranthus annuus[1]	.	.	.	.	.	.	.	.	.	.	.	.	.	.	.	.	.	.	.	.	.
*Stellaria media	.	.	.	.	.	.	.	5	.	.	.	.	.	.	.	.	.	.	.	.	.
*Stellaria graminea	.	.	.	.	.	.	.	.	.	.	.	.	.	.	.	.	.	.	.	.	.
*Raphanus raphanistrum	1	1	.	.	.	.	.	.	.	.	.	.	.	.	.	.	.	.	.	.	.
*Stachys arvenis	.	.	.	.	.	.	.	.	.	1	.	.	.	.	.	.	.	.	.	.	.
*Prunella vulgaris	.	.	.	.	.	.	.	.	.	.	.	.	.	.	.	.	.	.	.	.	.
*Galium aparine	.	.	.	.	.	.	.	.	.	.	.	.	.	.	.	.	.	.	.	.	.
*Urtica dioica	.	cf.1	.	.	.	.	.	.	.	.	.	.	.	.	.	.	1	.	.	.	.
*Solanum nigrum	.	.	.	.	.	.	2	.	.	.	.	.	.	.	.	.	.	.	.	.	.
*Capsella bursa-pastoris	.	.	.	.	.	.	.	.	.	.	.	.	.	.	.	.	.	.	.	.	.
Vicia	.	.	.	.	.	.	.	.	½	.	.	.	.	1	.	.	.	.	.	.	.
*Viola	.	.	.	.	.	.	.	.	.	.	.	.	.	.	.	.	.	.	.	.	.
*Plantago lanceolata	.	.	.	.	.	.	1	.	.	.	.	.	.	.	.	1	.	.	.	.	1
*Ranunculus repens	.	.	.	.	.	.	1	.	.	.	.	.	.	.	.	.	.	.	.	.	.
Lotus	.	.	.	.	.	.	.	.	.	.	.	.	.	.	.	.	.	2	.	.	.
*Trifolium repens	3	.	.	.	1	.	2	.	.	.	.	2	.	.	.	.	.	.	.	.	.
*Medicago lupulina	.	.	.	.	.	.	.	.	.	.	.	.	.	.	.	.	.	.	.	.	1
*Festuca rubra	.	.	.	.	.	.	.	.	.	.	.	.	.	.	.	.	.	.	.	.	.
Phleum	.	.	.	.	.	.	.	.	.	.	.	.	.	.	.	.	.	.	.	.	.
Poa	.	.	.	.	.	.	.	.	.	.	.	.	1	.	.	.	.	.	.	.	.
*Potentilla cf. erecta	.	.	.	.	.	.	.	.	.	.	.	.	.	.	.	.	.	.	.	.	.
Calluna vulgaris[2]	.	.	.	.	.	.	.	.	.	.	.	.	.	.	.	.	.	.	.	.	.
Valeriana officinalis	.	.	.	.	.	.	.	.	.	.	.	.	.	.	.	.	.	.	.	.	.
Carex nigra-type	.	.	.	.	.	.	.	.	.	.	.	.	.	.	.	.	.	.	.	.	.
Carex serotina-type	.	.	.	.	.	.	.	.	.	.	.	.	.	.	.	.	.	.	.	.	.
Carex rostrata/vesicaria	.	.	.	.	.	.	.	.	.	.	.	.	.	.	.	.	.	.	.	.	.
Eleocharis palustris	.	.	.	.	.	.	1	.	.	.	.	.	.	.	.	.	.	.	.	.	.
Oenanthe aquatica	.	.	.	.	.	.	.	.	.	.	.	.	.	.	.	.	.	.	.	.	.
Cenococcum geophilum[3]	22	1375	.	1	4	4	32	550	7	.	.	50	.	2	.	.	.	.	.	.	6

[1] Calyces   [2] Leafed stem fragments   [3] Sclerotia

952	957	959	999	159	848	849	856	857	980	168	612	654	702	717	912	995	1004	Sample number
IV	IV	IV	IV	IV?	IV?	IV?	IV?	IV?	IV?	?	?	?	?	?	?	?	?	Phase
.	1	.	12	3	.	.	2	1	.	3	.	3	1	1	.	.	1	Panicùm miliaceum
.	.	.	3	.	.	.	.	.	.	.	.	.	.	.	.	.	.	Avena fatua/sativa
5	1	.	10	.	.	.	1	.	.	.	.	.	.	.	.	.	10	Secale cereale
.	.	.	.	.	.	.	2	1	.	.	.	.	.	.	.	1	2	Hordeum vulgare
.	.	.	.	.	.	.	.	.	.	.	.	.	.	.	.	.	.	Triticum dicoccum
.	.	.	.	.	.	.	.	.	.	.	.	.	.	.	.	.	.	Triticum spikelet forks
.	.	.	.	.	.	.	.	.	.	.	.	.	.	.	.	.	.	Triticum glume bases
.	.	.	.	.	.	.	.	.	.	.	.	.	.	.	.	.	.	Camelina sativa
.	.	.	.	.	.	.	.	.	.	.	.	.	.	.	.	.	.	Linum usitatissimum
.	.	.	.	.	.	.	.	.	.	.	.	.	.	.	1	.	.	Vicia faba var. minor
.	.	.	.	.	.	.	.	.	.	.	.	.	.	.	+	.	.	Corylus avellana
.	.	.	.	.	.	.	.	.	.	.	.	.	.	.	.	.	.	Pyrus malus
.	.	.	.	.	.	.	.	.	.	.	.	.	.	.	.	.	.	Rubus idaeus
.	.	.	.	.	.	.	.	.	.	.	.	.	.	.	.	.	.	Humulus lupulus
1	.	.	1½	.	.	.	.	.	.	.	.	.	.	.	.	.	½	*Bromus mollis/secalinus
.	.	.	.	.	.	.	1	.	.	.	.	.	.	.	.	.	.	*Echinochloa crus-galli
.	.	.	.	.	.	.	.	.	.	.	.	.	.	.	.	.	.	*Setaria viridis/italica
.	.	1	.	.	.	.	.	.	.	.	.	.	.	.	.	.	.	*Digitaria ischaemum
1	.	.	23	1	.	.	1	2	.	8	.	.	.	.	.	.	9	*Spergula arvensis
.	.	1	2	.	cf.1	.	3	1	.	1	.	.	.	.	1	.	.	*Chenopodium album
.	1	.	.	.	.	.	.	.	.	.	.	.	.	.	.	.	.	*Chen. rubrum/glaucum
.	.	.	.	.	.	.	.	.	.	.	.	.	.	.	.	.	.	*Chenopodium ficifolium
.	1	.	.	.	.	.	1	.	.	.	.	.	.	.	.	.	.	Atriplex hastata/patula
1	.	1	22	.	.	1	.	.	.	12	.	.	.	3	.	.	8	*Rumex acetosella
.	.	.	1	.	.	.	.	.	.	.	.	.	.	.	.	.	.	*Rumex crispus
.	.	.	4	.	½	.	.	.	.	.	.	1	2	.	.	.	.	*Polygonum lapathifolium
.	.	.	.	.	.	.	.	.	.	.	.	.	.	.	.	.	.	*Polygonum persicaria
.	.	.	.	.	.	.	.	.	.	.	.	.	.	.	.	.	.	*Polygonum aviculare
.	.	.	.	.	.	.	1	.	.	1	.	.	.	.	.	.	.	*Polygonum convolvulus
.	.	.	.	.	.	.	.	.	.	.	.	.	.	.	.	.	.	*Polygonum hydropiper
.	.	.	.	.	.	.	.	.	.	.	.	.	.	.	.	.	.	*Scleranthus annuus[1]
.	.	.	.	.	.	.	.	.	.	.	.	.	.	.	.	.	.	*Stellaria media
.	.	.	.	.	.	.	.	.	.	.	.	.	.	1	.	.	.	*Stellaria graminea
.	.	.	.	.	.	.	.	.	.	.	.	.	.	.	.	.	.	*Raphanus raphanistrum
.	.	.	.	.	.	1	.	.	.	.	.	.	.	.	.	.	.	*Stachys arvenis
.	.	.	.	.	.	.	.	.	.	.	.	.	.	.	.	.	.	*Prunella vulgaris
.	.	.	.	.	.	.	.	.	.	.	.	.	.	.	.	.	.	*Galium aparine
.	.	.	.	.	.	.	.	.	.	.	.	.	.	.	.	.	.	*Urtica dioica
.	.	.	1	.	.	.	.	.	.	1	.	.	.	.	.	.	.	*Solanum nigrum
.	.	.	.	.	.	.	.	.	.	.	.	.	.	.	.	.	.	*Capsella bursa-pastoris
.	.	.	.	.	.	.	.	.	.	.	.	.	.	.	.	.	.	Vicia
.	.	.	.	.	.	.	.	.	.	.	.	.	.	.	.	.	.	*Viola
.	.	1	.	.	.	.	.	.	.	.	.	.	.	.	.	.	1	*Plantago lanceolata
.	.	.	.	.	.	.	.	.	.	.	.	.	.	.	.	.	.	*Ranunculus repens
.	.	1	.	.	.	.	.	.	.	.	.	.	.	.	.	.	1	Lotus
.	.	1	1	.	.	.	.	.	.	.	.	.	.	.	.	.	.	*Trifolium repens
.	.	.	.	.	.	.	.	.	.	.	.	.	.	.	.	.	.	*Medicago lupulina
.	.	.	.	.	.	.	.	.	.	.	.	.	.	.	.	.	.	*Festuca rubra
.	.	.	.	.	.	.	.	.	.	.	.	.	.	.	.	.	.	Phleum
.	.	.	.	.	.	.	.	.	.	.	.	.	.	.	.	.	.	Poa
1	.	.	.	.	.	.	.	.	.	.	.	.	.	.	.	.	.	*Potentilla cf. erecta
.	.	.	.	.	.	.	.	.	.	.	.	.	.	.	.	.	.	Calluna vulgaris[2]
.	.	.	.	.	.	.	.	.	.	.	.	.	.	.	.	.	.	Valeriana officinalis
.	.	.	.	.	.	.	.	.	.	.	.	.	.	.	.	.	.	Carex nigra-type
.	.	.	.	.	.	.	.	.	.	.	.	.	.	.	.	.	.	Carex serotina-type
.	.	.	.	.	.	.	.	.	.	.	.	.	.	.	.	.	.	Carex rostrata/vesicaria
.	.	.	.	.	.	.	.	.	.	.	.	.	.	.	.	.	.	Eleocharis palustris
.	.	.	.	.	.	.	.	.	.	.	.	.	.	.	.	.	.	Oenanthe aquatica
.	1	16	1	.	.	.	.	5	3	.	180	1	12	10	.	4	.	Cenococcum geophilum[3]

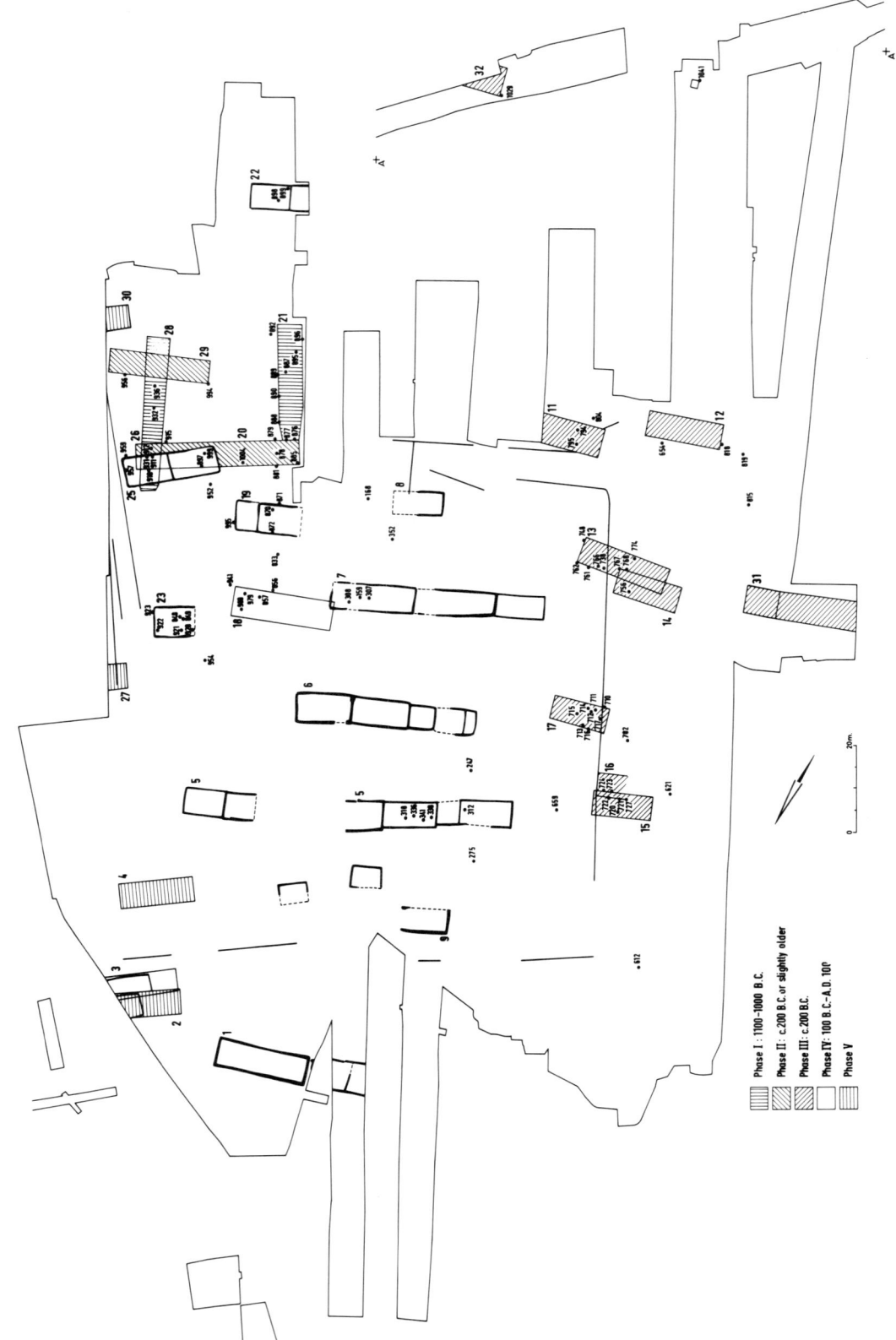

Fig. 2. Survey of the Noordbarge settlement showing the location of the houses and of the samples examined for charred plant remains. Five phases of habitation are distinguished. Phase I: houses 21 and 28; phase II: houses 20 (=26) and 29; phase III: houses 11-17, 31 and 32; phase IV: houses 1, 5-9, 19, 22, 23 and 25; phase V: houses 2, 4, 20 and 27.

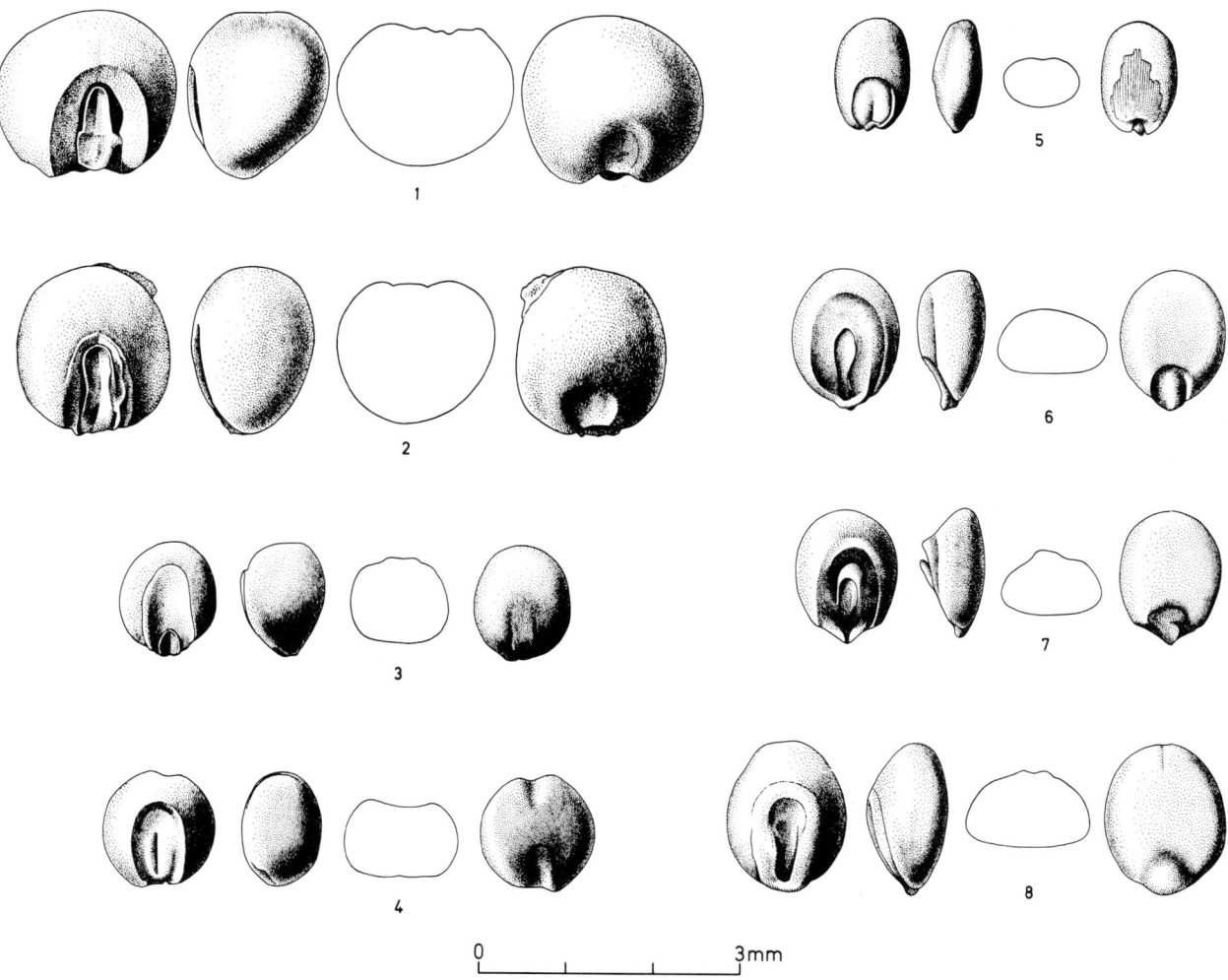

Fig. 3. Millet-type grains from Noordbarge. Sample numbers are in brackets. 1-2: *Panicum miliaceum* (336); 3-4: *Setara* cf. *viridis* (815); 5: *Digitaria ischaemum* (815); 6-7: *Echinochloa crus-galli* (738); 8: *Echinochloa crus-galli* (815).

the fill of pits, post-holes and such-like. There are no indications of storage pits. Samples 336 and 1041, which yielded rather great numbers of crop-plant grains, are both from the fill of post-holes. The vegetable remains in the pits and post-holes are there in secondary position. They must have been present in the soil which was shovelled into the pit or the post-hole.

The charred seed contents of most, if not of all the Noordbarge samples must have been of mixed origin. Some seeds may have formed part of the fully processed crop ready for storage or for the preparation of food,

others may have originated from the residue left after crop processing, such as threshing and seed cleaning. If the plant remains in a sample represent already one particular crop-processing activity, if, for instance, they are from the tail-corn left after sieving or basket-winnowing, this may be difficult to demonstrate. Moreover, considerable numbers of seeds of wild plants did not arrive in the site together with the harvested crop, but they were brought in, intentionally or unintentionally, as a result of other activities of the inhabitants of the site. Consequently, in evaluating the palaeobotanical data one should be aware

Table 3. Dimensions in mm and index values for *Hordeum vulgare* from Noordbarge 1004
(N = 72).

	L	B	T	100L:B	100T:B
min.	4.4	2.3	1.7	158	71
aver.	5.51	3.05	2.45	181	80
max.	6.5	3.8	3.1	219	92

Table 4. Mean dimensions in mm and index values for *Hordeum vulgare* from various sites. E.I.A. = Early Iron Age; R.I.A = Roman Iron Age; Med. = Medieval.

	L	B	T	100L:B	100T:B	period
Angelsloo 84	4.91	2.31	1.69	214	73	E.I.A.
Gees 14a	5.42	2.83	2.20	193	78	E.I.A.
Dalfsen 436	5.46	2.78	2.16	197	78	R.I.A.
Wijster 1239	5.58	2.89	2.19	193	76	R.I.A.
Noordbarge 1004	5.51	3.05	2.45	181	80	R.I.A.
Gasselte 418	5.56	2.93	2.23	191	76	Med.
Gasselte 428	5.92	2.87	2.21	206	77	Med.
Gasselte 544	5.45	2.78	2.14	197	77	Med.

Table 5. Dimensions in mm and index values for *Triticum dicoccum* from Noordbarge.

		L	B	T	100L:B	100T:B
Sample 1041	min.	5.3	2.7	2.5	165	85
(N = 7)	aver.	5.47	2.97	2.66	186	90
	max.	6.2	3.2	2.8	223	97
Sample 898	min.	4.6	2.2	1.7	171	77
(N = 8)	aver.	4.85	2.52	2.29	193	90
	max.	5.1	3.0	2.6	215	103

of the usually heterogeneous origin of the plant remains in the samples examined.

3. CROP PLANTS AND POSSIBLE CROP PLANTS

3.1. Panicum miliaceum (fig. 3: 1-2)

Of the crop-plant species, *Panicum miliaceum* (broomcorn millet) shows the highest presence percentage. It was found in 44 of the 81 samples which yielded charred seeds. Sample 336 is a nearly pure *Panicum* sample, with only a minor admixture of other seeds. The preservation of the millet in this sample was not particularly good, so that only a rather small number of grains was suitable for measuring (table 7). At least during phases III and IV, in the period from 200 B.C. to A.D. 100, *Panicum miliaceum* must have been cultivated at Noordbarge. The samples from phases I and II which yielded a few *Panicum* grains have been attributed to these periods with some reserve, so that there is no firm evidence for the presence of this crop plant in Middle Bronze Age and Early Iron Age Noordbarge.

The economic importance of *Panicum* at Noordbarge during phases III and IV, that is to say, its relative proportion in the crop-plant production, cannot be determined. One may, however, assume that it must have been a common crop plant. From the charred seed finds published so far it appears that from the Middle Bronze Age on, *Panicum* was present in the north of the Netherlands (Van Zeist, 1970). As in the samples concerned millet constitutes only an insignificant admixture to barley or wheat, is was not clear whether this species was grown intentionally. On the other hand, grain impressions in pottery suggest that millet must have been a crop plant in its own right. This conclusion is confirmed by the Noordbarge evidence.

From the above it is, once again, evident that isolated, large charred seed finds may give a picture of prehistoric crop-plant husbandry which is incomplete and which to some extent may even be misleading. In this respect the analysis of a great number of samples originating from various features in the settlement must provide more reliable information although each individual sample may be rather poor in seeds and fruits. Due to lack of material for comparison the role of broomcorn millet in the economy of Late Bronze Age and Iron Age farmers on the sandy soils in the north of the Netherlands is still obscure. There are no indications for the cultivation of *Panicum miliaceum* in Iron Age sites in the coastal area of the north of the Netherlands. Broomcorn millet is reported for Jemgumkloster, on the Ems river in Northwest Germany, in samples from the last century B.C. and from the beginning of our era (Behre, 1972). In contrast to most other coastal settlement sites, Jemgumkloster was situated in a fresh-water environment. According to Knörzer (1971) broomcorn millet must have been an important crop plant in Early Iron Age Rhineland; it is represented there in various sites by considerable numbers of grains.

3.2. Hordeum vulgare

The barley of Noordbarge is of the hulled type. Grains of the naked variety have not been found. *Hordeum* is represented in a relatively great number of samples, but generally only by a few grains. An exception in this respect forms sample 1041, from the fill of a granary post-hole, which yielded nearly 2000 barley caryopses. This is a predominantly barley sample (more than 90% *Hordeum*) with an admixture of other cereal grains. Unfortunately, the preservation is poor, the grains being more or less deformed due to puffing. As a consequence only a relatively small number of caryopses could be selected for measuring (table 3), but also of these grains the dimensions may more or less seriously have been affected by the carbonization. In addition to the grains, samples 794 and 1041 yielded 1 and 3 rachis internodes, respectively.

In table 4 the mean dimensions and index values of charred barley grains from various sites on sandy soils in the north of the Netherlands are shown. From this table it appears that the mean dimensions of the Noordbarge grains do not differ markedly from those of most other barley samples from prehistoric and early-historical sites. The barley from Angelsloo is notably small. This is also true for the *Triticum dicoccum* from the same sample, suggesting that at least during the season in which the crop plants of this Angelsloo sample were grown the conditions for plant cultivation must have been rather poor. The barley from Gasselte sample no. 428 is conspicuously long, but in other samples from this site the grains are smaller (Van Zeist & Palfenier-Vegter, 1979, table 7). One may conclude that the Noordbarge barley is of average size.

The comparatively low mean L:B index value of the Noordbarge barley may have been the effect of the carbonization (decrease in length, increase in width). The same explanation may apply for the relatively high average T:B index value.

3.3. Triticum dicoccum

Emmer wheat is represented in a much smaller number of samples than barley. On the other hand, there can be no doubt that it was cultivated by the Noordbarge farmers. It is unlikely that *Triticum dicoccum* could maintain itself as an admixture to other crop-plant species. The charred seed evidence for the north of the Netherlands shows that the role of emmer wheat diminished in Late Bronze Age and Iron Age times. This wheat species gave way, first to barley and subsequently also to oats and rye. In medieval sites in the north of the Netherlands, *Triticum dicoccum* has disappeared altogether. The decreasing economic importance of emmer wheat in late prehistoric times is also reflected at Noordbarge.

From samples 1041 (phase III) and 898 (phase IV) small numbers of emmer wheat grains have been measured (table 5). The grains of sample 898 are distinctly smaller than those of sample 1041. The measured specimens from the latter sample are conspicuously large for emmer wheat from sandy soils (cf. Van Zeist, 1968 [1970], table 64).

Sample 898, with predominantly emmer wheat, yielded a *Triticum monococcum*-type grain fragment.

3.4. Secale cereale

The presence of rye grains at Noordbarge has been discussed in a previous paper (Van Zeist, 1976). The question whether the rye grains in this site point to the intentional cultivation of this species or whether it may have occurred only as an admixture to other crop plants has been treated rather extensively. Particularly samples 307 and 308, in which *Secale* grains occur in fairly large numbers, were considered as conclusive evidence of rye cultivation. The present author sees no reason to change this conclusion. Indeed, current investigations of charred plant remains from the multi-period settlement site of Peeloo, north of Assen (Harsema, 1979; Kooi, 1980), yielded *Secale* grains from Roman Iron Age and early medieval features. Consequently, for the north of the Netherlands rye cultivation in the first centuries A.D. is now attested for two sites lying c. 40 km apart.

As for the beginning of rye cultivation at Noordbarge, the following should be mentioned. In the 1976 paper, sample 1004 which yielded c. 10 rye grains, has been attributed to phase II, suggesting that during the period concerned *Secale* was grown here. A re-examination of the settlement plan has led to the conclusion that the post-hole of sample 1004 should not be assigned to any particular building phase. Consequently, in table 2 this sample is shown among the group of samples of questionable periodization.

Sample 856 (with one rye grain) should not be attributed to phase III, as was done erroneously in the 1976 paper, but to phase IV, be it with some reserve (table 2). As for sample 738, also with one rye grain, the following should be remarked. The sample is from the fill of a pit inside house no. 13 and it is reasonable to assume that this pit belongs to the house concerned and that consequently it must be attributed to phase III. However, posts of a fence of phase IV have been dug in the fill of this pit. For that reason the fill of the pit must have been contaminated with material from phase IV. It will be clear that the single rye grain in sample 738 could possibly date from phase IV and that it cannot be considered as evidence of rye cultivation during phase III.

The conclusion that only for phase IV is there firm evidence of rye cultivation at Noordbarge is not really affected by the fact that the largest rye sample (no. 307), from the fill of a post-hole, has been attributed to phase IV with some reserve. The post concerned was no structural element of house no. 7. On the other hand, there are no other buildings of which this post could have formed part.

The dimensions and index values of rye grains from sample 307 are shown in table 6.

3.5. Avena fatua/sativa

The oat grains in the Noordbarge samples pose some problems. In an earlier paper (Van Zeist, 1976) it was suggested that *Avena sativa* formed part of the crop-plant assortment of the Noordbarge farmers. However, after a reconsideration of the factual evidence the present author feels obliged to take a more sceptical point of view. The numbers of *Avena* fruits are in themselves no indication of the intentional growing of oats, that is of *A. sativa*. In none of the samples is it by far the dominant cereal grain type. The greatest number of oat caryopses (c. 25 specimens) occurs in sample 1041, but this is a fairly large barley sample the proportion of *Avena* being only 1.2%.

Altogether two *Avena* flower bases have been recovered. One of them, in sample 774,

shows the oval articulation scar characteristic of *A. fatua* (wild oat). The other flower base, in sample 1041, has been rather seriously damaged which makes a reliable identification next to impossible. Most probably the flower base is of *A. sativa*. It will be clear that this damaged flower base cannot be considered as firm evidence of oat cultivation. On the other hand, it is well possible that *A. sativa* was grown by the Noordbarge farmers.

3.6. Linum usitatissimum

Sample 815 yielded about 15 seeds of *Linum usitatissimum*. All seeds (fig. 4:8) are more or less seriously deformed as a result of the carbonization. Five specimens have been measured although the dimensions may deviate rather considerably from those of the seeds before carbonization: 3.5 x 1.7, 3.0 x 1.5, 3.3 x 1.4, 3.0 x 1.8, 3.0 x 1.8 mm. Although flax is represented by a small number of seeds in only one sample, one may assume that this crop plant was cultivated by the Noordbarge farmers, at least at some stage of the habitation of the site. It is unlikely that linseed could have occurred as a weed in the fields. This crop plant cannot maintain itself for some time without being cultivated intentionally. One could hypothesize that the linseeds had been obtained by trade, *e.g.* from the inhabitants of the coastal area where *Linum* was cultivated rather widely. However, this suggestion must be regarded as rather unlikely.

3.7. Camelina sativa

Camelina sativa seeds (fig. 4:3) occur in fairly considerable numbers in samples 738 and 815 from phase III. As usual the oleaginous seeds of this species have more or less seriously been affected by the carbonization. For 26 specimens from sample 815 length and breadth have been determined: 1.52 (1.4-1.7) x 0.96 (0.8-1.1) mm.

The *Camelina* seeds at Noordbarge do not necessarily imply that this species was grown intentionally. Gold-of-pleasure could have

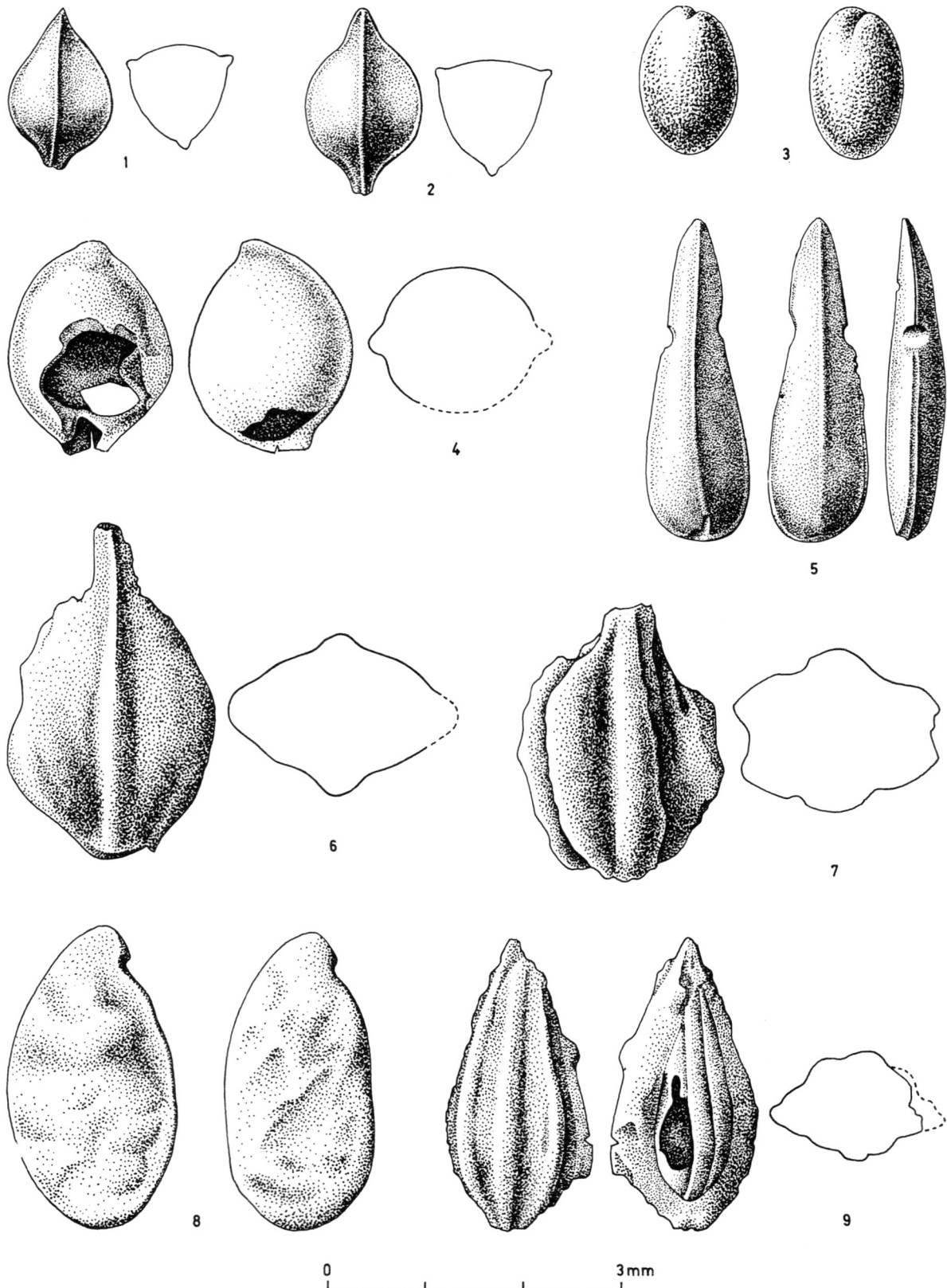

Fig. 4. Charred seeds from Noordbarge. 1: *Rumex crispus* (872); 2: *Rumex crispus* (247); 3: *Camelina sativa* (815); 4: *Humulus lupulus* (910); 5: *Valeriana officinalis,* inner fruit (738); 6-7: unidentified seed (1041); 8: *Linum usitatissimum* (815); 9: *Oenanthe aquatica* (766).

occurred as a weed in linseed fields. In this respect it should be remembered that the *Linum* seeds discussed above occurred in one of the two samples with *Camelina* (sample 815). The *Camelina* seeds could consequently have formed part of the residue left after the cleaning of the linseed crop.

The cultivation of *Camelina sativa* could be demonstrated convincingly for pre-Roman and Roman Iron Age settlement sites in the coastal area of the north of the Netherlands (Van Zeist, 1974) and of Northwest Germany (Körber-Grohne, 1967). Moreover, in various Iron Age sites in the Rhineland area *Camelina sativa* is well represented (Knörzer, 1978; 1980). It seems justified to assume that this species was grown by the Noordbarge farmers because of its oil-rich seeds. The question whether it was cultivated only during phase III or also during phases II and IV must remain undecided. In this connection it should be kept in mind that in general *Camelina* seems to be poorly represented in the charred seed record.

4. WILD PLANT SPECIES

4.1. Wild millet-type fruits

Wild millet-type fruits, which include grains of *Echinochloa crus-galli, Digitaria ischaemum* and *Setaria* cf. *viridis*, have been found in various samples. In only two samples, nos. 738 and 815, do these grains occur in somewhat greater numbers.

4.1.1. *Setaria cf. viridis* (fig. 3:3-4)

Setaria grains are characterized by a long and rather narrow radicle shield. The species identification of the charred fruits, somewhat greater numbers of which were found in samples 738 and 815, proved to be difficult. Is *Setaria viridis*, which could have occurred as a weed in the fields, concerned here, or must the charred grains be attributed to *S. italica*, implying that Italian millet was a crop plant of the Noordbarge farmers? In this connection it should be mentioned that according to Knörzer

(1971) *Setaria italica* was grown in the Rhineland area in Iron Age times.

The charred *Setaria* grains from Noordbarge show more resemblance to those of modern *S. italica* than of *S. viridis*. However, to what extent may the carbonization have changed the shape of the grains? To gain some insight into the possible effect of carbonization on the shape of millet-type fruits, 50 modern dehusked grains of *Panicum miliaceum* were measured (table 7). A comparison with the dimensions of the charred *Panicum* grains from sample 336 is not meaningful because the original size of the Noordbarge specimens may have differed rather considerably from that of the modern ones. On the other hand, the index values are probably more indicative of possible changes in the dimensions as a result of carbonization. The lower mean L:B index value in the charred grains indicates that relative to the breadth of the grains the length has decreased. The higher mean T:B index value suggests that the thickness of the grains increased relative to the width. The carbonization resulted in a decrease in length, an increase in thickness and probably also in some increase in width. The charred grains are more compact than the non-carbonized modern ones.

Further, 25 modern dehusked grains of both *Setaria italica* and *S. viridis* were measured for comparison with 8 charred *Setaria* grains from sample 815 (table 7). The latter were the only charred *Setaria* fruits suitable for measurement. Modern naked grains of *S. italica* and *S. viridis* from various proveniences present in the seed reference collection of the Biologisch-Archaeologisch Instituut differ quite distinctly from each other. The fruits of *S. viridis* are more slender than those of *S. italica*. This finds expression in the L:B index values (table 7) which do not even show an overlap (89-113 for *S. italica* and 138-169 for *S. viridis*). The grains of the cultivated *S. italica* are, on average, shorter than those of its wild ancestor *S. viridis*, but, on the other hand, they are noticeably thicker.

Table 6. Dimensions in mm and index values of *Secale cereale* from Noordbarge 307 (N = 32).

	L	B	T	100L:B	100T:B
min.	3.9	1.6	1.5	167	70
aver.	4.99	2.14	1.95	235	91
max.	6.2	2.5	2.4	350	108

Table 7. Dimensions in mm and index values of *Panicum miliaceum* and *Setaria* species.

		L	B	T	100L:B	100T:B
Panicum miliaceum	min.	2.0	1.7	1.3	108	71
modern (N = 50)	aver.	2.20	1.86	1.41	118	76
	max.	2.3	2.0	1.6	132	83
Panicum miliaceum	min.	1.5	1.4	1.2	88	73
Noordbarge 336	aver.	1.79	1.72	1.43	105	86
(N = 52)	max.	2.1	2.1	1.7	121	95
Setaria italica	min.	1.4	1.2	0.8	89	67
modern (N = 25)	aver.	1.51	1.48	1.07	102	72
	max.	1.6	1.6	1.1	113	81
Setaria viridis	min.	1.6	1.0	0.6	138	60
modern (N = 25)	aver.	1.72	1.16	0.75	149	65
	max.	1.8	1.3	0.9	169	73
Setaria	min.	1.1	1.0	0.8	100	63
Noordbarge 815	aver.	1.22	1.12	0.83	109	75
(N = 8)	max.	1.3	1.3	0.9	115	85

Table 8. Dimensions in mm of *Echinochloa crus-galli* from various sites.

	Length	Breadth	Thickness
Noordbarge 738 (N = 14)	1.33 (1.1-1.6)	1.10 (1.0-1.3)	0.69 (0.6-0.9)
Rheydt[1] (N = 10)	1.30 (1.2-1.4)	1.03 (0.9-1.1)	0.58 (0.5-0.7)
Neuss[2] (N = 10)	1.64 (1.5-1.7)	1.36 (1.2-1.5)	0.69 (0.6-0.8)
Gasselte[3] (N = 11)	1.64 (1.3-2.0)	1.31 (1.1-1.5)	

[1] Knörzer 1971 (Hallstatt)   [2] Knörzer 1970 (Roman)   [3] Van Zeist & Palfenier-Vegter 1979 (Medieval)

Which conclusions may be drawn from a comparison of the index values of the charred *Setaria* grains with those of the modern ones, taking into consideration the differences in the index values between modern and carbonized prehistoric *Panicum miliaceum*? The *Panicum* grains suggest an increase in the T:B index values of about 13% as a result of carbonization. A similar increase in the T/B ratio of *Setaria* grains would result in mean T:B index values of about 81 and 73 for *S. italica* and *S. viridis*, respectively, after carbonization. The mean T:B index value of 75 established for the charred *Setaria* from sample 815 could point to *S. viridis*. This suggestion is not confirmed by the L:B index values of the charred *Setaria*. In the charred *Panicum* grains the mean L:B index value is about 11% lower than that of the modern grains. For *Setaria italica* and *S. viridis* a similar decrease would result in mean L:B index values of about 91 and 133, respectively. The mean L:B index value of 109 obtained for the carbonized *Setaria* from Noordbarge does not conform more or less to either of the two calculated values.

It is self-evident that the effect of the carbonization on the shape of the grains is not necessarily the same in *Panicum* and *Setaria* and that, moreover, not too much weight should be attached to the measurements of only 8 charred *Setaria* grains. From the above it will be obvious that a satisfactory species identification is not possible. *Setaria viridis* seems to be the most likely candidate. It should, however, be mentioned that Knörzer (1971) attributes *Setaria* grains from Iron Age (Hallstatt C/D) Nettesheim in Rhineland, of about the same dimensions as the Noordbarge specimens (1.27 x 1.20 x 0.82 mm for 10 caryopses), to *S. italica*.

4.1.2. *Echinochloa crus-galli*

The grains of *Echinochloa crus-galli* are characterized by the flat ventral side and by the large radicle shield extending over about 3/4 of the domed dorsal side (fig. 3:6-8). In general the grains of *Echinochloa* can easily be recognized, but with some grains from samples 738 and 815 it was difficult to make a distinction between *Setaria* and *Echinochloa*. The dimensions (and index values of carbonized grains from Noordbarge and from a few other sites are shown in table 8.

The Noordbarge grains are of about the same size as those from Iron Age (Hallstatt) Rheydt (Rhineland), but are smaller than those from Roman Neuss and medieval Gasselte. The variations in the size of the *Echinochloa* grains are probably to be ascribed to local soil conditions.

4.1.3. *Digitaria ischaemum*

The caryopses of *Digitaria (Panicum) ischaemum* are more slender than those of *Echinochloa*, whereas the radicle shield extends over only 1/3 to 2/5 of the dorsal side (fig. 3:5). This species is scarcely represented at Noordbarge; only sample 815 yielded a somewhat greater number of *Digitaria* grains. Six specimens from this sample measure 1.21(1.1-1.3) x 0.76(0.7-0.8) x 0.48(0.3-0.6) mm. These dimensions agree with those obtained for seven *Digitaria ischaemum* grains from Iron Age (Hallstatt C/D) Rommerskirchen in Rhineland (Knörzer, 1971): 1.18 x 0.69 x 0.49 mm.

4.1.4. *Wild millets as food plants?*

Knörzer (1971) suggests that the grains of *Setaria* and *Echinochloa* served as food for prehistoric man. He assumes that these species occurred in the broomcorn millet fields, leaving undecided whether or not in Iron Age times intentional mixtures of wild and domesticated millets were grown. The wild millet-type grains from Noordbarge give no indication as to their possible economic role. The somewhat greater numbers in samples 738 and 815 could point to the harvesting of these grains for human consumption,

but they can also be interpreted as the residue of crop cleaning. The charred seeds and
fruits in the Noordbarge samples are very
probably of mixed origin, that is to say,
that they are derived from various domestic
activities (2.2.).

4.2. Other potential wild food plants

A few nutshell fragments of *Corylus avellana* indicate that hazelnuts were collected,
but no speculations can be made as for
their possible role in the diet of the inhabitants of the site. The picking of wild fruits
is attested by one *Rubus* fruitstone and by
half an apple pip. The find of a charred seed
of *Humulus lupulus* (fig. 4:4) is certainly
interesting, but it does not yet suggest that
hop was used for beer brewing.

The seeds of various weeds from fields
and other disturbed habitats represented at
Noordbarge are assumed to have served as
food for prehistoric man. Thus, Knörzer
(1977) makes a strong point of the intentional harvesting of the caryopses of *Bromus
secalinus*, a common weed in grain fields,
by prehistoric man. The fairly great number
of *Bromus* grains in sample 341 could indicate that they were consumed by the Noordbarge people. However, the *Bromus* caryopses in this sample could also represent
the residue of crop cleaning, implying that
the brome-grass grains had been removed
from the crop together with other weed
seeds. Five *Bromus* grains from sample 341
measure: 3.8 x 1.4 x 1.1, 3.9 x 1.7 x 1.0, 4.6
x 1.5 x 1.2, 4.6 x 1.6 x 1.3 and 4.6 x 1.4 x
1.3 mm.

Chenopodium album, Spergula arvensis
and *Polygonum lapathifolium/persicaria* are
regularly reported in palaeo-ethnobotanical
literature as potential wild food plants. Particularly for Iron Age sites in Jutland there is
firm evidence of the harvesting of the seeds
of these field weeds for human consumption.
Chenopodium album occurs in a great number
of seeds in sample 815, while *Spergula, Polygonum* and also *Rumex acetosella* are well

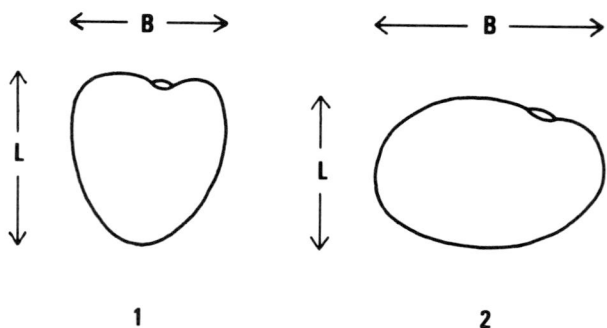

Fig. 5. Position of the measurements in seeds of *Trifolium*
(1) and *Lotus* (2).

represented in various samples. But, again,
the presence of greater quantities of these
seeds can be explained as an indication of
intentional gathering as well as in terms of
residues of crop cleaning.

4.3. Lotus and Trifolium repens

Two types of small leguminous seeds are represented at Noordbarge by more than only
one specimen, viz. those of *Trifolium repens*
and *Lotus*. On the basis of the shape, the seeds
of *Trifolium repens* can be distinguished from
those of the other *Trifolium* species which
could come into consideration for this site (cf.
Knörzer, 1970, pp. 80-82).

The seeds of *Lotus corniculatus* and *L. uliginosus* differ from each other only in size. The
dimensions of 25 modern seeds of both *Lotus*
species are shown in table 9. As is demonstrated by the measurements *Lotus corniculatus*
seeds are appreciably larger than those of
L. uliginosus. There is hardly any overlap. As
for the dimensions, the length is taken between
the side with the hilum and the opposite end.
The breadth is perpendicular to the length
axis (fig. 5). The carbonized *Lotus* seeds from
Noordbarge are smaller than those of modern
L. uliginosus (table 9). The carbonization must
have resulted in a more or less considerable
decrease in size. This is also clear from the
comparison of the dimensions of modern and
charred seeds of *Trifolium repens*.

Table 9. Dimensions in mm for modern and charred seeds of *Lotus* and *Trifolium repens*.

		L	B
Lotus	min.	0.7	0.9
Noordbarge (N = 9)	aver.	0.76	0.95
	max.	0.8	1.0
Lotus uliginosus	min.	0.8	0.9
modern (N = 25)	aver.	0.97	1.08
	max.	1.0	1.3
Lotus corniculatus	min.	1.1	1.3
modern (N = 25)	aver.	1.23	1.40
	max.	1.0	1.6
Trifolium repens	min.	1.0	0.9
modern (N = 25)	aver.	1.17	0.96
	max.	1.3	1.0
Trifolium repens	min.	0.8	0.6
Noordbarge (N = 6)	aver.	0.89	0.72
	max.	1.0	0.8

Without any information on the degree of shrinkage of *Lotus* seeds in carbonization the Noordbarge specimens cannot be identified to the species level. One may wonder whether the size reduction in the *Trifolium repens* seeds could provide a clue in this respect. Length and breadth of the Noordbarge *Trifolium* seeds are, on average, about 25% smaller than those of the modern seeds. A 25% reduction of both dimensions in the modern *Lotus* seeds results in mean values of 0.73 x 0.81 mm for *L. uliginosus* and 0.93 x 1.05 mm for *L. corniculatus*. A reduced average length of 0.73 mm for *L. uliginosus* corresponds well with the value of 0.76 mm obtained for the Noordbarge seeds. On the other hand, the average breadth of 0.95 mm for the latter seeds is notably larger than that for *L. uliginosus* after a 25% reduction (0.85 mm). The reduced dimensions of *L. corniculatus* (0.93 x 1.05 mm) are both distinctly larger than those of the Noordbarge specimens. Consequently, the dimensions would plead more in favour of *L. uliginosus* than of *L. corniculatus*. However, in view of the many uncertainties one should rather refrain from attributing the Noordbarge *Lotus* seeds to either of the two species which come into consideration here.

4.4. Cenococcum geophilum

Sclerotia of *Cenococcum geophilum* are quite regularly present at Noordbarge, in some samples even in very great numbers. These sclerotia are always somewhat problematic. They are at most listed in reports on charred seed finds, but generally without further comment. Natho (1957) published a rather detailed report on *Cenococcum* from Wahlitz. For the superimposed culture layers there — from the Neolithic (Rössen culture) to medieval times — he established increasing numbers of *Cenococcum* sclerotia with decreasing depth. Natho does not attribute this to possible

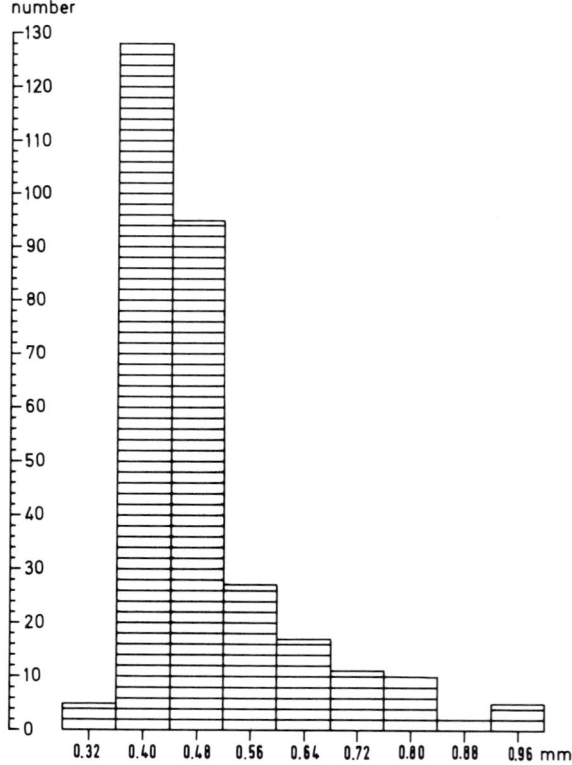

number

Fig. 6. Frequency distribution of the greatest dimension of 300 sclerotia of *Cenococcum geophilum* from sample 722.

secondary deposition (infiltration from the upper layers), but he assumes that in more remote times the growing conditions for *Cenococcum geophilum* were not favourable. This fungus species is reported to live in the humus layer of forest and heath-land and also in peaty soils (Natho, 1957).

The Noordbarge evidence suggests that carbonized *Cenococcum* sclerotia were not evenly distributed in the upper soil layers. Thus, in the fill of post-holes of houses 15 and 16 (samples 621, 720, 721, 722, 723, 724, 727) great numbers of *Cenococcum* sclerotia were found, whereas the fill of post-holes of houses 11, 12, 13, 14 and 17, equally attributed to phase III, yielded usually no or only few sclerotia (samples 711, 712, 713, 714, 715, 716, 748, 756, 762, 766, 767, 768, 794, 804, 818). The present author prefers to refrain from speculations to explain the local differences in the distribution of carbonized *Ceno-*

coccum sclerotia, but he only wishes to draw attention to this phenomenon.

For 300 sclerotia from sample 722 the greatest dimension has been determined: 0.3-1.0 (aver. 0.49) mm. The frequency distribution of the dimensions is shown in fig. 6. The Gasselte sclerotia are small; the absence of specimens larger than 1.0 mm is striking (cf. Natho, 1957).

5. SOME REMARKS ON THE DISTRIBUTION OF THE PLANT REMAINS

It has been suggested (2.2.) that the charred seeds and fruits in the fill of pits and post-holes must have originated from the top soil in the vicinity of these features. This opinion leads to some considerations regarding the presence of man on the terrain prior to the construction of the houses. The seeds and fruits must for the greater part have arrived in the settlement area as the result of human activities. Without the interference of man no crop-plant seeds and only small numbers of weed seeds were to be expected in the area concerned.

In the samples from phase I, dated to the Middle Bronze Age, seeds occur in only small numbers. Apart from a few broomcorn-millet grains in sample 892, attributed to phase I with some reserve, no other crop-plant fruits have been recovered from phase I samples. Samples 857 and 895, not listed in table 2, did not yield any seeds at all. Apparently the Bronze Age settlement had been built on a terrain on which domestic activities which, among other things, result in carbonized seeds and fruits, had up to then been of only limited extent. It was not a terrain that had been used intensively for some time prior to the construction of the settlement.

The same applies to the samples from phase II. In addition to the samples listed in table 2, 4 samples from phase II (house no. 20) were devoid of seeds and fruits. It looks as if settlement phase II was built on a terrain which only recently had been occupied. In this connection it should be remembered that there is a hiatus of 800 to 1000 years between phases I and II (see 1.)

In contrast, the plant remains of phase III samples point to a longer activity of man in the settlement area before the construction of the houses. In many places carbonized seeds and fruits of crop plants and weeds from fields were present on or near the surface of the soil. This is no surprise because the phase III houses were built on the same terrain as those of phase II and no interruption in the habitation had taken place.

However, some caution has to be observed in drawing the kind of conclusions presented above. The Noordbarge evidence indicates that even on a terrain that has been in use already for a longer period carbonized seeds and fruits may locally have been scarce or absent. Thus, in samples 920 to 923, from house no. 23 (phase IV), crop plants are not represented at all and other species only poorly. This fact invalidates to some degree the assumption that settlement phases I and II had been erected on a terrain that only just recently had been taken into exploitation. In fact the phase I samples examined for plant remains are all from the same house (no. 21). For phase II, 8 out of the 12 samples (including the samples not listed in table 2) are from the same house (no. 20). One cannot exclude the possibility that it is by accident that the houses of phases I and II which were sampled for palaeobotanical examination yielded only small numbers of seeds and fruits and that those samples are not representative for the whole of the settlement phases concerned.

6. THE ECOLOGICAL SIGNIFICANCE OF THE WILD PLANTS

Nearly all wild plants represented at Noordbarge grow on dry to moderately moist soils. Species from marshy habitats, such as *Eleocharis* and *Carex rostrata/vesicaria*, are only scarcely represented, which is rather surprising. One may assume that not far to the east of

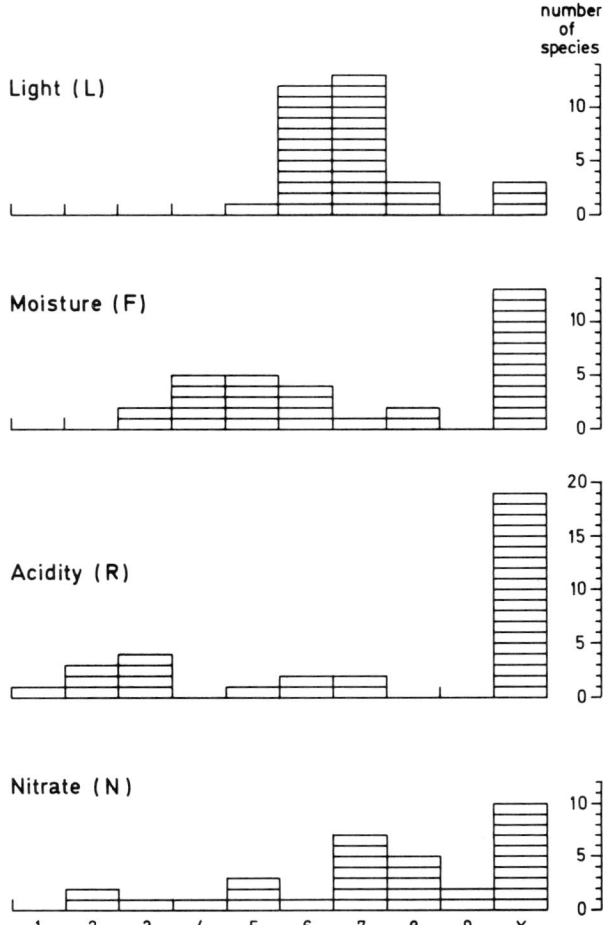

Fig. 7. So-called palaeoethnobotanical eco-diagrams: frequency distribution graphs of ecological indicator values. For explanation see text (section 6).

the site a rather large body of water, viz. the Bargermeer (Lake of Barge) was present. The reconstructed maximum extension of the Bargermeer is shown in fig. 1. To the west, north and east, the lake was bordered by the higher sandy soils. The 20 m contour line must have corresponded approximately with the highest lake level. Raised bog constituted the border of the lake to the south. At its maximum size the lake must have extended to less than 500 m from the locality of the Noordbarge site. It is likely that at the time of the settlement the lake had not yet reached its maximum extension, so that it was somewhat further away from the site than 500 m. However, the lake shore, and consequently marshy vegetations, must have been situated at a rather short distance from the settlement.

As for the other wild plant species, the majority of them are weeds from fields, waste places and road sides. In previous papers it has been attempted to determine the vegetation types, *i.e.* the syntaxonomic units in the sense of the Zürich-Montpellier School of Phytosociology, in the vicinity of the settlement. This reconstruction of the vegetation is based upon the phytosociological affinity of the species represented in a particular site. In some cases, for instance for sites from the coastal area in which non-carbonized plant remains are well preserved, this may permit a rather detailed reconstruction of the vegetation and hence of the environment in the vicinity of the site. If the vegetable remains are predominantly of species from disturbed habitats this method is somewhat less satisfactory, particularly because it is hardly justified to determine the prehistoric field-weed vegetation types below the highest syntaxonomic units (see *e.g.* Van Zeist & Palfenier-Vegter, 1979, pp. 290-91). For that reason in this paper another approach to evaluate the ecological information provided by the species demonstrated for Noordbarge will be presented, viz. that of the indicator values (*Zeigerwerte*) of plants introduced by Ellenberg (1974). This method has already been applied to archaeological plant remains by Willerding (1978; 1980).

The so-called ecological behaviour of a plant is expressed in its reaction to six environmental factors, viz. three climatic factors: light (L), temperature (T), continentality (K); and three soil factors: moisture (F), acidity (R), nitrogen content (N). Nine degrees of behaviour (1-9) with regard to these factors are distinguished. Thus, L 1 is a full-shadow plant, L 5 a half-shadow plant, and L 9 a full-light species. R 1 grows on very acid soils, R 5 on moderately acid soils, and R 9 on neutral soils. N 3 is found on soils poor in nitrogen, N 8 is a nitrogen indicator plant. Indifferent behaviour with regard to an environmental factor is indicated with an X.

In the graphs of fig. 7 the ecological behaviour is expressed of 32 wild plants which could be identified to the species level with a fair degree of certainty (in table 2 indicated by an asterisk). Marsh plants are not included in this group of species. For four of Ellenberg's environmental factors the frequency distributions of the indicator values of the Noordbarge species are shown. From these graphs the following conclusions can be drawn.

By far the majority of the plants may be considered as moderately light demanding. This will be no surprise because they are species from fields and other open habitats. Plants from neutral to basic soils are in the minority. Most of the Noordbarge species are either characteristic of acid soils or indifferent to soil reaction. As for the moisture content, moderately dry soils (M4-M6) must have prevailed. A rather great number of species prefer soils which are rich in nitrogen (N7-N9). This could indicate that the fields were rather rich in nitrate suggesting that some kind of fertilization (litter from the forest, animal manure, vegetable debris) was practised. On the other hand, it is also possible that the species concerned were particularly found in and near the settlement, in places where refuse was dumped.

7. REFERENCES

Behre, K.-E., 1972. Kultur- und Wildpflanzenreste aus der Marschgrabung Jemgumkloster/Ems (um Christi Geburt). *Neue Ausgrabungen und Forschungen in Niedersachsen 7*, pp. 164-184.
Ellenberg, H., 1974. *Zeigerwerte der Gefässpflanzen Mitteleuropas* (= Scripta Geobotanica Bd. 9).
Harsema, O.H., 1976. Noordbarge. In: Archeologisch Nieuws. *Bulletin van de Kon. Ned. Oudheidk. Bond 75*, pp. 52-55.
Harsema, O.H., 1979. Peeloo, gem. Assen. In: Kroniek van opgravingen en vondsten in Drenthe in 1977. *Nieuwe Drentse Volksalmanak 96*, pp. 150-151.
Knörzer, K.-H., 1970. *Römerzeitliche Pflanzenfunde aus Neuss* (=Limesforschungen Bd. 10: Novaesium IV). Berlin.
Knörzer, K.-H., 1971. Eisenzeitliche Pflanzenfunde im Rheinland. *Bonner Jahrbücher 171*, pp. 40-58.
Knörzer, K.-H., 1977. Pflanzliche Grossreste des bandkeramischen Siedlungsplatzes Langweiler 9. In: R. Kuper, H. Löhr, J. Lüning, P. Stehli & A. Zimmermann (eds.), *Der bandkeramische Siedlungsplatz Langweiler 9, Ge-*

meinde Aldenhoven, Kreis Düren (= Rheinische Ausgrabungen Bd. 18). Bonn, pp. 279-304.

Knörzer, K.-H., 1978. Entwicklung und Ausbreitung des Leindotters (*Camelina sativa* s. 1.). *Ber. Deutsch. Bot. Ges.* 91, pp. 187-195.

Knörzer, K.-H., 1980. Neue metallzeitliche Pflanzenfunde im Rheinland. *Archaeo-Physika* 7, pp. 25-34.

Kooi, P.B., 1979. *Pre-Roman urnfields in the North of the Netherlands*. Groningen.

Kooi, P.B., 1980. Peeloo, gem. Assen. In: Kroniek van opgravingen en vondsten in Drenthe in 1978. *Nieuwe Drentse Volksalmanak* 97, p. 182.

Körber-Grohne, U., 1967. *Geobotanische Untersuchungen auf der Feddersen Wierde*. Wiesbaden (Textband & Tafelband).

Natho, G., 1957. Cenococcum geophilum Fr. in Wahlitz. *Beiträge zur Frühgeschichte der Landwirtschaft* 3, pp. 161-169.

Willerding, U., 1978. Paläo-ethnobotanische Befunde an mittelalterliche Pflanzenreste aus Süd-Niedersachsen, Nord-Hessen und dem östlichen Westfalen. *Ber. Deutsch. Bot. Ges.* 91, pp. 129-160.

Willerding, U., 1980. Anbaufrüchte der Eisenzeit und des frühen Mittelalters, ihre Anbauformen, Standortsverhältnisse und Erntemethoden. In: *Untersuchungen zur eisenzeitlichen und frühmittelalterlichen Flur in Mitteleuropa und ihrer Nutzung*, Tl. II (=Abhandl. d. Akad. d. Wiss. in Göttingen, phil.-hist. Klasse, 3. Folge, Nr. 116). Göttingen, pp. 126-191.

Zeist, W. van, 1968 (1970). Prehistoric and early historic food plants in the Netherlands. *Palaeohistoria* 14, pp. 42-173.

Zeist, W. van, 1974. Palaeobotanical studies of settlement sites in the coastal area of the Netherlands. *Palaeohistoria* 16, pp. 223-371.

Zeist, W. van, 1976. Two early rye finds from the Netherlands. *Acta Botanica Neerlandica* 25, pp. 71-79.

Zeist, W. van & R.M. Palfenier-Vegter, 1979. Agriculture in medieval Gasselte. *Palaeohistoria* 21, pp. 267-299.